T0181275

Einstein Studies

Editors: Don Howard   John Stachel

Published under the sponsorship
of the Center for Einstein Studies,
Boston University

More information about this series at http://www.springer.com/series/4890

Dennis Lehmkuhl · Gregor Schiemann
Erhard Scholz

Editors

Towards a Theory
of Spacetime Theories

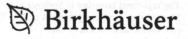 Birkhäuser

Editors
Dennis Lehmkuhl
Einstein Papers Project and HSS Division
California Institute of Technology
Pasadena, CA
USA

Gregor Schiemann
Theory and History of Science, Department
 of Philosophy, School of Human and
 Social Sciences
University of Wuppertal
Wuppertal
Germany

Erhard Scholz
History of Mathematics (em.), Department
 of Mathematics
University of Wuppertal
Wuppertal
Germany

The Einstein Studies series is published under the sponsorship of the Center for Einstein Studies, Boston University.

ISSN 2381-5833   ISSN 2381-5841   (electronic)
Einstein Studies
ISBN 978-1-4939-7998-1   ISBN 978-1-4939-3210-8   (eBook)
DOI 10.1007/978-1-4939-3210-8

Mathematics Subject Classification (2010): 83–02, 83C05, 83C10, 83C40, 83C45, 83D05

Printed on acid-free paper

This book is published under the trade name Birkhäuser
The registered company is Springer Science+Business Media LLC
The registered company address is: 233 Spring Street, New York, NY 10013, U.S.A.
(www.birkhauser-science.com)

Preface

This book is the outgrowth of a conference organized at the *Interdisciplinary Centre for Science and Technology Studies (IZWT)* of the University of Wuppertal, Germany, in 2010, July 21–23. Around that time the editors of the present volume had the pleasure of a close interdisciplinary cooperation for several years. We thank the centre for giving us the opportunity to organize the conference. We also express our gratitude to the Fritz Thyssen Foundation and the University of Wuppertal for their financial support. Two of us would like to use this opportunity to acknowledge that the bulk of work for preparing the conference was generously taken over by D. L., who also acted as the main editor of this book. However, most of all, we want to thank all contributors to the conference and to this book.

Wuppertal/Pasadena   Dennis Lehmkuhl
June 2016   Gregor Schiemann
           Erhard Scholz

Contents

Introduction: Towards A Theory of Spacetime Theories

Dennis Lehmkuhl

The title of this book—Towards a Theory of Spacetime Theories—is an attempt at false modesty. Or, rather, maybe: an act of an unreasonable raising of the chin in the face of a task supposedly impossible to master. After all, we do not even have a comprehensive map of the solution space of general relativity; by far the most established and most investigated spacetime theory; how then are we supposed to draw a map of the space of spacetime theories in which general relativity itself is but one little point? It seems a daunting and impossible task. And still, we cannot afford not to take it on.

General Relativity (GR) brought with it the idea that spacetime is not just a static but a dynamical playing field in its own right; according to GR, spacetime itself is *dynamical*. This is a wonderful insight, and in many ways we still have to come to grips with its implications. Much of the current literature in philosophy of physics wrestles with the right interpretation of Einstein's 1915 theory. But GR is not the only candidate for the 'right' theory of spacetime and gravitation out there. Indeed, in recent years it has come under pressure. We need to understand the rivals of GR, and try to play a part in getting an objective picture of how to decide between these theories.

Even if we were absolutely sure that GR is the right theory of spacetime and gravity, even if we were only concerned with *its* interpretation rather than with that of its outlandish rivals, we would still *need to* look at said rivals. For you cannot understand how *special* GR is as a theory, as a point in the space of possible spacetime theories, without looking at least at its immediate neighbourhood in this space. Is GR the only serious theory of gravity in which special relativity is locally valid? Or is this a feature shared by a wide array of post-1915 theories and we should stop seeing it as something noteworthy about GR in particular? How about the notorious 'geometrization' of the gravitational field and the unification of gravity and inertia?

D. Lehmkuhl (✉)
Einstein Papers Project and HSS Division, California Institute of Technology,
Pasadena, CA, USA
e-mail: lehmkuhl@caltech.edu

© Springer Science+Business Media, LLC 2017
D. Lehmkuhl et al. (eds.), *Towards a Theory of Spacetime Theories*,
Einstein Studies 13, DOI 10.1007/978-1-4939-3210-8_1

Is GR the only theory achieving these goals? Are the field equations of the theories in GR's neighbourhood equally non-linear, or are there different degrees or kinds of non-linearity in a sense to be specified? How exactly are the field equations and the equations of motion of particles related to one another, in GR and in theories close to GR? How about the shape of the solution space of GR's field equations? Is it a special characteristic of GR that there is a *unique* exterior static and spherically symmetric solution to the field equations, or is this result (known as Birkhoff's theorem) germane not to GR itself but to the wider neighbourhood around and including GR in the space of spacetime theories? If so, how big is this neighbourhood? And how about gravitational waves: how special are the gravitational waves allowed for by GR as compared to the gravitational waves allowed for by other theories?

1   Which Space of Theories?

We need a map. Maybe a map of the whole space of spacetime theories is too much to ask for at the moment, but we can aim at a map of at least the immediate neighbourhood of GR in the space of theories.

Here, however, we already hit the first obstacle. When I first envisaged the workshop on which this book is based, I had no doubt that the space GR lives in is the space of spacetime theories, and that this was the space of which we had to draw a map. But what distinguishes the space of spacetime theories from the space of field theories? Maxwell–Lorentz electrodynamics is a member of the space of field theories but not a member of the space of spacetime theories. General Relativity, on the other hand, is arguably both a spacetime theory and a field theory. But which neighbourhood are we looking at then: GR's neighbourhood in the space of field theories, or its neighbourhood in the space of spacetime theories? If we look at the former neighbourhood, then special relativistic electrodynamics is arguably a very close neighbour to GR, if we look at the latter, it is arguably not a neighbour at all. But why? What makes a theory a *spacetime* theory?

We will not start with an answer to this question. Instead, one of the hopes for this book is that by comparing different spacetime theories, we will come closer to an inkling of what they have in common, in contrast to a field theory that is supposed to be about an abstract space (like a fibre-bundle formulation of electrodynamics), rather than about physical spacetime.

2   A Comparison of Different Theories of Spacetime Theories

There are three kinds of (meta)-theories of spacetime theories; three ways to draw a map. The first kind is the most ambitious: try to develop a framework so general and encompassing that it can capture all the properties of each spacetime theory (or a major subset of all spacetime theories), and allow us to compare them with one another. The first meta-theory of this camp was never declared as one, even though it is implicit in the works of Weyl, Eddington, Einstein, Schouten, etc. The idea is to focus on the underlying geometry on which general relativity is based—pseudo-Riemannian geometry—and to find new geometries that are generalisations of GR's geometry on which, in turn, more general physical theories can be based. Examples include Weyl geometry, in which the independently defined metric and the connection are semi-compatible rather than compatible like in GR, or affine theory, in which only the affine connection is fundamental. Further generalisations can be obtained by allowing the metric and/or the connection to be asymmetrical, or for them not to be compatible with one another at all. Depending on which generalised geometry one chooses, one will have a torsion tensor arising from the connection in addition to the curvature tensor, or obtain more than one curvature tensor.[1]

The big drawback of the above approach is that it focuses merely on the fundamental mathematical objects of the different theories, but not so much on the dynamics. Indeed, Eddington published on his affine approach, in which a generalised affine connection is the only fundamental variable, without ever giving field equations, let alone details on how the affine connection, whose Ricci tensor was to represent the unified gravitational and electromagnetic fields, coupled to ordinary matter.[2] Furthermore, depending on how restrictive your notion of 'geometry' is, the approach might keep you from investigating generalisations of GR in which mathematical objects occur which do not have a 'clear geometrical meaning'.[3] Finally, the geometrical approach easily leads us to believe that in order to get a theory that is in an interesting sense different from GR we need to move away from pseudo-Riemannian geometry; and indeed this was the leading approach between the 1920s and the 1940s.

However, in the 1950s and 1960s, the renaissance of gravitational research, more and more alternatives to general relativity were found that, like GR, were based on pseudo-Riemannian geometry. Their main difference to GR lay in new gravitational fields in addition to the metric tensor (and not necessarily related to geometry), in different field equations, and different coupling structures. The first formalism aiming to capture this plethora of gravitational theories was 'Dicke's formalism', due to Robert Dicke. The main aim of the approach was to deliver a framework that was theoretically as unrestrictive and prejudiced as possible, so as to provide a framework in which different gravitational theories could be judged against new data (more precise measurements in the earth–moon system, the solar system, and in earthbound free-fall experiments) without preferring one theory over other theories. Thus, Dicke's formalism only demanded that the following minimal postulates are fulfilled in any theory of gravity: (1) that spacetime is represented by a 4-dimensional

---

[1] See Goenner [9, 10] especially, but also van Dongen [19], Vizgin [20] and Bergia [2].

[2] See Goenner [9], section 4.3. Einstein's main engagement with Eddington's affine approach was to deliver the field equations for the affine connection, see especially the recently published Volumes 13 and 14 of the Collected Papers of Albert Einstein.

[3] Einstein's own notion of 'geometry' was very unrestrictive. Indeed, he argued that an arbitrary vector v^μ is not more nor less 'geometrical' than a metric tensor $g_{\mu\nu}$. See [13] for details.

manifold, and (2) that every theory of gravity to be categorised is expressed in generally covariant form. On top of this, the following constraints were imposed: (i) gravity must be associated with one or more tensor fields; and (ii) the theory must have a Lagrangian formulation. It was explicitly *not* demanded that (a) any kind of equivalence principle holds, nor (b) that freely falling particles move on geodesics.[4]

Dicke's framework was designed to compare gravitational experiments, especially in the solar system regime. In order to complement and build on this, Thorne et al. [17] constructed an overarching system designed to compare the conceptual and mathematical intricacies of different theories of gravity. I will call their framework the TTL-framework for short. The authors start by giving separate definitions of 'spacetime theory' and 'mathematical representation of a spacetime theory', distinguish between 'kinematically possible trajectories' and 'dynamically possible trajectories', three different kinds of variables that can appear in a spacetime theory (confined, absolute, and dynamical variables), and they give abstract definitions of the distinction between boundary conditions, prior geometric constraints, decomposition equations, and dynamical laws. With this toolbox, they can succinctly and completely characterise any theory that is a 'spacetime theory' according to their definition of a spacetime theory: "any theory that possesses a mathematical representation constructed from a 4-dimensional spacetime manifold and from geometric objects defined on that manifold."[5] Thorne et al. go on to define a gravitation theory as a special kind of spacetime theory, namely any spacetime theory that essentially predicts Kepler's laws for a binary star system.[6] A metric theory of gravity is, in turn, a special case of gravitation theory that has a particular mathematical representation in which (I) spacetime is endowed with a metric tensor; (II) the world lines of test bodies are geodesics of said metric; and (III) the Einstein equivalence principle (as defined within TTL) is satisfied; in particular, all non-gravitational laws in any freely falling frame reduce to their counterparts in special relativity. Much could be said about this definition of a metric theory of gravity, especially with regard to the question of how the three conditions relate to one another. Note in particular that some of these conditions will be theorems in one metric theory and assumptions in another; I will come back to the special case of the second assumption further below.

The second kind of metatheoretical approach to spacetime theories, theories of gravitation in particular, aims not to capture *all* of the properties of a given spacetime theory but only how it behaves in a certain limit. One may ask why one would go for this kind of approach if one could also go for the first approach listed above. The answer is mostly practical. First, most of our empirical data tells us about the

---

[4]See Dicke [5] and Will [21], p. 10, for a summary of Dicke's framework.

[5]The notion of 'geometric object' they draw on here is that of Trautman [18]; which is designed to "include nearly all the entities needed in geometry and physics". They are less explicit about what they mean by a '4-dimensional spacetime manifold' rather than a '4-dimensional manifold'; I would argue that a spacetime manifold needs at least conformal structure (an equivalence class of metrics defined on it) in order to be called a spacetime manifold, for only then can we distinguish between spatial and temporal dimensions.

[6]See Thorne et al. [17], p. 18, for qualifications regarding what kind of binary star system they are talking about, and how close the theory has to come to Kepler's laws.

properties of comparatively weak gravitational fields. If we want to know which theory of gravity fares best in the light of this empirical data, then it is easiest to look at the limit where gravitational fields are relatively weak and where non-linear effects are typically negligible. But also, secondly, it makes it much easier to compare gravitational theories almost at a glance, and to see in which respects they are alike or not alike (in the limit).

The most prominent example in this second kind of approach is the parametrised post-Newtonian (PPN) framework. It takes two tentative conclusions of Dicke's work/framework as a starting point and postulates (1) that there exists a metric tensor $g_{\mu\nu}$ of signature -2 defined on the 4-dimensional manifold, (2) that this metric is 'read' by rods and clocks, and (3) that every material system is associated with an energy-momentum tensor, whose covariant divergence vanishes, $\nabla^\mu T_{\mu\nu} = 0$. Here, the covariant derivative ∇^μ is compatible with the metric $g_{\mu\nu}$ defined in the first postulate.

Postulates 1 and 3 restrict the kinds of allowed gravitational theories significantly, to a set of theories that is in important respects very much like GR. In addition to the metric $g_{\mu\nu}$, it allows for further fields associated with gravity, which can couple to both the metric and the matter fields in various ways. But just like in GR, rods and clocks are associated with *one* metric, and free particles follow the geodesics of the Levi-Civita connection compatible with said metric. However, these postulates do not yet constrain the different theories of gravity to a limiting case. This is achieved by two further constraints, namely (i) that the gravitational sources are weak in the sense that in the Newtonian limit of the respective theory the Newtonian potential is $<10^{-6}$; and (ii) that gravitational sources (except electromagnetic fields) move slowly with respect to one another, in the sense that $v^2 < 10^{-7}$.[7]

The PPN framework has been incredibly succesfull. It was originally designed primarily to be applied to the solar system regime in order to compare how different theories of gravity handle, say, Mercury's perihelion or the relative motion of Earth and Sun. A particularly interesting difference found between different theories of gravity as compared in the PPN framework is the so-called Nordvedt effect, which some alternative theories of gravity (like Jordan–Brans–Dicke theory) predict while others (including GR) do not. It is often referred to as a test of the strong equivalence principle which holds in GR but not in all alternative theories of gravity.[8] However, since its inception in the 1960s, the PPN framework has also been applied to compare the predictions of different gravitational theories outside of the solar system regime; for example, their predictions with respect to how binary pulsars behave.

---

[7]I am using geometrized units here, in which the speed of light is 1, and the Newtonian potential dimensionless. Cf. Will [21], p. 87.

[8]The Nordvedt effect *would* obtain if the ratio of inertial and gravitational mass would be different from 1 for sufficiently large, sufficiently self-gravitating bodies. Thus it would show that while test bodies move on geodesics (the weak equivalence principle), not all massive gravitating bodies do, even if their spin is neglected. The most important experimental realisation were lunar laser tests of the Earth–Moon system in the 1960s, in which no Nordvedt effect was discovered. However, it remains possible that more massive bodies (black holes in particular) would exhibit a Nordvedt effect. See Nordtvedt [15] and Will [21], section 8.1, for details.

Other limiting frameworks have been proposed, designed to compare the limits of gravitational theories in different regimes. The parametrised Post-Friedmannian (PPF) framework is particularly interesting, for it allows us to compare different theories of gravity which have the Friedmann solution in its solution space.[9] In effect, the PPF framework is designed to judge how different theories handle the cosmological data that has been garnered since the discovery of the cosmic microwave background, which brought about the widespread conviction that the whole history of the universe we actually live in can be represented by the Friedmann solution.[10]

The articles in this book go into another direction than these grand metatheoretic approaches. Rather than drawing a map in broad strokes, they focus on particularly rich regions in the space of spacetime theories. Indeed, most of them compare our most successful theory of gravity, GR, with one or two other theories of gravity. The general idea is that by getting to know our own home city, and by comparing it to our immediate neighbours, we will be better prepared for the grand journey still ahead of us. By comparing GR in all its details to particular spacetime theories, we hope to get a better idea about what is special about GR, to see patterns in the immediate neighbourhood of GR that might be much more difficult to see on a map with a grand scale.

3 The Case Studies Contained in this Book

The first article in this book is by James Weatherall; it compares GR with geometrized Newtonian gravitation theory (also known as Newton–Cartan theory). In particular, Weatherall compares the role of geodesic theorems in the two theories; the possibility to *derive* the geodesic motion of matter within these theories. In the context of GR alone it has sometimes been claimed (especially by Brown [4]) that the existence of a theorem that tells us that matter moves on geodesics *explains* inertial motion of matter rather than presuming it. In drawing on his own proof that a similar theorem exists in geometrized Newtonian gravity, Jim discusses the similarities and differences of the respective theorems within the two theories, and uses the result to reflect on the different senses in which one could take inertial motion to be explained. The article shows clearly the direct benefits of a comparative analysis of two theories of gravity: it is only through the comparison with geometrized Newtonian gravity that the different ways in which we might interpret the sentence 'GR explains inertial motion' come into focus.

While Weatherall's chapter zooms in on a feature of GR that many have taken as being at the core of GR (the geodesic theorem), Erik Curiel's chapter focuses

---

[9]See Ferreira [8] for details on the PPF framework.

[10]Of course, the conviction that the Friedmann solution adequately represents the era that we currently live in goes back to Hubble's discovery of the redshift of galaxies. But only the discovery of the Cosmic Microwave Background gave convincing evidence of the big bang theory, i.e. the idea that the Friedmann solution applies to the beginning of the universe too; indeed that there was a beginning in the first place. See Smeenk [16] for details.

on something that many have taken to be at the periphery of GR. 'A primer on energy conditions' gives a comprehensive overview of the different conditions that can be imposed on the energy-momentum tensor, the source term of the Einstein field equations, and discusses different possible interpretation for each of them. In doing so, Curiel shows that there is almost nothing one can do with GR without imposing energy conditions; they are anything but at the periphery of the theory. But as Curiel shows, the conditions and their interpretations stretch way beyond GR, towards the question of what it means to be a spacetime theory. In his final section, 'Constraints on the character of spacetime theories', Curiel has moved from what was only apparently the periphery to the very heart of GR and spacetime theories more generally: from his analysis of energy conditions he draws conclusions about features that any spacetime theory must have, especially with regard to the relation between the stress-energy of matter and the (local and global) structure of spacetime.

Where Weatherall focused on the role of geodesic theorems and Curiel on the role of energy conditions in spacetime theorems, Oliver Pooley focuses on a property of GR that has too often been discussed without detailed comparison as to how the property features in other spacetime theories: the alleged background independence of GR. Pooley discusses different attempts of defining the notion of background independence that result in different versions of GR, and compares them with different versions of the special relativistic theory of a scalar field defined on a flat background. In doing so he disentangles different conceptions of background independence, diffeomorphism invariance, and dynamicality of spacetime.

Friedrich Hehl investigates whether GR can be brought into the form of a gauge theory akin to the standard model of elementary particle physics. Building on work by Utiyama, Sciama and Kibble, he develops Riemann–Cartan theory on the basis of the translation group of Minkowski spacetime; the resulting theory has non-vanishing curvature and non-vanishing torsion. It contains General Relativity (vanishing torsion, non-vanishing curvature) and teleparallel theory (non-vanishing torsion, vanishing curvature) as special cases, and shows how the empirical equivalence of the latter two theories can be understood and what it demands. This empirical equivalence has recently been the topic of philosophical discussion [11]; Hehl's article sheds new light on this equivalence by embedding both theories into the more general framework of Riemann–Cartan theory. Further philosophical work could use his results to question the long-standing position that the diffeomorphism group is to be regarded as the gauge group of GR.

Erhard Scholz looks at a different geometric generalisation of GR, pioneered by Hermann Weyl. Scholz chooses a particular variant of Weyl geometry, one that is integrable and not intrinsically linked to electromagnetism, and shows how looking at both GR and Jordan-Brans-Dicke theory uncovers new relationships between the two theories. He then relates this framework to electroweak theory, and discusses the relationship between the gravitational scalar field of Weyl geometry and the scalar Higgs field of the standard model. Scholz then derives new cosmological models in this framework, and sheds new light on the cosmological models of GR.

Claus Beisbart's chapter differs from most other chapters in this volume in that it casts the net more widely: Beisbart starts from a general framework for a 'theory of theories', the semantic or model-theoretic conception of what it is to be a theory, and applies it to the case of spacetime theories. This allows Beisbart to reflect on the question of what makes a theory a *spacetime* theory, and he can propose several possible answers drawing on the semantic conception of theories. In the end, he applies the framework to compare GR and Brans-Dicke theory in particular. He argues that GR 'is not a theory of its own, but rather a relationship between theories', and that the familiar claim that Brans–Dicke theory reduces to GR in a limit is problematic when the limiting relationship is spelt out using the semantic framework.

David Wallace investigates how exactly the (special) relativity principle and the equivalence principle are related in the context of GR. He starts by summarising the often-voiced position that both principles are true in sufficiently small regions of spacetime, namely in regions where curvature is negligible. He challenges this claim by introducing a thought experiment he terms *Galileo's black hole*: a system where curvature is *not* negligible yet the relativity and equivalence principles still hold. He diagnoses that if a general relativistic system is *isolated*, its metric at sufficiently large distances is the same as the metric of any system at sufficiently small distances: it is the Minkowski metric with the Poincaré group as its symmetry group. For such systems, he argues, the relativity and equivalence principles hold even though curvature/gravity is not negligible. Thus, he shows what exactly makes the two principles hold both locally and at large distances from isolated bodies.

Both the relativity and the equivalence principle are specific symmetry principles. The general characterisation of different kinds of symmetries is the topic of Adán Sus' paper. He carefully distinguishes between global and local symmetries, investigates how global conservation laws arise even in the context of theories with local symmetries, and which types of symmetries exactly have direct empirical significance. He points to the precise relationship between different types of symmetries and different types of conserved currents, and their interpretation, in answering these questions.

While the previous chapters have all investigated and compared different classical (non-quantum) spacetime theories, the last two chapters in the volume turn to the question of what spacetime (and spacetime theories) are in the context of approaches to quantum gravity. Claus Kiefer investigates the different concepts of time in GR and in quantum theory, and analyses the extent to which these differences present an obstacle for the construction of a quantum theory of gravity. Kiefer argues that one of the two requirements that *any* theory of quantum gravity must fulfil is the recovery of GR in a classical limit.

Christian Wüthrich investigates how one of the most promising approaches for the correct theory of quantum gravity, Loop Quantum Gravity (LQG), deals with this problem. As he points out, many approaches to quantum gravity start from the assumption that the world does not contain spacetime as part of its fundamental structure, but as something that has to be regained in the classical limit. Wüthrich investigates how exactly this might take place in LQG. In discussing different interpretational options, one thing about Wüthrich's analysis is particularly interesting in the context of this volume: the (re)-emergence of spacetime in a classical limit seems to uncover a rather different limiting relationship between LQG and GR as compared to, say, GR and special relativity (compare Pooley's chapter), or GR and Newtonian gravity (compare Weatherall's chapter).

4 Outlook

It might seem strange to finish the introduction to a book with a few words about what could come after the book. But in the end, this book is supposed to be a step (or a couple of steps) *towards* a theory of spacetime theories. Thus, it seems appropriate to say something about what the next steps could be.

I contrasted the chapters in this book with the grand schemes of Dicke's framework, the Thorne–Lee–Lightman framework, the PPN- and the PPF framework. While these schemes try to cast their net widely, to cover as many spacetime theories as possible, the chapters in this book instead focus on a detailed comparison of certain pairs or triples of spacetime theories. A natural next step could be to investigate which lessons can be drawn for modified grand-scheme frameworks; whether the results of the pairwise comparison of spacetime theories allow for 'bottom-up results' that directly impinge on the details of the overarching frameworks. They might even motivate the construction of a new overarching framework to complement the existing ones.

Another avenue would be to explore something that neither the grand schemes nor the pairwise comparisons of this volumes have focused on so far, although both endeavours have touched on it. Both approaches (grand frameworks and pairwise comparisons) have focused on comparing the field equations, symmetry groups and fundamental (geometric) objects of the different theories. The solutions to these equations were not the focus of these investigations. However, much if not all of the actual predictive work of spacetime theories is achieved by solutions to the field equations. We see this, for example, by the fact that when Einstein [7] predicted the perihelion of Mercury, he did not even have the final field equations of GR; but the (approximation to) the Schwarzschild solution he used in making the prediction was a solution both to the field equations he used in that paper and to the (soon-to-be-called) Einstein equations of the final theory. In defining the solution he used to model the gravitational field of the Sun that Mercury is subject to, Einstein demanded that (i) the field is spherically symmetric; (ii) the field is static; and (iii) that it is asymptotically flat.

Birkhoff [3] soon showed that the Schwarzschild solution is the *unique* exterior spherically symmetric solution that is also static and asymptotically flat. In other words, it is unnecessary to demand staticity and asymptotic flatness as independent assumptions; if the solution is demanded to be spherically symmetric, then one gets the other two characteristics Einstein assumed 'for free'. This is a rather striking property of the solution space of Einstein's field equations. One way of learning more about how special (or ordinary) GR is in its immediate neighbourhood in the 'space of spacetime theories' would be to gauge whether (a counterpart of) Birkhoff's theorem holds in other spacetime theories; and if so which other features these theories have in common with GR.[11]

Investigating the subspace of spherically symmetric solutions is particularly important in every theory of gravity, for most astrophysical bodies (stars, black holes, planets) are approximately spherically symmetric. An equally important solution subspace is that of gravitational wave solutions. We know that in GR gravitational waves have two modes of polarisation, two degrees of freedom. Is this typical of this solution subspace in most theories of gravity? Interestingly, it is not: Eardley et al. [6] showed that the most general gravitational wave solution in a metric theory of gravity (as defined in the TTL-framework) has six possible polarisation modes. GR, as a special case of a metric theory of gravity, allows for only two of those, Brans–Dicke theory allows for three.

The first gravitational wave has only just been detected experimentally [1]. However, the two LIGO detectors at Hanford and Louisiana were arranged in such a way that the likelihood of detecting a gravitational wave hitting the Earth from an arbitrary direction was to be maximised; not to distinguish, say, a GR-wave from a Brans–Dicke wave. This will change once more the gravitational wave interferometers go online.

There is so much left to be done towards an encompassing theory of spacetime theories. But we are getting there. Step by Step.

References

1. BP Abbott, R Abbott, TD Abbott, MR Abernathy, F Acernese, K Ackley, C Adams, T Adams, P Addesso, RX Adhikari, et al. Observation of gravitational waves from a binary black hole merger. *Physical Review Letters*, 116(6): 061102, 2016.
2. Silvio Bergia. Attempts at unified field theories (1919–1955). alleged failure and intrinsic validation/refutation criteria. In John Earman, Michel Janssen, and John Norton, editors, *The Attraction of Gravitation. New Studies in the History of General Relativity*, volume 5 of *Einstein Studies*. Birkhäuser, 1993.
3. G.D. Birkhoff. *Relativity and modern physics*. 1923.
4. Harvey R. Brown. *Physical Relativity. Space-time structure from a dynamical perspective*. Oxford University Press, USA, Revised edition 2007.
5. Robert H Dicke. Experimental relativity. *Relativity, Groups and Topology. Relativité, Groupes et Topologie*, pages 165–313, 1964.
6. Douglas M Eardley, David L Lee, and Alan P Lightman. Gravitational-wave observations as a tool for testing relativistic gravity. *Physical Review D*, 8(10): 3308, 1973.
7. Albert Einstein. Erklärung der Perihelbewegung des merkur aus der allgemeinen Relativitätstheorie. *Königliche Preussische Akademie der Wisenschaften (Berlin)*, Reprinted as Vol.6, Doc. 24 CPAE 1915.

---

[11] Just a few preliminary results on this: A counterpart to Birkhoff's theorem exists for the Einstein–Maxwell solutions: the unique spherically symmetric solution to the Einstein–Maxwell solutions is the (likewise static and asymptotically flat) Reissner–Nordström solution. There are some candidates for Birkhoff counterparts in some scalar-tensor theories, but typically extra assumptions are necessary to prove uniqueness [12]. There are also generalisations of Birkhoff's theorem for some but not all theories with torsion [14], and for Lovelock gravity [22].

8. Timothy Clifton, Pedro G Ferreira, Antonio Padilla, and Constantinos Skordis. Modified gravity and cosmology. *Physics Reports*, 513: 1–189, 2012

9. Hubert F. M. Goenner. On the history of unified field theories. *Living Rev. Relativity, 7, :2004, 2. [Online Article]: cited [13.2.2004]*, http://www.livingreviews.org/lrr-2004-2, 2004.

10. Hubert F.M. Goenner. On the history of unified field theories. part ii.(ca. 1930–ca. 1965). *Living Rev. Relativity*, 17(5), 2014.

11. Eleanor Knox. Newton–Cartan theory and teleparallel gravity: The force of a formulation. *Studies in History and Philosophy of Science Part B: Studies in History and Philosophy of Modern Physics*, 42(4): 264–275, 2011.

12. KD Krori and D Nandy. Birkhoff's theorem and scalar-tensor theories of gravitation. *Journal of Physics A: Mathematical and General*, 10(6): 993, 1977.

13. Dennis Lehmkuhl. Why Einstein did not believe that general relativity geometrizes gravity. *Studies in History and Philosophy of Science Part B: Studies in History and Philosophy of Modern Physics*, 46: 316–326, 2014.

14. Donald E Neville. Birkhoff theorems for r+ r 2 gravity theories with torsion. *Physical Review D*, 21(10): 2770, 1980.

15. Kenneth Nordtvedt. Equivalence principle for massive bodies. ii. theory. *Physical Review*, 169(5): 1017–1025, 1968.

16. Chris Smeenk. Einstein's role in the creation of relativistic cosmology. In Michel Janssen and Christoph Lehner, editors, *The Cambridge Companion to Einstein*. Cambridge University Press, 2015.

17. Kip. S. Thorne, David L. Lee, and Alan P. Lightman. Foundations for a theory of gravitation theories. *Physical Review D*, 7:3563–3578, 1973.

18. Andrzej Trautman. Foundations and current problems of general relativity (notes by graham dixon, petros florides and gerald lemmer). 1965.

19. Jeroen van Dongen. *Einstein's Unification*. Cambridge University Press, Cambridge, 2010.

20. Vladimir Vizgin. *Unified Field Theories in the first third of the 20th century*. Birkhäuser, 1994.

21. Clifford M. Will. *Theory and Experiment in Gravitational Physics*. Cambridge University Press, 1993.

22. Robin Zegers. Birkhoff's theorem in lovelock gravity. *Journal of mathematical physics*, 46(7): 072502, 2005.

Inertial Motion, Explanation, and the Foundations of Classical Spacetime Theories

James Owen Weatherall

Abstract I begin by reviewing some recent work on the status of the geodesic principle in general relativity and the geometrized formulation of Newtonian gravitation. I then turn to the question of whether either of these theories might be said to "explain" inertial motion. I argue that there is a sense in which *both* theories may be understood to explain inertial motion, but that the sense of "explain" is rather different from what one might have expected. This sense of explanation is connected with a view of theories—I call it the "puzzleball view"—on which the foundations of a physical theory are best understood as a network of mutually interdependent principles and assumptions.

1 Introduction

There is a very old question in the philosophy of space and time, concerning how and why bodies move in the particular way that they do in the absence of any external forces. The question originates with Aristotle, and indeed, the puzzle is particularly acute when one thinks of it as the ancients might have. Given some external influence on a body, it might seem clear why that body moves in one fashion rather than another: the external influence forces it to do so. But when there are no forces present, what does the work of picking one possible state of motion over any other? Consider planetary motion: there are no apparent forces acting on planets, and yet they proceed along fixed trajectories. Why these orbits rather than others? In Aristotelian terms, what determines the "natural motions" of a body?

This manuscript was prepared in 2012 and has not been significantly revised since then. I still hold the philosophical views defended here, but have not attempted to update the manuscript in light of more recent work by myself or others.

J.O. Weatherall (✉)
Department of Logic and Philosophy of Science, University of California,
Irvine, CA 92697-5100, USA
e-mail: weatherj@uci.edu

© Springer Science+Business Media, LLC 2017
D. Lehmkuhl et al. (eds.), *Towards a Theory of Spacetime Theories*,
Einstein Studies 13, DOI 10.1007/978-1-4939-3210-8_2

The modern answer to the question originates with Galileo and Descartes, but finds its canonical form in Newton's first law of motion, which states that in the absence of external forces, a body will move in a straight line at constant velocity. This "law of inertia," as Newton called it, is preserved, *mutatis mutandis*, in general relativity, where inertial motion is governed by the *geodesic principle*. The geodesic principle states that in the absence of external forces, the possible trajectories through four-dimensional spacetime of a massive test point particle will be timelike geodesics— i.e., bodies will move along "locally straightest" lines without acceleration.

In standard presentations of general relativity, the geodesic principle is stated as a postulate (cf. [27, 33, 36, 54]), much like Newton's first law.[1] However, shortly after Einstein presented the theory, he and others began to suspect that one could equally well conceive of the geodesic principle as a theorem, at least in the presence of other standard assumptions of relativity theory [17, 19, 20]. This shift from geodesic-principle-as-postulate to geodesic-principle-as-theorem has led to a widespread and deeply influential view that general relativity has a special explanatory virtue that distinguishes it from other theories of space and time. In the words of Harvey Brown, general relativity "... is the first in the long line of dynamical theories ... that *explains* inertial motion" [4, pg. 163]. In other words, it may be that Newtonian physics answers the "how" part of Aristotle's question, but there is a sense in which only general relativity answers the "why" part.

Although Einstein's early attempts to prove the geodesic principle were not unambiguously successful, more recent efforts have shown that there is a precise sense in which the geodesic principle may be understood as a theorem of general relativity [23].[2] However, it turns out that relativity is not unique in this regard. Geometrized Newtonian gravitation (sometimes, Newton–Cartan theory) is a reformulation of

---

[1]For a detailed and enlightening discussion of the status of the first law of motion in standard Newtonian gravitation, see Earman and Friedman [14].

[2]There have been several steps along the way to proving the geodesic principle as a rigorous theorem of general relativity. The most significant early attempt was the work of Einstein and Grommer [19] and Einstein et al. [20], with subsequent work due to Mathisson [34, 35] (see also Sauer and Trautman [44]), Taub [50], Thomas [51], and Newman and Posadas [11, 38, 39]. Many of these are described and criticized briefly in Geroch and Jang [23]; for more expansive discussions, see Blanchet [2] and Damour [10]. This history of Einstein's efforts in this domain is described by Havas [26] and Kennefick [28]. There are currently two approaches to the problem that are widely recognized as successful: the one developed by Geroch and Jang [see also 18], which will be my focus in the present paper, and one developed by Sternberg [47] and Souriau [45], among others, which models a massive test point particle as an order-zero distribution with support along a curve. One can then show that if the distribution is (weakly) conserved, the curve must be a geodesic. Note, however, that although the Geroch–Jang approach and the Sternberg–Souriau approach are *prima facie* different, there is a sense in which they turn out to be equivalent [24]. It is worth observing that, although modern attempts to derive equations of motion in general relativity may be thought of as addressing the same problem that Einstein and his contemporaries sought to address, the theorems have a significantly different form. To give an example, Einstein et al. [20] claimed to show that the geodesic principle followed from the vacuum form of Einstein's equation; the Geroch–Jang theorem, meanwhile, makes no explicit reference to Einstein's equation, and, as we will see below, if it is related to Einstein's equation at all, it is because the theorem assumes that matter is represented by a divergence-free energy-momentum field—an assumption that may be thought to follow from Einstein's equation with sources, but not the vacuum form of the equation. And so, while I take the

Newtonian gravitation due to Cartan [5, 6] and Friedrichs [22] that shares many of the qualitative features of general relativity. In geometrized Newtonian gravitation one represents space and time as a four-dimensional spacetime manifold, the curvature of which depends dynamically on the distribution of matter on the manifold. Gravitational influences, meanwhile, are not understood as forces, as in traditional formulations of Newtonian gravitation; rather, they are a manifestation of the curvature of spacetime. And in particular, inertial motion is governed by the geodesic principle: in the absence of external (nongravitational) forces, bodies move along the geodesics of (curved) spacetime. Recently, I have shown that the geodesic principle can be understood as a theorem of geometrized Newtonian gravitation [55]. Mathematically, the Newtonian theorem is nearly identical to the Geroch–Jang theorem. Moreover, as I have argued elsewhere, when the background assumptions needed to prove these theorems are examined in the contexts of each theory, one can reasonably conclude that the geodesic principle has essentially the same status in both cases, though in neither theory is the situation as simple as one might have hoped [56].

One consequence of this recent work is that Einstein and others' idea that the status of the geodesic principle in general relativity distinguishes the theory from other theories of space and time seems more difficult to hold on to. But it also raises a related issue. When one attends carefully to the details of these theorems, several complications arise concerning the strength and status of the assumptions necessary for proving them. Given these complications, one might reasonably ask, do either of these theories *explain* inertial motion? It is this second question that I will take up in the present paper.[3]

(Footnote 2 continued)
Geroch–Jang theorem to provide a kind of answer to a problem Einstein recognized, it may be that the form of the answer is sufficiently different from what Einstein expected that Einstein would not have found it satisfactory. I am grateful to an anonymous referee for emphasizing this last point to me.

[3]The recent literature on whether and in what sense general relativity and Newtonian gravitation explain inertial motion originates with Brown [4]. Brown is not especially concerned to give an "account" of the sense of explanation he has in mind, in the sense of providing necessary or sufficient conditions for when some argument, theorem, etc., is an explanation (nor, I should say, am I!), though the idea is that the geodesic principle is explained in general relativity because there is a sense in which it is a consequence of the central dynamical principle of the theory, Einstein's equation. Sus [48] has expanded on this view, calling the form of explanation at issue "dynamical explanation," and further defending Brown's claim that general relativity is distinguished from other spacetime theories with regard to the explanation it provides of inertial motion. Malament [32] and I [56, 57], meanwhile, have pointed out that the geodesic principle does *not* follow merely from Einstein's equation, and that a strong energy condition is also required; moreover, as I note above, a theorem remarkably similar to the one that holds in the relativistic case also holds in geometrized Newtonian gravitation. But these latter discussions largely set aside the question of what sense of explanation is at issue, if any. More recently, Tamir [49] has pointed out that in general relativity, at least, the geodesic principle is false for realistic matter. He then considers almost-geodesic motion as a kind of universal phenomenon in the sense of Batterman [1]. From this latter perspective, these theorems provide explanations in the sense of showing how certain behavior can be expected to arise approximately for a wide variety of substances. The remarks in the present paper are of a rather different character than (most of) this earlier work, and so I will not engage with it closely in the text.

I will begin with a brief overview of geometrized Newtonian gravitation, after which I will review the relevant theorems concerning the geodesic principle in that theory and general relativity. I will focus on the subtle ways in which the theorems differ, and on the complications that arise when one tries to interpret them. Once this background material has been laid out, I will turn to the question at hand. The starting point for this discussion will be to observe that on one way of thinking about explanation in scientific theories, the answer to the question is "no": neither of these theories explains inertial motion, at least if the assumptions going into the theorems have the character I describe. I want to resist this view, however, because I think it takes for granted that one can make clear distinctions between "levels" or "tiers" of fundamentality of the central principles of a theory. Careful analysis of the geodesic principle theorems, meanwhile, suggests that there is another way of thinking about how the principles of a theory fit together. The alternative view I will develop—I will call it the "puzzleball view" or, perhaps more precisely, the "puzzleball conjecture"— holds that the foundations of physical theories, or at least *these* physical theories, are best conceived as a network of mutually interdependent principles, rather than as a collection of independent and explanatorily fundamental "axioms" or "postulates." On this view, one way to provide a satisfactory explanation of a central principle of a theory, such as the geodesic principle in general relativity or geometrized Newtonian gravitation, would be to exhibit its dependence on the other central principles of the theory, i.e., to show how the principle-to-be-explained is a consequence of the other central principles and basic assumptions of the theory. And this is precisely what the theorems I will describe do. And so, I will argue that there *is* a sense in which both theories explain inertial motion, though some care is required to say what is meant by "explain" in this context.

I should be clear from the start: the language of explanation is a convenient one, but I am not ultimately interested in the semantics of the word "explain." The goal is not to argue whether one thing or another is *really* an explanation. The dialectic, rather, is as follows. Many people have suggested that general relativity provides an important kind of insight with regard to inertial motion, something to be valued and sought after in our physical theories. One might call this thing an "explanation," or not. The point, though, is that when one looks in detail at just what one gets in relativity theory (and in geometrized Newtonian gravitation), it seems to work in a different way than one might have initially guessed it would. One response to this observation would be to say that we have not actually gotten what we were promised—or, in the language above, that general relativity does *not* explain inertial motion. But another response is to try to better understand what we *do* get. My principal thesis is that if one takes this second path, an alternative picture emerges of how the foundations of theories work. And on this alternative picture, general relativity and geometrized Newtonian gravitation both do provide an important and very useful kind of insight into inertial motion, and more, there are clear reasons why one should value and seek out this sort of insight. Indeed, one might even think that what we ultimately get is what we should have wanted in the first place. I am inclined to use the word "explanation" for this sort of insight, but fully recognize that this usage may seem nonstandard or incorrect to some readers.

2 Overview of Geometrized Newtonian Gravitation

Geometrized Newtonian gravitation is best understood as a translation of Newtonian gravitation into the language of general relativity, a way of making Newtonian physics look as much like general relativity as possible, for the purposes of addressing comparative questions about the two theories.[4] The result is a theory that is strikingly similar in many qualitative respects to general relativity, but which differs in certain crucial details. Recall that in general relativity, a *relativistic spacetime* is an ordered pair (M, g_{ab}), where M is a smooth four-dimensional manifold and g_{ab} is a smooth Lorentzian metric on the manifold. In geometrized Newtonian gravitation, meanwhile, one similarly starts with a smooth four-dimensional manifold M, but one endows this manifold with a different metric structure. Specifically, one defines *two* (degenerate) metrics. One, a *temporal metric* t_{ab}, has signature $(1, 0, 0, 0)$. It is used to assign temporal lengths to vectors on M: the temporal length of a vector ξ^a at a point p is $(t_{ab}\xi^a\xi^b)^{1/2}$. Vectors with nonzero temporal length are called *timelike*; otherwise, they are called *spacelike*. The second metric is a *spatial metric* h^{ab}, with signature $(0, 1, 1, 1)$. In general one requires that these two metrics satisfy an orthogonality condition, $h^{ab}t_{bc} = 0$. It is important that the temporal metric is written with covariant indices and the spatial metric with contravariant indices: since both metrics have degenerate signatures, they are not invertible, and so in general one cannot use either to raise or lower indices. In particular, this means that the spatial metric cannot be used to assign spatial lengths to vectors directly. Instead, one uses the following indirect method. Given a spacelike vector ξ^a, one can show that there always exists a (nonunique) covector u_a such that $\xi^a = h^{ab}u_b$. One then defines the spatial length of ξ^a to be $(h^{ab}u_au_b)^{1/2}$, which can be shown to be independent of the choice of u_a.

Given a Lorentzian metric g_{ab} on a manifold M, there always exists a unique covariant derivative operator ∇ that is compatible with g_{ab} in the sense that $\nabla_a g_{bc} = 0$. This does not hold for the degenerate Newtonian metrics. Instead, there are an uncountably infinite collection of derivative operators that satisfy the compatibility conditions $\nabla_a t_{bc} = 0$ and $\nabla_a h^{bc} = 0$. This means that to identify a model of geometrized Newtonian gravitation, one needs to specify a derivative operator in addition to the metric field. Thus, we define a *classical spacetime* as an ordered quadruple $(M, t_{ab}, h^{ab}, \nabla)$, where M, t_{ab}, h^{ab}, and ∇ are as described, the metrics satisfy the orthogonality condition, and the metrics and derivative operator satisfy the compatibility conditions. A classical spacetime is the analog of a relativistic spacetime. Note that the signature of t_{ab} guarantees that at any point p, one can find a covector t_a such that $t_{ab} = t_a t_b$; in cases where such a field can be defined globally, we call the associated spacetime *temporally orientable*. In what follows, we will always restrict attention to temporally orientable spacetimes, and will replace t_{ab} with t_a whenever we specify a classical spacetime.

---

[4]This brief overview of geometrized Newtonian gravitation is neither systematic nor complete. The best available treatment of the subject is given in Malament [33]; see also Trautman [52]. My notation and conventions here follow Malament's.

In both theories, timelike curves—curves whose tangent vector field is always timelike—represent the possible trajectories of point particles (and idealized observers). And as in general relativity, matter fields in geometrized Newtonian gravitation are represented by a smooth symmetric rank-2 field T^{ab} (with contravariant indices). In general relativity, this field is called the *energy-momentum tensor*; in geometrized Newtonian gravitation, it is called the *mass-momentum tensor*. The reason for the difference concerns the interpretations of the fields. In relativity theory, the four-momentum density of a matter field with energy-momentum tensor T^{ab} is only defined relative to some observer's state of motion: given an observer whose worldline has (timelike) tangent field ξ^a, the four-momentum density P^a as determined by the observer is given by $P^a = T^{ab}\xi_b$. When P^a is timelike or null, one can define the mass density ρ of the field at a point, relative to the observer, as the length of P^a. Moreover, the four-momentum field can be further decomposed (relative to ξ^a) as $P^a = E\xi^a + p^a$, where $E = P^n\xi_n$ is the relative energy density as determined by the observer, and $p^a = P^n(\delta^a{}_n - \xi^a\xi_n)$ is the relative three-momentum density. Thus, the field T^{ab} encodes the relative mass, relative energy, and relative momentum densities as determined by any observer. In geometrized Newtonian gravitation, meanwhile, *all* observers make the same determination of the four-momentum density of a matter field at a point: for any observer, P^a is given by $P^a = T^{ab}t_b$. Given a particular observer whose worldline has tangent field ξ^a, though, one can decompose P^a as $P^a = \rho\xi^a + p^a$, where $\rho = P^a t_a (= T^{ab}t_a t_b)$ is the (observer-independent) mass density associated with the matter field, and where $p^a = P^n(\delta^a{}_n - \xi^a t_n)$ is the relative three-momentum density of the matter field as determined by the observer. Thus in geometrized Newtonian gravitation, T^{ab} encodes the (absolute) mass density of a matter field, as well as its momentum relative to any observer.[5]

It is standard in both theories to limit attention to matter fields that satisfy several additional constraints. In particular, in both cases one assumes that matter fields satisfy the *conservation condition*, which states that their energy/mass-momentum fields are divergence free (i.e., $\nabla_a T^{ab} = \mathbf{0}$). One also usually requires that such fields satisfy various *energy conditions*. In geometrized Newtonian gravitation, only one such condition is standard: it is the so-called *mass condition*.

> **Mass condition**: A mass-momentum field satisfies the mass condition if, at every point, either $T^{ab} = \mathbf{0}$ or $T^{ab}t_a t_b > 0$.

Since $T^{ab}t_a t_b = \rho$ is the mass density, this assumption states that whenever the mass-momentum tensor is nonvanishing, the associated matter field has positive mass. The situation is more complicated in general relativity, where there are several energy conditions that one may consider. I will mention a few because they are of particular interest for present purposes. One, called the *weak energy condition*, is (at least *prima*

---

[5]Note that is general relativity, one makes a distinction between the mass and energy densities relative to a given observer, where relative mass density is the length of the four-momentum density determined by an observer at a point ($\rho = (P^a P_a)^{1/2}$) and relative energy density is $E = T^{ab}\xi_a\xi_b = P^a\xi_a$, where ξ^a is the tangent field to the observer's worldline. In geometrized Newtonian gravitation, this distinction collapses.

facie) quite similar to the mass condition. It states that the energy density of a matter field as determined by any observer is always nonnegative.

Weak energy condition: An energy-momentum field satisfies the weak energy condition if, given any timelike vector ξ^a at a point, $T^{ab}\xi_a\xi_b \geq 0$.

It is also common to consider stronger conditions. For instance, there are the *dominant energy condition* and the *strengthened dominant energy condition*:

Dominant Energy Condition: An energy-momentum field satisfies the dominant energy condition if, given any timelike vector ξ_a at a point, $T^{ab}\xi_a\xi_b \geq 0$ and $T^{ab}\xi_a$ is timelike or null.

Strengthened Dominant Energy Condition: An energy-momentum field satisfies the strengthened dominant energy condition if, give any timelike covector ξ_a at a point, $T^{ab}\xi_a\xi_b \geq 0$ and either $T^{ab} = 0$ or $T^{ab}\xi_a$ is timelike.

If these two conditions obtain for some matter field, then not only do all observers take the field to have nonnegative energy density, they also take its four-momentum to be causal or timelike (respectively). In other words, these latter conditions capture a sense in which matter must propagate at or below the speed of light.

The curvature of a classical spacetime is defined in the standard way: given a derivative operator ∇, the *Riemann curvature tensor* $R^a{}_{bcd}$ is the unique tensor field such that for any vector field ξ^a, $R^a{}_{bcd}\xi^b = -2\nabla_{[c}\nabla_{d]}\xi^a$. The *Ricci curvature tensor*, meanwhile, is given by $R_{ab} = R^n{}_{abn}$. In both contexts, one says that a spacetime is *flat* if $R^a{}_{bcd} = \mathbf{0}$; in geometrized Newtonian gravitation, one also says that a (possibly curved) spacetime is *spatially flat* if $R^{abcd} = R^a{}_{mno}h^{bm}h^{cn}h^{do} = \mathbf{0}$ or, equivalently, $R_{mn}h^{ma}h^{nb} = \mathbf{0}$. Given these ingredients, one can state the sense in which in geometrized Newtonian gravitation, the curvature of spacetime depends on the distribution of matter: namely, the central dynamical principle of the theory, the *geometrized Poisson equation*, states that $R_{ab} = 4\pi\rho t_a t_b$, where ρ is the mass density defined above. This expression explicitly relates the Ricci curvature of spacetime to the distribution of matter. It is the Newtonian analog of Einstein's equation, $R_{ab} = 8\pi(T_{ab} - \frac{1}{2}Tg_{ab})$, where $T = T^{ab}g_{ab}$, or equivalently $8\pi T_{ab} = R_{ab} - \frac{1}{2}Rg_{ab}$, where $R = R_{ab}g^{ab}$.

There are a few points to emphasize here concerning the geometrized Poisson equation. For one, if the geometrized Poisson equation holds of a classical spacetime for some mass-momentum tensor T^{ab}, then the classical spacetime is spatially flat, since $R_{nm}h^{na}h^{mb} = 4\pi\rho t_n t_m h^{ma}h^{nb} = \mathbf{0}$. This fact is a way of recovering a familiar feature of Newtonian gravitation, namely that *space* is always flat, even though in the geometrized theory *spacetime* may be curved. Second, in general relativity one can freely think of both the metric and the derivative operator as (systemically related) dynamical variables in the theory. In geometrized Newtonian gravitation, this is not the case: instead, the metrical structure of a classical spacetime is fixed, and only the derivative operator (or more specifically, the Ricci curvature, which is defined in terms of the derivative operator) is a dynamic variable. Finally, there is a sense in which, given some matter distribution, the geometrized Poisson equation "fixes" a derivative operator on a classical spacetime, but one has to be careful, as one can typically only recover a unique derivative operator satisfying the geometrized Poisson equation for a given matter distribution in the presence of additional boundary conditions or other assumptions.

The geometrized Poisson equation provides the sense in which in geometrized Newtonian gravitation, spacetime is curved in the presence of matter; the sense in which gravitational effects may be understood as a manifestation of this curvature is just the same as in general relativity. That is, a derivative operator allows one to define a class of geometrically privileged curves, the *geodesics* of the spacetime, which consist of all curves whose tangent fields ξ^a satisfy $\xi^n \nabla_n \xi^a = \mathbf{0}$ everywhere. I have already said that the timelike curves of a spacetime represent the possible trajectories for massive particles; the timelike geodesics, meanwhile, represent the possible *unaccelerated* trajectories of particles in both theories. The geodesic principle then connects these geometrically privileged curves with force-free motion. Thus, in geometrized Newtonian gravitation, as in general relativity, the distribution of matter throughout space and time affects the possible trajectories of massive point particles not by causing such particles to accelerate, but rather by dynamically determining a collection of unaccelerated curves.

These features of geometrized Newtonian gravitation provide the sense in which the theory is qualitatively similar to general relativity. But one might wonder what undergirds the implicit claim that geometrized Newtonian gravitation is in some sense *Newtonian*. One sense in which the theory is Newtonian is immediate: the degenerate metric structure of a classical spacetime captures the implicit geometry of space and time in ordinary Newtonian gravitation, where one has a temporally ordered succession of flat three-dimensional manifolds representing space at various times (cf. [46]). But there is more to say. In standard formulations of Newtonian gravitation, spacetime is flat. Gravitation is a force mediated by a gravitational potential, which is related to the distribution of matter by Poisson's equation. In the present four-dimensional geometrical language, this can be expressed as follows. We begin with a classical spacetime (M, t_a, h^{ab}, ∇) as before, but now we require that ∇ is flat, i.e., $R^a{}_{bcd} = \mathbf{0}$. We again represent matter by its mass-momentum field T^{ab}, defined just as above, but we also define a scalar field φ, which is the gravitational potential. Poisson's equation is written as $\nabla^a \nabla_a \varphi = 4\pi\rho$ where the index on ∇^a is raised using h^{ab}, and where $\rho = T^{ab} t_a t_b$. And now the acceleration of a massive test point particle in the presence of a gravitational potential φ is given by $\xi^n \nabla_n \xi^a = -\nabla^a \varphi$, where ξ^a is the tangent to the particle's trajectory. In other words, in standard Newtonian gravitation matter accelerates in the presence of mass.

It turns out that standard Newtonian gravitation (thus understood) and geometrized Newtonian gravitation are systematically related [33, ch. 4.2]. Specifically, given a classical spacetime (M, t_a, h^{ab}, ∇) with ∇ flat, a smooth mass density ρ, and a smooth gravitational potential φ satisfying $\nabla^a \nabla_a \varphi = 4\pi\rho$, there always exists a unique derivative operator $\tilde{\nabla}$ such that $(M, t_a, h^{ab}, \tilde{\nabla})$ is a classical spacetime, $\tilde{R}_{ab} = 4\pi\rho t_a t_b$, and such that for any timelike vector field ξ^a, $\xi^n \nabla_n \xi^a = -\nabla^a \varphi$ if and only if $\xi^n \tilde{\nabla}_n \xi^a = \mathbf{0}$. In other words, given a model of standard Newtonian gravitation, there is always a model of geometrized Newtonian gravitation with precisely the same mass density and allowed trajectories. Additionally, the derivative operator $\tilde{\nabla}$ will always satisfy two curvature conditions: $\tilde{R}^{ab}{}_{cd} = \mathbf{0}$ and $\tilde{R}^a{}_b{}^c{}_d = \tilde{R}^c{}_d{}^a{}_b$. This result

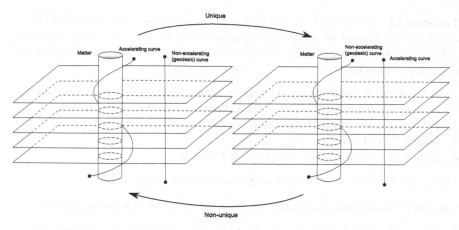

Fig. 1 In general, it is possible to translate between geometrized Newtonian gravitation and standard Newtonian gravitation, as depicted in this figure. On the left is a model of standard Newtonian gravitation: one has a matter field represented by the world tube of some body, such as the sun, and a curve orbiting this body, representing, say, a small planet. This curve corresponds to an allowed trajectory insofar as it is accelerating by the appropriate amount. On the right is the corresponding model of geometrized Newtonian gravitation. One has precisely the same matter distribution, and the same allowed trajectory (i.e., the same orbit), but now we understand this trajectory to be allowed by the theory because it is a geodesic of a curved derivative operator, with curvature determined by the matter distribution. Note that both theories have the same metrical structure, represented here by a succession of flat slices representing space at various times

is known as the Trautman geometrization lemma; it provides the sense in which one can always translate from standard Newtonian gravitation into the geometrized theory. One can also prove a corresponding recovery lemma (also due to Trautman), allowing for translations back: namely, given a classical spacetime $(M, t_a, h^{ab}, \tilde{\nabla})$ and smooth mass density ρ satisfying $\tilde{R}_{ab} = 4\pi\rho t_a t_b$, if $\tilde{R}^{ab}{}_{cd} = 0$ and $\tilde{R}^a{}_b{}^c{}_d = \tilde{R}^c{}_d{}^a{}_b$ then at least locally there always exists a flat derivative operator ∇ and a gravitational potential φ such that (M, t_a, h^{ab}, ∇) is a classical spacetime, $\nabla^a \nabla_a \varphi = 4\pi\rho$, and again for any timelike vector field ξ^a, $\xi^n \nabla_n \xi^a = -\nabla^a \varphi$ if and only if $\xi^n \tilde{\nabla}_n \xi^a = 0$. Note that this recovery result only holds in the presence of the two additional curvature conditions stated above; moreover, in general the translation from geometrized Newtonian gravitation to standard Newtonian gravitation will *not* be unique. (See Fig. 1.)

3   The Geodesic Principle as a Theorem

With the background of the previous section in place, I can now state the precise sense in which the geodesic principle may be understood as a theorem in general relativity and geometrized Newtonian gravitation. I will begin by stating both theorems, and then double back to the question of how one should interpret them.

Theorem 3.1 *[23][6]Let (M, g_{ab}) be a relativistic spacetime, and suppose M is oriented. Let $\gamma : I \to M$ be a smooth imbedded curve. Suppose that given any open subset O of M containing $\gamma[I]$, there exists a smooth symmetric field T^{ab} with the following properties.*

1. *T^{ab} satisfies the* strengthened dominant energy condition, *i.e., given any timelike covector ξ_a at a point, $T^{ab}\xi_a\xi_b \geq 0$ and either $T^{ab} = \mathbf{0}$ or $T^{ab}\xi_a$ is timelike;*
2. *T^{ab} satisfies the* conservation condition, *i.e., $\nabla_a T^{ab} = \mathbf{0}$;*
3. *$supp(T^{ab}) \subset O$; and*
4. *there is at least one point in O at which $T^{ab} \neq \mathbf{0}$.*

Then γ is a timelike curve that can be reparametrized as a geodesic.

One can prove an almost identical theorem in geometrized Newtonian gravitation.[7]

Theorem 3.2 *[55] Let $(M, t_{ab}, h^{ab}, \nabla)$ be a classical spacetime, and suppose that M is oriented. Suppose also that $R^{ab}{}_{cd} = \mathbf{0}$. Let $\gamma : I \to M$ be a smooth imbedded curve. Suppose that given any open subset O of M containing $\gamma[I]$, there exists a smooth symmetric field T^{ab} with the following properties.*

1. *T^{ab} satisfies the* mass condition, *i.e., whenever $T^{ab} \neq \mathbf{0}$, $T^{ab}t_a t_b > 0$;*
2. *T^{ab} satisfies the* conservation condition, *i.e., $\nabla_a T^{ab} = \mathbf{0}$;*
3. *$supp(T^{ab}) \subset O$; and*
4. *there is at least one point in O at which $T^{ab} \neq \mathbf{0}$.*

Then γ is a timelike curve that can be reparametrized as a geodesic.

As a first remark, it may not be obvious that either of these theorems should be understood to capture the geodesic principle at all, at least in a natural way. A principal difficulty in trying to derive the geodesic principle as a theorem concerns a kind of ontological mismatch between the geodesic principle and the rest of general relativity: namely, general relativity is a field theory, whereas the geodesic principle is a statement concerning point particles. One strategy for dealing with this problem is to try to model massive point particles as "small" bits of extended matter, and then show that under sufficiently general assumptions, the world tubes of such small bits of matter will contain timelike geodesics. But this turns out to be false in general—geodesic motion only obtains in the idealized limit where the world tube of a body collapses to a curve, in which case one can no longer represent matter as a smooth

---

[6]This particular statement of the theorem is heavily indebted to Malament [33, Prop. 2.5.2].

[7]Note that the following theorem may be understood to include inertial motion in *standard* (i.e., non-geometrized) Newtonian gravitation as a special case where the derivative operator associated with the classical spacetime in the proposition happens to be flat. So the present result may be taken to show that Newton's first law can be thought of as a theorem, too. In the case of standard Newtonian gravitation, however, gravitational interactions are conceived as forces and correspond to failures of the mass-momentum tensor to be conserved (relative to the fixed background choice of flat derivative operator), so strictly fewer physical situations correspond to inertial motion. For this reason, it is more interesting to focus on the geometrized theory, since the result is both stronger in that case and more directly analogous to the Geroch–Jang theorem.

field on spacetime.[8] The Geroch–Jang strategy, meanwhile, is different. Instead of starting with some matter and asking what kind of trajectory it follows, one starts with a curve and asks under what circumstances that curve can be understood as a trajectory for arbitrarily small bits of extended matter. Both theorems then state that the only curves along which arbitrarily small bits of matter can be constructed are timelike geodesics.

Importantly, one represents a "small bit of matter" by a smooth symmetric rank 2 tensor field with support in some neighborhood of the curve. But a curve is not understood as a possible trajectory for a free massive test point particle if one can construct *any* smooth symmetric rank 2 tensor field in arbitrarily small neighborhoods of the curve—rather, one limits attention to fields that satisfy additional constraints. The claim, then, is that these theorems capture the geodesic principle in both theories insofar as the additional constraints on the matter fields adequately capture what we intend by "free massive test matter." This means that the interpretation of the theorems turns on the status of these conditions. And so, for a comparative study of the status of the geodesic principle in each theory, one wants to compare the status of each of these assumptions relative to their respective theories.

Two of the assumptions can be set aside immediately: in both theorems, assumptions (3) and (4) play the role of setting up the limiting process implicit in the theorems. Assumption (3) limits attention to matter fields that vanish outside one's chosen neighborhood of the curve (which captures the sense in which one is considering arbitrarily small bits of matter propagating along the curve), and assumption (4) indicates that the matter field must be nonvanishing somewhere along curve, ruling out the trivial case. These assumptions are identical in both cases, and neither is troublingly strong.

There is also an obvious difference that can be safely ignored. In the Newtonian theorem, we place an additional constraint on the curvature, namely $R^{ab}{}_{cd} = \mathbf{0}$. This is precisely the curvature condition needed to prove the Trautman recovery theorem, allowing one to translate from a model of geometrized Newtonian gravitation to a model of standard Newtonian gravitation. For this reason, the curvature condition is naturally interpreted as a restriction to models of geometrized Newtonian gravitation that are *Newtonian*, in the sense that they admit translations back to models of standard Newtonian gravitation. This presumably is the case of greatest interest, and so I am inclined to think of the assumption as benign.[9] Moreover, there is good reason to think that this assumption can be dropped, though to my knowledge, proving as much is still an open (and perhaps interesting) problem.

The most striking difference between the two theorems concerns the respective assumptions (1).[10] In the Newtonian theorem, this is the mass condition, i.e., that whenever the mass-momentum field is nonvanishing, the mass density determined by any observer must be positive. This is the standard energy condition in geometrized

---

[8]This point is emphasized by Tamir [49].

[9]For another view on this matter, see Sus [48].

[10]For an enlightening and much more detailed discussion of energy conditions in general relativity, including the Strengthened Dominant Energy Condition, see Curiel [9].

Newtonian gravitation, and more, it is natural in this context, as it captures the sense in which the bits of matter being represented are massive. In the Geroch–Jang theorem, meanwhile, one requires the strengthened dominant energy condition, which states that (a) all observers must assign nonnegative energy density to the matter field (the weak energy condition) and (b) that if $T^{ab} \neq \mathbf{0}$, then the four-momentum assigned to the matter field by any observer must be timelike. It seems natural to think that the *weak* energy condition, (a), is playing the role played by the mass condition in the Newtonian case: namely, it captures the sense in which the small bits of matter are massive, by requiring that they always have nonnegative mass. But from this perspective, the second part of the condition, (b), is a strong additional requirement. In Newtonian gravitation, it would seem, one needs only to assume that mass is always positive to get timelike geodesic propagation, whereas in general relativity, one also needs to make an assumption about the timelike propagation of energy-momentum.[11]

However, the situation is not quite so simple as this. Although the mass condition appears to be nothing more than an assumption about positive mass, it, too, contains an implicit assumption about timelike propagation. To see this, consider a different (nonstandard) Newtonian energy condition, which I will call the *weakened mass condition*.

> **Weakened Mass Condition**: A mass-momentum field T^{ab} satisfies the weakened mass condition if at every point, $T^{ab} t_a t_b \geq 0$.

The weakened mass condition has a good claim on being the Newtonian analog of the weak energy condition and might similarly be understood as the claim that mass/energy density is always nonnegative. But it is *strictly weaker* than the mass condition, since the weakened mass condition may be satisfied by mass-momentum fields that are spacelike, in the sense that $T^{ab} \neq \mathbf{0}$ but $T^{ab} t_a t_b = 0$ (for example, consider $T^{ab} = u^a u^b$, with u^a a spacelike vector field). In other words, the mass condition amounts to the weakened mass condition plus the additional assumption that T^{ab} is timelike. We can make this explicit by defining an equivalent condition, the *modified mass condition*.

> **Modified Mass Condition**: A mass-momentum field T^{ab} satisfies the modified mass condition if at every point, $T^{ab} t_a t_b \geq 0$ and either $T^{ab} = \mathbf{0}$ or $T^{ab} t_a$ is timelike.

The modified mass condition is equivalent to the mass condition, but would appear to be the natural translation of the strengthened dominant energy condition.[12]

---

[11] This is precisely how I present the situation in [56, 57]. However, I now think matters are still more complicated than I indicate there, as I explain in the text. Still, the principal morals of those previous discussions are unchanged by these additional considerations.

[12] To see that the mass condition and modified mass condition are equivalent, consider the following. Fix a classical spacetime (M, t_a, h^{ab}, ∇) and a mass-momentum field T^{ab} on M. First suppose T^{ab} satisfies the mass condition. Then at every point, either $T^{ab} = \mathbf{0}$ or $T^{ab} t_a t_b > 0$. If $T^{ab} = \mathbf{0}$ everywhere, then it also satisfies the modified mass condition, so suppose there is a point p such that $T^{ab} \neq \mathbf{0}$. Thus at p, $T^{ab} t_a t_b > 0$. It follows that the temporal length of $T^{ab} t_a$ is positive, and thus that the vector $T^{ab} t_a$ must be timelike at p. So T^{ab} satisfies the modified mass condition.

Table 1 This table summarizes the relationship between the various energy conditions discussed in the text. Single arrows represent "apparently natural translations"; double arrows represent logical implications

Geometrized Newtonian gravitation		General relativity
Modified mass condition	$\longleftrightarrow$	Strengthened dom. energy condition
$\Updownarrow$		$\Downarrow$
Mass condition	$\longleftrightarrow$	Strengthened weak energy condition
$\Downarrow$		$\Downarrow$
Weakened mass condition	$\longleftrightarrow$	Weak energy condition

Returning to general relativity, one can also consider the *strengthened weak energy condition.*

Strengthened Weak Energy Condition: An energy-momentum field satisfies the strengthened weak energy condition if, give any timelike vector ξ^a at a point, either $T^{ab} = \mathbf{0}$ or $T^{ab}\xi_a\xi_b > 0$.

This condition seems like the natural translation of the (standard) mass condition, but it is strictly weaker than the strengthened dominant energy condition and strictly stronger than the weak energy condition![13]

This situation is summarized in Table 1. There are thus two ways of thinking about the relationship between the energy conditions used in these theorems, depending on which "natural translations" one emphasizes. On one way of thinking, the mass condition is essentially the same as the strengthened weak energy condition. From this point of view, then, the strengthened dominant energy condition in the Geroch–Jang theorem is a strictly stronger assumption than the corresponding assumption in its Newtonian counterpart, Theorem 3.2. More, one might be inclined to think that one gets something additional for free in the Newtonian case, since the mass condition turns out to imply the (apparently) stronger modified mass condition, whereas the strengthened weak energy condition does not imply the strengthened dominant energy condition. Meanwhile, on the other way of thinking about things, one argues that the strengthened dominant energy condition is essentially the same as the modified mass condition, which is fully equivalent to the mass condition. And so one concludes that the energy conditions required by the two theorems are essentially the same.

(Footnote 12 continued)

Now suppose T^{ab} satisfies the modified mass condition. Once again, if T^{ab} vanishes everywhere, it automatically satisfies the mass condition, so suppose there is a point p such that $T^{ab} \neq \mathbf{0}$. At that point, we know $T^{ab}t_a$ is timelike, and thus that $T^{ab}t_at_b \neq 0$. But since $T^{ab}t_at_b \geq 0$, it follows that $T^{ab}t_at_b > 0$. Thus T^{ab} satisfies the mass condition.

[13]The strengthened weak energy condition is also strictly weaker than the (strict) dominant energy condition, and so Prop. 4 of Weatherall [57] implies that the strengthened weak energy condition is not strong enough to prove the Geroch–Jang theorem.

There is also another possibility, which is to say that one cannot perform simple translations between the energy conditions in these two theories at all. I am inclined to endorse this last option, though this raises new questions about how one should compare the theorems. There are a few things to say. First, irrespective of how one tries (or does not try) to translate these conditions, there are still two senses in which the strengthened dominant energy condition is arguably stronger than the mass condition. One is that the timelike propagation clause of the strengthened dominant energy condition can be understood as the assumption that the instantaneous speed of matter, relative to any observer, must be strictly less than the speed of light. The corresponding clause of the (modified) mass condition, meanwhile, amounts to the assumption that matter cannot propagate at *infinite* speed relative to any observer. And the assumption that a number must be less than a fixed finite value is stronger than the assumption that it must be finite, but not bounded.

The second, more significant sense in which the strengthened dominant energy condition is stronger is that the *only* way in which matter in Newtonian gravitation can be "massive" (i.e., have positive mass as determined by some observer) is if it satisfies the mass condition. Matter that satisfies the weakened mass condition but not the mass condition will necessarily have zero mass. And so one might argue that the mass condition is necessary to capture what is meant by "massive" in the context of Newtonian gravitation. In general relatively, meanwhile, matter can be "massive" in two senses, *without* satisfying the strengthened dominant energy condition: it can be massive in the sense that it has positive energy density (i.e., it satisfies the weak energy condition), and it can be massive in the sense that some observers will assign it positive mass density (i.e., the relative four-momentum density as determined by some observers is timelike). This second sense trades on an important distinction between *some* observers assigning positive mass density and *all* observers assigning positive mass density. One might have thought that in order for a matter field to be massive, it would be sufficient if some observers—say, co-moving observers making determination of "rest mass density," when that makes sense—determine that the field has positive mass density. But the strengthened dominant energy condition requires considerably more than this. In geometrized Newtonian gravitation, meanwhile, all of these distinctions collapse. If anyone determines a matter field has positive mass, then everyone does.

A final remark is that, understood within the context of the respective theories, the strengthened dominant energy condition is a more surprising assumption to have to make than the mass condition. One often thinks of relativity theory as *forbidding* superluminal propagation of energy-momentum, in the sense that somehow the geometric structure of the theory renders superluminal energy-momentum incoherent. But here, at least, it seems that we need to rule out superluminal propagation of energy-momentum as an additional assumption in order to derive the geodesic principle. This point can be made precise by asking whether one can drop or weaken the energy condition in the Geroch–Jang theorem and still derive the geodesic principle. And the answer is "no." If one drops the energy condition altogether, it is possible to construct bits of matter that propagate along *any* timelike curve [32]. And if one weakens the energy condition to the weak energy condition or the dominant energy condition, one can construct bits of matter that propagate along spacelike or null curves, respectively [57]. To be sure, in the Newtonian case the mass condition is similarly necessary (the considerations offered in Weatherall [57] can be adapted to show that the weakened mass condition is not enough to get timelike geodesic motion in geometrized Newtonian gravitation), but this does not seem as striking, since one does not expect Newtonian gravitation to imply restrictions on the propagation of matter (or more specifically, mass-momentum), even if it is standard to assume that matter cannot propagate instantaneously in the theory.

This leaves the conservation condition, assumption (2) in both theorems. The statement of the assumption is identical in both cases, namely that the tensor fields representing matter must be divergence-free. And in both theories, this assumption is a way of capturing that the bits of matter must be *free* in the sense of noninteracting. This interpretation is justified because in both theories there is a standard background assumption that at every point of spacetime, *total* energy/mass-momentum must be divergence free, and more, that a particular energy/mass-momentum field fails to be divergence free at a point just in case it is interacting with some other such field at that point. And so, to say that a particular field satisfies the conservation condition everywhere is to say that that field cannot be exchanging energy/mass-momentum with any other fields.

So far, it would seem that these assumptions have precisely the same status in both theorems. But this is too quick. Although the assumptions are equally natural ways of capturing the desired sense of "free" in both cases, they only have that interpretation in the presence of the background assumption regarding the local conservation of total energy/mass-momentum. And there is an argument to be made that this background assumption has a different status in general relativity than in geometrized Newtonian gravitation. In general relativity, Einstein's equation implies the conservation condition, at least for total source matter. This is because the equation can be written as $8\pi T^{ab} = R^{ab} - \frac{1}{2}g^{ab}R$, and it is a brute geometrical fact (known as Bianchi's identity) that the right-hand side of this equation is always divergence free. Thus, the left-hand side must also be divergence-free.

The geometrized Poisson equation, however, does *not* imply the conservation condition. And so, if one has Einstein's equation lurking in the background, one might be inclined to say that the background assumption that matter is conserved comes for free in general relativity, whereas it is an additional brute assumption in geometrized Newtonian gravitation. There is an important caveat here—the argument that Einstein's equation implies the conservation condition only applies for source matter, whereas the geodesic principle is supposed to govern *test* matter, i.e., matter that may be neglected as a source in Einstein's equation. Nonetheless, one might think that the conservation condition has a special status—even, to anticipate the discussion in the next section, a privileged explanatory status—in general relativity because of its relation to Einstein's equation.[14]

---

[14]I should emphasize: one does not *need* to think of the conservation condition as having a different status in general relativity than in geometrized Newtonian gravitation. For instance, I have elsewhere argued that one can think of the conservation condition as a meta-principle, in the sense that the

In the next section, I will turn to the question of whether either of these theorems should count as explanations of inertial motion. But before I do so, it will be helpful to sum up the discussion in the present section. I have now made precise the sense in which one can prove the geodesic principle as a theorem of both general relativity and geometrized Newtonian gravitation. But, as I hope has become clear, interpreting and comparing these theorems is quite subtle. It is not quite right to say that the theorems have the same interpretation or significance: on the one hand, there is arguably a sense in which the conservation condition, necessary for both theorems, has a different and perhaps privileged status in general relativity; and on the other hand, there are several senses in which the energy condition required for the Geroch–Jang theorem is stronger than the condition required for the Newtonian theorem, both in absolute terms and relative to the respective theories. Despite these differences, however, there is at least one important sense in which the status of the geodesic principle is strikingly similar in both theories. In both cases, one *can* prove the geodesic principle as a theorem. But to do so, one needs to make strong assumptions about the nature of matter. The status of these assumptions will play a central role in what follows.

4 Explaining Inertial Motion?

General relativity and geometrized Newtonian gravitation, like any physical theory, involve a number of basic assumptions and central principles. For instance, general relativity begins with some background assumptions about matter and geometry: space and time are represented by a four-dimensional, possibly curved Lorentzian manifold; matter is represented by its energy-momentum tensor, a smooth symmetric rank two tensor field on spacetime. One then adds some additional assumptions, as principles indicating how to interpret and use the theory. One may stipulate that total energy-momentum at a point must satisfy the conservation condition. One assumes that matter fields satisfy various possible energy conditions, that idealized clocks measure proper time along their trajectories, and that free massive test point particles traverse timelike geodesics. We postulate a dynamical relationship between the geometrical structure of spacetime and the energy-momentum field, and so on. Some of these assumptions involve stipulating kinematical structure; others involve basic constraints and dynamical relationships; others still tell us how to extract empirical content from the theory. All of them have some claim to centrality or fundamentality in the theory.

But they are not necessarily independent. For instance, as I mention above, the conservation condition may be understood as a consequence of Einstein's equation,

---

(Footnote 14 continued)

assumption that matter is conserved is expected to hold true in a wide variety of theories, and that from this perspective the status of the assumption is much the same in both general relativity and geometrized Newtonian gravitation [56]. (Of course, the assumption that a matter field is divergence free is not exactly the same as the assumption that total mass or energy is constant over time, but it does deserve to be called the relativistic version of traditional conservation principles.)

at least for source matter. And so, at least in some contexts, one might want to think of the conservation condition as somehow subordinate to Einstein's equation. One might even be inclined to say that it is Einstein's equation that really deserves to be called the "fundamental principle," while the conservation condition has some other, less fundamental status—or, in other words, that Einstein's equation *explains* why matter is locally conserved. We might even say that this is what it *means* to say that something like the conservation condition is explained by a theory: it can be derived from more fundamental principles in the theory.

From this point of view, one might have thought that when Einstein, Eddington, and others have claimed that general relativity explains inertial motion, in the sense that one can prove the geodesic principle as a theorem, the claim would have been analogous to what I have just said about the conservation condition: namely, one can take some collection of other principles of the theory and use them to derive the geodesic principle. One might then think that the geodesic principle has the same subordinate status as the conservation condition. It may be central to the theory, but not truly fundamental. The fundamental principles are the ones that go into proving the geodesic principle. On this view, one thinks of the foundations of general relativity as a two-tiered system. On the top tier are the truly fundamental principles; on the lower tier are the other central principles that can be derived from the top-tier principles. Initially, perhaps, one thought that the geodesic principle and conservation condition were top-tier principles; but the Geroch–Jang theorem and Bianchi's identity show that they are really second-tier principles.[15]

Thinking this way can lead to problems, however. The main moral of the last section was that although one can prove the geodesic principle as a theorem in both general relativity and geometrized Newtonian gravitation, to do so requires strong assumptions about the nature of matter. And so, if we want to move the geodesic principle to the lower tier, it would seem that we need to understand these assumptions as top-tier principles. But this raises a question: why should we think of *these* principles as the truly fundamental ones? Or more specifically, why should we think of the conservation condition and the respective energy conditions as more fundamental than the geodesic principle itself?

If one were committed to the idea that the geodesic principle is a second-tier principle in one or both of these theories, perhaps one would be willing to include the assumptions needed to prove the geodesic principle among the truly fundamental principles of that theory. But it is hard to see how this is an appealing move on independent grounds. Even if one were to argue that dynamical principles such as Einstein's equation and the geometrized Poisson equation are clearly more fundamental than the geodesic principle, it remains the case that the strong energy condition needed to prove the Geroch–Jang theorem is entirely independent of Einstein's equation. (And neither the conservation condition nor the mass condition follows from the geometrized Poisson equation.)

---

[15]Indeed, it seems Einstein originally *did* think of the conservation condition as a top-tier principle, in the sense that he thought it was an independent assumption that any realistic field equation would need to be compatible with. See Earman and Glymour [15, 16].

More, there is a sense in which one can draw all of the inferential arrows in the opposite direction, at least in one important case. Consider an energy/mass-momentum field of the form $T^{ab} = \rho \xi^a \xi^b$, for some smooth scalar field ρ and smooth vector field ξ^a. An energy/mass-momentum field of this form is the natural way of representing a matter field composed of mutually noninteracting massive point particles (at least when ρ is nonnegative). And so, since the geodesic principle governs the behavior of free massive test point particles, we can use it to derive features of this matter field: specifically, the geodesic principle implies that the flow lines of the field, which represent the trajectories of each speck of dust, must be timelike geodesics. These flow lines are just the integral curves of ξ^a, and so it follows that ξ^a must be timelike and geodesic (i.e., $\xi^n \nabla_n \xi^a = \mathbf{0}$). But if ξ^a is timelike, then T^{ab} satisfies the strengthened dominant energy condition (or respectively, the mass condition in geometrized Newtonian gravitation). And if it is geodesic, then T^{ab} is divergence free. Thus the geodesic principle allows us to derive that matter fields consisting of noninteracting massive test point particles satisfy precisely the two conditions we need to assume in order to prove the geodesic principle.[16]

So perhaps we should not be so quick to declare the conservation condition and energy conditions top-tier in either theory. At the very least, it is not perfectly clear that these assumptions are more fundamental than the geodesic principle. But thinking in this way might lead one to conclude that *neither* of the geodesic principle theorems has much explanatory significance, since (the intuition might go) explanations always proceed from more fundamental or basic facts to less fundamental facts. Here, meanwhile, the arrows of fundamentality are muddled. And this would mean that not only is general relativity not special with regard to its explanation of inertial motion—it does not explain inertial motion at all!

5   The Puzzleball Conjecture

I do not find the argument I offer in the previous section compelling. It rests on a basic intuition: to explain something like the geodesic principle, one must begin with some truly fundamental principles and then provide an argument for why the principle-to-be-explained must follow from these more fundamental ones. This intuition takes for granted that we can make sense of a distinction between different tiers of fundamentality among the central principles of a theory like general relativity or geometrized Newtonian gravitation. And I think that this is a mistake—or at least, that there is a more compelling way of thinking about things.

Consider what the geodesic principle theorems *do* accomplish. In both of these cases, the theorems show how in the presence of other basic assumptions of the respective theories, the geodesic principle follows. Or in other words, they show that given that one is committed to the rest of (say) general relativity, one must

---

[16]Of course, the present argument does not imply that the conservation condition and energy conditions hold for *all* matter—just for the type of matter directly governed by the geodesic principle.

Fig. 2 One way of thinking about the foundations of physical theories would have it that some of the central principles of a theory have a distinguished status as the "truly fundamental" principles. An alternative view, which I describe and advocate here, is that the foundations of a theory are better thought of as a network of mutually interdependent principles, interlocking like the pieces of a spherical puzzle. On this view, one would tend to expect that any of the central principles of a theory should be derivable from the rest of the theory with that principle removed, much like the overall shape of a puzzleball constrains the shape of any individual piece

also be committed to the geodesic principle. One cannot freely change the geodesic principle without also changing the rest of general relativity: one cannot "fiddle" with the theory by (merely) replacing the geodesic principle with the assumption that free massive test point particles traverse some other class of curves—uniformly accelerating curves, say, or spacelike curves. The geodesic principle is not modular, in the sense that one cannot construct a collection of perfectly good theories that differ only in how they treat inertial motion. More, the theorems clarify precisely how it is that the geodesic principle "fits in" among the other central principles of general relativity.

It seems to me that these reflections suggest a proposal. Instead of thinking of the foundations of a physical theory as consisting of a collection of essentially independent postulates from which the rest of the theory is derived, one might instead think of the foundations of a theory as consisting of a network of mutually interdependent principles—a collection of interlocking pieces, as in the spherical puzzle in Fig. 2.[17] The idea is that, as with the geodesic principle, one should generally expect

---

[17]Feynman [21] makes a distinction between two ways of understanding physical theories that is similar to the one I make here. On the "Greek" view of theories, one begins with a collection of fixed fundamental axioms or postulates. Feynman does not like this way of thinking about theories. Instead, he endorses the "Babylonian" view, on which one observes that the principles of a theory are more richly connected: perhaps it is sometimes convenient to take certain principles of a theory as axioms and others as theorems, but one needs to recognize that in other cases one might want to switch this around and think of your theorems as the axioms, and use them to prove your former axioms. He then observes that "If all these various theorems are interconnected by reasoning there is no real way to say 'These are the most fundamental axioms,' because if you were told something different instead you could also run the reasoning the other way. It is like a bridge with lots of

that many of the central principles of a physical theory may be proven as theorems, given the rest of the theory. Trying to make a distinction between the top-tier principles and the second-tier principles of a theory is not fruitful, then, since most, or even all, of the principles can be understood equally well as either postulate or theorem, and indeed, in different contexts it may well be desirable to think of them in different ways. Importantly, theories are not modular, in the sense described above. We cannot simply replace any given principle with some other one, at least not without changing the rest of the theory in possibly dramatic ways. And theorems such as the Geroch–Jang theorem and its Newtonian counterpart are of interest because they exhibit the details of these interdependencies. They show just how the pieces interlock.

To be sure, nothing I have said thus far should count as an apodictic argument for the view I have described (call it the "puzzleball view"). Nor will I give such an argument—indeed, I am not sure what an argument for the claim that there are no truly fundamental principles of general relativity would look like. Instead, I merely offer the view as an alternative way to think of the kinds of interrelations between the principles of physical theories on display with theorems like the Geroch–Jang theorem. Perhaps the proposal is best conceived as a conjecture, albeit one with some compelling early evidence, for the following reason: while the senses in which, for instance, the conservation condition and the geodesic principle follow from other standard assumptions of general relativity are now established, the senses in which other principles, such as Einstein's equation or various energy conditions, are derivable from or constrained by the rest of the theory are less clear.[18] And so we have the skeleton of a mathematical question: are *all* of the central principles of relativity theory and geometrized Newtonian gravitation (or other theories still) indeed mutually interderivable in the way that I have suggested? Most or many of them? Or are the geodesic principle and conservation condition anomalies?

Some important work has already been done on this topic: Dixon [12] has shown a sense in which the geometrized Poisson equation is the unique dynamical principle compatible with a collection of natural assumptions in Newtonian gravitation; similarly, Sachs and Wu [43] and Reyes [42] have shown that there is a sense in which the (vacuum form of) Einstein's equation can be derived from (in effect) the geodesic principle, among other assumptions, and Lovelock [30, 31], Navarro and Sancho [37], and Curiel [8] have argued for various senses in which the Einstein tensor is the unique tensor that can appear on the left-hand side of Einstein's equation, even in the non-vacuum case.

Meanwhile, Duval and Künzle [13] and Christian [7] have argued that even though the conservation condition in geometrized Newtonian gravitation does not follow from the geometrized Poisson equation, one can nonetheless derive it from other principles, at least if one considers Lagrangian formulations of the theory. One might

---

(Footnote 17 continued)

members, and it is overconnected; if pieces have dropped out you can reconnect it another way" (pg. 46). The view I describe here is firmly in Feynman's Babylonian tradition. I am grateful to Bill Wimsatt for pointing out this connection.

[18]Indeed, it would seem that there are no known nontrivial derivations of energy conditions from other central principles of relativity theory. See Curiel [9].

even understand Newton's argument for universal gravitation as a kind of heuristic argument that the inverse square law of gravitation is the unique dynamical principle compatible with certain other central principles of standard Newtonian gravitation (including generalized empirical facts, such as Kepler's Harmonic Law). Results such as these provide tentative evidence for my basic hypothesis, that one should expect many or all of the central principles of these spacetime theories to be mutually interdependent.

But I do not think these results are yet conclusive. Specifically, what has not been done is to systematically study such results in order to try to characterize, in the way that has now been done for the geodesic principle theorems, (1) just what the assumptions going into these theorems are, (2) how natural the assumptions are in the contexts of the relevant theories, and (3) how these assumptions, in turn, depend on the other central principles of the theories (if they do). And the attractiveness of the proposal presented here turns on the answers to these questions. It is only after a project of this form has been carried out that one can fully evaluate whether the central principles of these theories are really as tightly intertwined as the puzzleball view would have it.

That said, if a careful study of this sort reveals that only some of the central principles of a theory are interconnected, it may still be fruitful to think about the foundations of theories in the way I propose here, since the discovery that some central principles of a theory (say, energy conditions in general relativity) are more peripheral than others need not imply that one can make sense of a unique or privileged collection of the most fundamental or basic principles. Much will depend on just what the structure of the situation turns out to be.

It is worth emphasizing that mapping out these kinds of relations among the central principles of a physical theory is of some independent interest, since understanding the extent to which the central principles of general relativity in particular are mutually interdependent could play an important role in the construction of future theories (and in some ways, it already has).[19] The reason has to do with the idea of "fiddling" with physical theories. There is a long tradition of attempting to modify general relativity with small changes: for instance, in Brans–Dicke theory, one modifies Einstein's equation to include an additional scalar field; in TeVeS gravitational theories, one also considers vector fields. In still other cases, one modifies general relativity by allowing derivative operators with torsion. In each of these examples (and many others), one makes what appears to be a local change in the central principles of general relativity.

---

[19]Feynman makes a related point about the practical importance of his Babylonian approach to theories. He writes, "If you have a structure that is only partly accurate, and something is going to fail, then if you write it with just the right axioms maybe only one axiom fails and the rest remain, you need only change one little thing. But if you write it with another set of axioms they may all collapse, because they all lean on that one thing that fails. We cannot tell ahead of time, without some intuition, which is the best way to write it so that we can find out the new situation. We must always keep all the alternative ways of looking at a thing in our heads; so physicists do Babylonian mathematics, and pay but little attention to the precise reasoning from fixed axioms" [21, pg. 54].

But these small changes can have dramatic consequences: for instance, in Einstein–Cartan theory, a modification of general relativity with torsion, the conservation condition does not generally hold. One does not have a geodesic principle, at least in the ordinary sense, since in general the collection of self-parallel curves picked out by the derivative operator do not agree with the collection of extremal curves picked out by the metric, and free massive test point particles need not propagate along either class of curves. Thus, apparently small tweaks can lead to a dramatically different theory, conceptually speaking. A clearer picture of just how the central principles of general relativity *do* fit together and constrain one another may provide important clarity into just what the consequences of these "small" modifications to the theory are, and more, may help guide us in the search for alternative theories of gravitation, by indicating which principles are more or less tightly connected to which others. Indeed, for this reason there is a sense in which the situation I describe above, where some principles are very tightly interlocking and others turn out to be more loosely connected (for instance, some principles play a role as assumptions in some theorems, but cannot be proved in complete generality themselves) is the most interesting from the practical perspective of mapping out the space of possible future theories.

In the next section, I will return to the question of explanation, now from the perspective of the present view. But before I do so, I want to clarify the puzzleball view slightly, as the language I have used to describe it may call to mind two other well-known ideas. It seems to me that the view I have described is distinct from both. First, note that the present proposal involves a picture of theories on which one emphasizes the ways in which the principles cohere with one another. This way of thinking may be reminiscent of coherentism in epistemology, a variety of anti-foundationalism that holds that to justify a belief is to show how it coheres with one's other beliefs (cf. [29]). But there is at least one major difference. Coherentism takes the coherence of one's beliefs to be a form of justification for those beliefs. Nothing about the puzzleball view should be taken to suggest that the justification for general relativity comes from the apparent fact that one can derive certain central principles from others— rather, the justification for the theory is based on its empirical successes. Or perhaps more precisely, our justification for general relativity is essentially independent of the relationship between the theory's central principles. To see the point most clearly, one might well expect both general relativity and geometrized Newtonian gravitation to be coherent, in the sense of having mutually interdependent central principles. But this does not imply that they are equally well justified—indeed, general relativity is better justified than geometrized Newtonian gravitation even if the pieces of geometrized Newtonian gravitation are more tightly interlocking.[20]

---

[20]The suggestion of a connection to coherentism raises a second, related issue. Even if we do not take the coherence of a body of beliefs as justification for any particular belief, one might nonetheless think of coherence as a virtue for a body of beliefs: all else being equal, one might tend to prefer to hold coherent beliefs than not. Should one say the same thing about physical theories? All else being equal, should one prefer a theory whose pieces interlock? I am not sure that anything in the body of the paper depends on this, but I am inclined to say "yes," for several reasons. First, as I argued above, when the central principles of theories are (partially) mutually interdependent,

Another view that the puzzleball view may be reminiscent of is some variety of Quinean holism (cf. [40]). Quine famously used the "web of belief" metaphor when arguing for the interdependencies of our scientific beliefs, and against the analytic/synthetic distinction. One might worry that the puzzleball picture above is just an alternative metaphor used to make a strikingly similar point—indeed, the claim that we cannot make a fruitful distinction between top-tier and second-tier principles sounds like an argument against an analytic/synthetic distinction, at least in the narrow domain of the foundations of certain physical theories. And perhaps it is right that I have recapitulated Quine here, though if it is, I think the point deserves to be made again since it is relevant for the present discussion of the geodesic principle. Still, while this article is not the occasion for detailed Quine exegesis, I will point to two ways in what I have proposed is *prima facie* different from Quine's holism, at least on the web-of-belief version.[21]

The first difference concerns just what the holism is supposed to be doing. Quine uses the interdependencies between beliefs as an argument for a radical form of conventionalism: when faced with evidence that conflicts with our beliefs, we have considerable leeway in choosing which parts of the web of beliefs to revise. Indeed, the web image is supposed to support a distinction between "central" or "core" beliefs and "peripheral" beliefs such that we can always accommodate challenges to our full collection of beliefs by modifying only the peripheral beliefs and leaving the core

---

(Footnote 20 continued)

the theory provides a guide for the building of future related theories in a way that may be helpful for scientific practice. Second, principles that are mutually interdependent are protected against claims of being *ad hoc*. A particular principle cannot be considered arbitrary or unmotivated if it is derivable, perhaps in multiple ways, from one's other principles. To put this point in a more experimentally oriented way, if the pieces of a theory are mutually interdependent, then testing any one principle can be understood as an implicit test of the other principles of a theory [25]. A third reason comes from Wimsatt [58], who argues that interderivability (or rather, multiple interderivability) is an indication of theoretic robustness and confers a kind of stability under theory change.

[21] My goal in the text is to distinguish the puzzleball view from web-of-belief holism. But this should not be taken to imply that Quine does not come much closer to the puzzleball view in other parts of his opus. For instance, Quine [41, Sec. V] distinguishes "legislative postulates" from "discursive postulates." "Legislative postulation," he writes, "institutes truth by convention ..." whereas "... discursive postulation is mere selection, from some preëxisting body of truths, of certain ones for use as a basis from which to derive others, initially known or unknown" [41, pg. 360]. He then goes on to argue that "conventionality is a passing trait, significant at the moving front of science but useless in classifying the sentences behind the lines. It is a trait of events and not of sentences." In other words, one might, when first developing a new scientific theory, begin with some bare, legislative postulates. But as the theory develops, these truths "... become integral to the corpus of truths; the artificiality of their origin does not linger as a localized quality, but suffuses the corpus. If a subsequent expositor singles out those once legislatively postulated truths again as postulates, this signifies nothing; he is engaged only in discursive postulation. He could as well choose his postulates from elsewhere in the corpus, and will if he think this serves his expository ends" [41, pg. 362]. The idea, I take it, is that once one has a well-developed scientific theory—such as general relativity—one often identifies postulates for the purposes of deriving new facts about the theory, but these are always discursive, and more, which facts or statements of the theory one will take to be the postulates in any given case will depend on one's purposes. This picture seems quite close to the puzzleball view, indeed. I am grateful to Pen Maddy for pointing out this connection to me.

beliefs intact. But this is precisely the opposite of what I have argued here, at least with regard to the foundations of spacetime theories. Instead, the idea is supposed to be that the foundations of physical theories are *not* modular, and that in general one has remarkably little latitude in how one revises a theory in light of new evidence. And this, I take it, is a desirable feature, since it provides a way out of the radical conventionalism I just described. Since the various principles of a physical theory constrain one another, we have very few degrees of freedom for enacting minor changes in theories in light of new evidence.

The second difference is related (and relates, too, to coherentism as described above). Quine's web of belief is supposed to be a (descriptive) metaphor for the sum total of one's beliefs. The view I have described here is much narrower in its scope. I do not claim that all of one's beliefs interlock in the way described; nor do I claim that scientific knowledge as a whole can be characterized by a puzzleball. The view does not even hold that particular scientific theories have this feature. The suggestion is that the *central principles* of some scientific theories are mutually interderivable, or in other words, that the foundations of some physical theories should be thought of in a certain way. I have been deliberately vague about just what is supposed to count as a central principle, in large part because I think that trying to list these principles in advance, even for well-understood theories such as general relativity or geometrized Newtonian gravitation, would be unproductive. In fact, one might expect that a full account of just what the central principles of a theory are may have to wait until one sees just what assumptions are necessary nodes when trying to map out the network of interconnected principles at the heart of a given physical theory. What I have done so far—and what I think can be done at this stage—is give examples of central principles of particular theories. And so one can say that among the central principles of general relativity, for instance, are things such as the conservation condition, Einstein's equation, the geodesic principle, and various energy conditions. But the point is that a claim about a collection of principles of this specific character is quite different from a claim about human knowledge quite broadly.

Note that this last point means that there is still a robust sense in which one can think of some parts of a theory as having a special "fundamental" status, even on the puzzleball view. Specifically, one might take *all* of the central principles of a theory to be fundamental. This leaves quite a bit of a theory as non-fundamental—for instance, particular predictions of a theory would not be among the central principles, and so these would not count as fundamental. If the puzzleball view is to be viewed as anti-foundational, then, it is only with regard to determinations of relative fundamentality *among* the central principles of a theory.[22]

---

[22]Feynman, and Wimsatt [58], argue that in cases where some principles can be proved in many different ways and others cannot be proved or can be proved from fewer starting cases, one can recover a different sense of "fundamental" principles, namely that the principles that can be proved in the most different ways should be understood as the most fundamental. Note that this turns the idea discussed above—where the most fundamental principles were the top-tier principles from which other principles would be derived, not the ones most often derived themselves—on its head. This idea is intriguing, but I mention it only to set it aside as it plays no role in the present discussion.

6 Explaining Inertial Motion, Redux

Now that I have described the puzzleball view in some detail, I can return to the question of principle interest in this paper: namely, is there a sense in which we should understand the Geroch–Jang theorem and its Newtonian counterpart as explanations? As a first remark, let me reiterate that in the context of the puzzleball view, it does not make sense to think of theories as based on "top-tier" principles and other, derived principles: in short, there is no way to make the distinction, at least among the central principles of the theory. None of the assumptions of a theory are distinguished as the truly basic or fundamental ones. And if this is right, then the kind of explanation that we apparently cannot get of the geodesic principle in general relativity and in geometrized Newtonian gravitation is uninteresting. No, we cannot derive the geodesic principle in either theory from more fundamental principles, but that is because it does not make sense to talk of unambiguously "more fundamental" principles in the first place.

Instead, what we can do is show how the geodesic principle in both of these theories fits into the rest of the puzzle (as it were). This, too, may be understood as an answer to the question, "Why do bodies move in the particular way that they do in the absence of an external force?" These theorems reveal that in the absence of an external force, in the context of their respective theories, bodies *must* move along timelike geodesics. In other words, the other basic assumptions of the theory constrain the motion of (small) bodies. Why timelike geodesic motion rather than any other? Because in general relativity, we understand matter to be conserved, and to be such that observers always attribute instantaneous subluminal velocities to it at every point. And it turns out that these assumptions, in the presence of the rest of the theory, imply that the only curves along which free massive test point particles can propagate are timelike geodesics. If we are committed to the rest of general relativity, then there is only one candidate principle for inertial motion.

So do general relativity and/or geometrized Newtonian gravitation explain inertial motion? Given the considerations just mentioned, I think the answer in both cases is "yes," so long as one understands "explain" in the right way. At the very least, these theorems provide deep insight and understanding into why bodies move in the particular way that they do in the absence of any external force—which is precisely what we were after when we asked the question. Moreover, the insight provided is that, in the context of the other central principles of the theories, the geodesic principle is *necessary*, the only principle governing inertial motion that is compatible with our other principles. It is a demonstration of precisely the ways in which the working parts of general relativity and geometrized Newtonian gravitation respectively constrain one another.

That said, this kind of explanation differs in some important ways from other kinds of explanations that one may be accustomed to thinking about. In particular, if the puzzleball view is correct, the kind of explanation I have just described need not be asymmetrical. That is, if general relativity might be said to explain inertial motion in the present sense by appealing to the fact that one can derive the geodesic principle from various other assumptions in the theory, one should *not* conclude that the geodesic principle cannot play a role in other derivations that should also count as explanatory—even derivations of the assumptions going into the Geroch–Jang theorem or its Newtonian counterpart. Indeed, one should expect that just as the geodesic principle is constrained by the other central assumptions of general relativity, so too are the conservation condition, the strengthened dominant energy condition, and even Einstein's equation constrained. And by the same reasons I have offered above for the view that one might justly call the Geroch–Jang theorem an explanation of inertial motion, one might also say that explanations can be given for the conservation condition or Einstein's equation, by showing how these principles of general relativity are derivable from the other central principles of the theory. In other words, general relativity explains inertial motion by appeal to Einstein's equation, but it may equally well explain Einstein's equation by appeal to the geodesic principle and other central assumptions of general relativity.

This observation may give some readers pause. There is, by now, a long tradition of philosophers of science worrying about the so-called "problem of explanatory asymmetry" (cf. [3]): intuitively, explanations appear to run in one direction and only one direction. The trajectory of a comet may explain why we see a bright light in the nighttime sky once every few hundred years, but a bright light in the nighttime sky cannot explain the trajectory of a comet; the height of a flagpole may explain the length of its shadow at sunset, but the length of the shadow does not explain the height of the flagpole. And so, many philosophers have argued, an account of explanation that allows symmetrical explanations—situations where A explains B and B explains A—is *prima facie* unacceptable.

A few remarks are in order. First, van Fraassen [53] has argued, I think correctly, that explanation should be understood as essentially pragmatic—and in particular that explanations should only be understood as responses to certain classes of question. To determine whether or not some particular response to a why question (say) should count as a satisfactory explanation depends on the context in which the question was asked and the particular demands of the questioner. While in some contexts we might want to say that a particular explanation runs in only one direction, there may well be other contexts in which the explanation would run in the other direction. If one is thinking in this way then the present example is simply a special case: if the question "why do bodies move in the particular way that they do in the absence of an external force?" is understood as "does general relativity require us to adopt the geodesic principle as the central principle governing inertial motion?", then one is rightly satisfied by a response along the lines of the Geroch–Jang theorem, even if in other contexts—i.e., in response to other questions—one might appeal to the geodesic principle to explain (say) Einstein's equation.

But there is also a more important point to make, here: I do not claim to be offering an "account of explanation," or anything like it. I have not suggested that a necessary or even sufficient condition for being an explanation is to show how the thing to be explained "fits in" with the rest of a physical theory, in the sense that it is derivable from other central principles of a theory. The point, rather, is to try to spell out the sense in which a particular class of theorems that show how the central principles of a spacetime theory fit together might be understood as explanatory—to say what, precisely, the theorems are doing, and why one might think of this as a kind of explanation, at least on the puzzleball view. Not all explanations work this way, nor do they need to in order for the story I have told here to be correct. And so, the fact that in *some* cases, we would want to say that if A explains B then B cannot explain A in no way undermines the claim that the present explanations simply do not work that way.

This point can be made most starkly by pointing to various other questions one can ask about inertial motion, even in general relativity, whose answers would be quite different from the Geroch–Jang theorem. Consider, for example, a question concerning a particular instance of inertial motion. Why, one might ask, does the perihelion of Mercury's orbit precess? One would answer this by appealing to some particular initial state of Mercury and features of spherically symmetric solutions to Einstein's equation to show that Mercury's orbit is the only allowed trajectory for a body with certain properties in a solar system like ours. The geodesic principle may play a role in this argument, insofar as one might idealize Mercury as a free massive test point particle, and Einstein's equation may play a role, insofar as one would want to consider a spacetime that is a solution to the equation, but the argument would have nothing to do with the Geroch–Jang theorem. And moreover, one would expect this sort of explanation to be asymmetric: Einstein's equation and the geodesic principle, along with some details concerning the state of the solar system and initial conditions for Mercury, explain the precession of the perihelion of Mercury; the precession of the perihelion of Mercury does not explain Einstein's equation or the geodesic principle.

But this is just the point. If the Geroch–Jang theorem and its Newtonian counterpart should be countenanced as explanations, it is only because they are satisfactory answers to particular questions, and they are only explanatory in the context of *those* demands for explanation. A question concerning the orbit of Mercury is quite different from a question concerning the nature of inertial motion generally. And these theorems answer only the most general version of the question: why *this* principle as opposed to any other? This is no mean task, but it is a specific one, and it needs to be treated with care.

Acknowledgements Versions of this work have been presented at CEA-Saclay (Paris), University of Wuppertal, University of Pittsburgh, University of Chicago, University of Texas at Austin, Columbia University, Brown University, New York University, University of Western Ontario (twice!), University of California–Berkeley, University of California–Irvine, Yale University, University of Southern California, Notre Dame University, Ludwig-Maximilians University, and at the 'New Directions' conference in Washington, DC. I am grateful to very helpful comments and discussion from all of these audiences, and particularly to (in no special order) John Manchak, Giovanni Valente, Craig Callender, Alexei Grinbaum, Harvey Brown, David Wallace, Chris Smeenk, Wayne Myrvold, Erik Curiel, Ryan Samaroo, John Norton, John Earman, Howard Stein, Mike Tamir, Bryan Roberts, Shelly Kagan, Sahotra Sarkar, Josh Dever, Bob Geroch, Bill Wimsatt, David Albert, Tim Maudlin, Sherri Roush, Josh Schechter, Chris Hill, Don Howard, Katherine Brading, Eleanor Knox, and Radin Dardashti. Special thanks are due to David Malament, Jeff Barrett, Kyle

Stanford, and Pen Maddy for many helpful discussions and comments on previous versions of this work. Thank you, too, to Dennis Lehmkuhl for organizing the 2010 workshop on which this volume is based, and for editing the volume.

References

1. Batterman, R., 2002. The Devil in the Details. Oxford University Press, New York.
2. Blanchet, L., 2000. Post-Newtonian gravitational radiation. In: Schmidt, B. (Ed.), Einstein's Field Equations and Their Physical Implications. Springer, Berlin, pp. 225–271.
3. Bromberger, S., 1966. Why-questions. In: Brody, B. A. (Ed.), Readings in the Philosophy of Science. Prentice Hall, Inc., Englewood Cliffs, pp. 66–84.
4. Brown, H., 2005. Physical Relativity. Oxford University Press, New York.
5. Cartan, E., 1923. Sur les variétés à connexion affine, et la théorie de la relativité généralisée (première partie). Annales scientifiques de l'École Normale Supérieure 40, 325–412.
6. Cartan, E., 1924. Sur les variétés à connexion affine, et la théorie de la relativité généralisée (première partie) (suite). Annales scientifiques de l'École Normale Supérieure 41, 1–25.
7. Christian, J., 1997. Exactly soluble sector of quantum gravity. Physical Review D 56 (8), 4844 –4877.
8. Curiel, E., 2012. On tensorial concomitants and the non-existence of a gravitational stress-energy tensor, available at: http://arxiv.org/abs/0908.3322v3.
9. Curiel, E., 2017. Towards a theory of spacetime theories. Birkauser, Boston, Ch. A Primer on Energy Conditions.
10. Damour, T., 1989. The problem of motion in Newtonian and Einsteinian gravity. In: Hawking, S. W., Israel, W. (Eds.), Three Hundred Years of Gravitation. Cambridge University Press, New York, pp. 128–198.
11. Dixon, W. G., 1964. A covariant multipole formalism for extended test bodies in general relativity. Il Nuovo Cimento 34 (2), 317–339.
12. Dixon, W. G., 1975. On the uniqueness of the Newtonian theory as a geometric theory of gravitation. Communications in Mathematical Physics 45, 167–182.
13. Duval, C., Künzle, H. P., 1978. Dynamics of continua and particles from general covariance of Newtonian gravitation theory. Reports on Mathematical Physics 13 (3).
14. Earman, J., Friedman, M., 1973. The meaning and status of Newton's law of inertia and the nature of gravitational forces. Philosophy of Science 40, 329.
15. Earman, J., Glymour, C., 1978. Einstein and Hilbert: Two months in the history of general relativity. Archive for History of Exact Sciences 19 (3), 291–308.
16. Earman, J., Glymour, C., 1978. Lost in the tensors: Einstein's struggles with covariance principles 1912–1916. Studies in the History and Philosophy of Science 9 (4), 251–278.
17. Eddington, A. S., 1924. The Mathematical Theory of Relativity. Cambridge University Press, Cambridge.
18. Ehlers, J., Geroch, R., 2004. Equation of motion of small bodies in relativity. Annals of Physics 309, 232–236.
19. Einstein, A., Grommer, J., 1927. Allgemeine Relativitätstheorie und Bewegungsgesetz. Verlag der Akademie der Wissenschaften, Berlin.
20. Einstein, A., Infeld, L., Hoffman, B., 1938. The gravitational equations and the problem of motion. Annals of Mathematics 39 (1), 65–100.
21. Feynman, R. C., 1967. The Character of Physical Law. Cornell University Press, Ithaca, NY.
22. Friedrichs, K. O., 1927. Eine invariante Formulierung des Newtonschen Gravitationsgesetzes und der Grenzüberganges vom Einsteinschen zum Newtonschen Gesetz. Mathematische Annalen 98, 566–575.
23. Geroch, R., Jang, P. S., 1975. Motion of a body in general relativity. Journal of Mathematical Physics 16 (1), 65.

24. Geroch, R., Weatherall, J. O., 2017. Equations of motion. Unpublished manuscript.
25. Harper, W. L., 2012. Isaac Newton's Scientific Method: Turning Data into Evidence about Gravity and Cosmology. Oxford University Press, New York.
26. Havas, P., 1989. The early history of the 'problem of motion' in general relativity. In: Howard, D., Stachel, J. (Eds.), Einstein and the History of General Relativity. Vol. 11 of Einstein Studies. Birkhäuser, Boston, pp. 234–276.
27. Hawking, S. W., Ellis, G. F. R., 1973. The Large Scale Structure of Space-time. Cambridge University Press, New York.
28. Kennefick, D., 2005. Einstein and the problem of motion: A small clue. In: Kox, A. J., Eisenstaedt, J. (Eds.), The Universe of General Relativity. Vol. 11 of Einstein Studies. Birkhäuser, Boson, pp. 109–124.
29. Kvanvig, J., 2007. Coherentist theories of epistemic justification. In: Zalta, E. N. (Ed.), The Stanford Encyclopedia of Philosophy. Stanford University, Stanford, CA, available at: http://plato.stanford.edu/entries/justep-coherence/.
30. Lovelock, D., 1971. The Einstein tensor and its generalizations. Journal of Mathematical Physics 12 (3), 498–501.
31. Lovelock, D., 1972. The four-dimensionality of space and the Einstein tensor. Journal of Mathematical Physics 13 (6), 874–876.
32. Malament, D., 2012. A remark about the "geodesic principle" in general relativity. In: Frappier, M., Brown, D. H., DiSalle, R. (Eds.), Analysis and Interpretation in the Exact Sciences: Essays in Honour of William Demopoulos. Springer, New York, pp. 245–252.
33. Malament, D. B., 2012. Topics in the Foundations of General Relativity and Newtonian Gravitation Theory. University of Chicago Press, Chicago.
34. Mathisson, M., 1931. Bewegungsproblem der feldphysik und elektronenkonstanten. Zeitschrift für Physik 69, 389Â408.
35. Mathisson, M., 1931. Die mechanik des materieteilchens in der allgemeinen relativitätstheorie. Zeitschrift für Physik 67, 826–844.
36. Misner, C. W., Thorne, K. S., Wheeler, J. A., 1973. Gravitation. W. H. Freeman.
37. Navarro, J., Sancho, J., 2008. On the naturalness of Einstein's equation. Journal of Geometry and Physics 58, 1007–1014.
38. Newman, E. T., Posadas, R., 1969. Motion and structure of singularities in general relativity. Physical Review 187 (5), 1784–1791.
39. Newman, E. T., Posadas, R., 1971. Motion and structure of singularities in general relativity, ii. Journal of Mathematical Physics 12 (11), 2319–2327.
40. Quine, W. V. O., 1951. Two dogmas of empiricism. The Philosophical Review 60, 20–43.
41. Quine, W. V. O., 1960. Carnap and logical truth. Synthese 12 (4), 350–374.
42. Reyes, G. E., 2009. A derivation of Einstein's vacuum field equations, available at: http://marieetgonzalo.files.wordpress.com/2009/12/a-derivation-of-einsteins-vacuum-field-equations1.pdf.
43. Sachs, R. K., Wu, H., 1973. General Relativity for Mathematicians. Springer-Verlag, New York.
44. Sauer, T., Trautman, A., 2008. Myron Matthison: What little we know of his life, http://arxiv.org/abs/802.2971.
45. Souriau, J.-M., 1974. Modèle de particule à spin dans le champ électromagnétique et gravitationnel. Annales de l'Institut Henri Poincaré Sec. A 20, 315.
46. Stein, H., 1967. Newtonian space-time. The Texas Quarterly 10, 174–200.
47. Sternberg, S., 2003. Semi-riemannian geometry and general relativity, available at: http://www.math.harvard.edu/~shlomo/.
48. Sus, A., 2011. On the explanation of inertia, unpublished.
49. Tamir, M., 2011. Proving the principle: Taking geodesic dynamics too seriously in Einstein's theory, available online at http://philsci-archive.pitt.edu/8779/.
50. Taub, A. H., 1962. On Thomas' result concerning the geodesic hypothesis. Proceedings of the National Academy of the USA 48 (9), 1570–1571.

51. Thomas, T. Y., 1962. On the geodesic hypothesis in the theory of gravitation. Proceedings of the National Academy of the USA 48 (9), 1567–1569.
52. Trautman, A., 1965. Foundations and current problem of general relativity. In: Deser, S., Ford, K. W. (Eds.), Lectures on General Relativity. Prentice-Hall, Englewood Cliffs, NJ, pp. 1–248.
53. van Fraassen, B., 1980. The Scientific Image. Clarendon Press, Oxford.
54. Wald, R. M., 1984. General Relativity. University of Chicago Press, Chicago.
55. Weatherall, J. O., 2011. The motion of a body in Newtonian theories. Journal of Mathematical Physics 52 (3), 032502.
56. Weatherall, J. O., 2011. On the status of the geodesic principle in Newtonian and relativistic physics. Studies in the History and Philosophy of Modern Physics 42 (4), 276–281.
57. Weatherall, J. O., 2012. A brief remark on energy conditions and the Geroch-Jang theorem. Foundations of Physics 42 (2), 209–214.
58. Wimsatt, W. C., 1981. Robustness, reliability, and overdetermination. In: Brewer, M., Collins, B. (Eds.), Scientific Inquiry in the Social Sciences. Jossey-Brass, San Francisco, pp. 123–162.

A Primer on Energy Conditions

Erik Curiel

Abstract An energy condition, in the context of a wide class of spacetime theories (including general relativity), is, crudely speaking, a relation one demands the stress-energy tensor of matter satisfy in order to try to capture the idea that "energy should be positive". The remarkable fact I will discuss in this paper is that such simple, general, almost trivial seeming propositions have profound and far-reaching import for our understanding of the structure of relativistic spacetimes. It is therefore especially surprising when one also learns that we have no clear understanding of the nature of these conditions, what theoretical status they have with respect to fundamental physics, what epistemic status they may have, when we should and should not expect them to be satisfied, and even in many cases how they and their consequences should be interpreted physically. Or so I shall argue, by a detailed analysis of the technical and conceptual character of all the standard conditions used in physics today, including examination of their consequences and the circumstances in which they are believed to be violated.

I thank Harvey Brown and David Malament for enjoyable conversations and debate that first put me on the path to a thorough investigation of energy conditions. I thank The Young Guns of the Spacetime Church of the Angle Brackets for suffering through a longer version of this paper (!) and giving me insightful help with a smile. Finally, I thank two anonymous referees (one of whom is Dennis Lehmkuhl) for giving me thorough and very helpful feedback, including the correction of technical infelicities and errors.

E. Curiel (✉)
Munich Center for Mathematical Philosophy, Ludwig-Maximilians-Universität,
Ludwigstraße 31, 80539 Munich, Germany
e-mail: erik@strangebeautiful.com

E. Curiel
Black Hole Initiative, Harvard University, 20 Garden Street, Cambridge, MA 02138, USA

© Springer Science+Business Media, LLC 2017
D. Lehmkuhl et al. (eds.), *Towards a Theory of Spacetime Theories*,
Einstein Studies 13, DOI 10.1007/978-1-4939-3210-8_3

1   The Character of Energy Conditions

An energy condition, in the context of a wide class of spacetime theories (including general relativity), is, crudely speaking, a relation one demands the stress-energy tensor of matter satisfy in order to try to capture the idea that "energy should be positive".[1] Perhaps the simplest example is the so-called weak energy condition: for any timelike vector ξ^a at any point of the spacetime manifold, the stress-energy tensor T_{ab} satisfies $T_{mn}\xi^m\xi^n \geq 0$. This has, *prima facie*, a simple physical interpretation: the (ordinary) energy density of the fields contributing to T_{ab}, as measured in a natural way by any observer (*e.g.*, using instruments at rest relative to that observer), is never negative. The remarkable fact I will discuss in this paper is that such simple, general, almost trivial seeming propositions have profound and far-reaching import for our understanding of the structure of relativistic spacetimes. It is therefore, especially, surprising when one also learns that we have no clear understanding of the nature of these conditions, what theoretical status they have *vis-à-vis* fundamental physics, what epistemic status they may have, when we should and should not expect them to be satisfied, and even in many cases how they and their consequences should be interpreted physically. Or so I shall argue.

Geroch and Horowitz [92, p. 260], in discussing the form of singularity theorems in general relativity, outline perhaps the most fundamental reason for the importance of energy conditions with the following pregnant observation:

> One would of course have to impose *some* restriction on the stress-energy of matter in order to obtain any singularity theorems, for with no restrictions Einstein's equation has no content. One might have thought, however, that only a detailed specification of the stress-energy at each point would suffice, *e.g.* that one might have to prove a separate theorem for each combination of the innumerable substances which could be introduced into spacetime. It is the energy condition which intervenes to make this subject simple. On the one hand it seems to be a physically reasonable condition on all types of classical matter, while on the other it is precisely the condition on the matter one needs for the singularity theorem.

---

[1]From hereon until §5, unless explicitly stated otherwise, the discussion should be understood to be restricted to the context of general relativity. Almost everything I say until then will in fact hold in a very wide class of spacetime theories, but the fixed context will greatly simplify the exposition. In general relativity, the fundamental theoretical unit, so to speak, is a spacetime model consisting of an ordered pair $(\mathscr{M}, g_{ab})$, where $\mathscr{M}$ is a four-dimensional, paracompact, Hausdorff, connected, differential manifold and g_{ab} is a pseudo-Riemannian metric on it of Lorentzian signature. 'T_{ab}' will always refer to the stress-energy tensor picked out in a spacetime model by the Einstein field equation, 'T' to the trace of T_{ab} ($T^n{}_n$), 'R_{ab}' to the Ricci tensor associated with the Riemann tensor $R^a{}_{bcd}$ associated with the unique torsion-free derivative operator ∇ associated with g_{ab}, 'R' to the trace of the Ricci tensor ($R^n{}_n$, the Gaussian scalar curvature), and 'G_{ab}' to the Einstein tensor ($R_{ab} - \frac{1}{2}Rg_{ab}$). For conventions about the metric signature and the exact definitions of these tensors, I follow Malament [133]. Unless otherwise explicitly noted, indicial lowercase Latin letters $(a, b, \ldots)$ designate abstract tensor-indices, indicial lower-case Greek letters $(\mu, \nu, \ldots)$ designate components with respect to a fixed coordinate system or tetrad of tangent vectors ($\mu \in \{0, 1, 2, 3\}$), and hatted indicial lower-case Greek letters ($\hat{\mu}, \hat{\nu}, \ldots$) designate the spacelike components ($\hat{\mu} \in \{1, 2, 3\}$) with respect to a fixed $1 + 3$ tetrad system. (For an exposition of the abstract index notation, see Penrose and Rindler [155], Wald [203], or Malament [133].)

I will return to this quote later, in §5, but for now the salient point is that a generic condition one imposes on the stress-energy tensor, "generic" in the sense that it can be formulated independently of the details of the internal structure of the tensor, which is to say independently of any quantitative or structural feature or idiosyncrasy of any particular matter fields, suffices to prove theorems of great depth and scope. Indeed, as Geroch and Horowitz suggest, without the possibility of relying on conditions of such a generic character, we would not have the extraordinarily general and far-reaching singularity theorems we do have. And it is not only singularity theorems that rely for their scope and power on these energy conditions—it is no exaggeration to say that the great renaissance in the study of general relativity itself that started in the 1950s with the work of Synge, Wheeler, Misner, Sachs, Bondi, Pirani, *et al.*, and the blossoming of the investigation of the global structure of relativistic spacetimes at the hands of Penrose, Hawking, Geroch, *et al.*, in the 1960's could not have happened without the formulation and use of such energy conditions.

What is perhaps even more remarkable is that many of the most profound results in the study of global structure—*e.g.*, the Hawking Area Theorem—do not depend on the Einstein field equation at all, but rather assume only a purely formal condition imposed on the Ricci tensor, which itself can be thought of as an "energy" condition if one invokes the Einstein field equation to provide a physical interpretation of the Ricci tensor. In a sense, therefore, energy conditions seem to reach down to and get a hold of a level of structure in our understanding of gravitation and relativistic spacetimes even more fundamental than the Einstein field equation itself. (I will discuss in §5 this idea of "levels of structure" in our understanding of general relativity in particular, and of gravitation and spacetime more generally.)

Now, most propositions of a fundamental character in general relativity admit of interpretation as either a postulate of the theory or as a derived consequence from some other propositions taken as postulates. That is to say, the theory allows one a great deal of freedom in what one will take as given and what one will demand a proof of. One can, for example, either assume the so-called Geodesic Principle from the start as a fundamental regulative principle of the theory, as, for example, in the exposition of Malament [133], or one can assume other propositions as fundamental, perhaps ones fixing the behavior of ideal clocks and rods, and derive the Geodesic Principle as a consequence of those propositions, as, for example, in the exposition of Eddington [64]. Which way one goes for any given proposition depends, in general, on the context one is working in, the aims of one's investigation, one's physical and philosophical intuitions and predilections, *etc.*[2]

This interpretive flexibility does not seem to hold, however, for energy conditions. I know of no substantive proposition that, starting from some set of other important "fundamental postulates", has as its consequence an energy condition. One either imposes an energy condition by fiat, or one shows that it holds for stress-energy tensors associated with particular forms of matter fields. One never imposes general

---

[2]See Weatherall [209, this volume] for an insightful discussion of a view of the foundations of spacetime theories, with particular regard to this issue, that I find sympathetic to my own views as I sketch them here.

conditions on other geometrical structures (*e.g.*, the Riemann tensor or the topology or the global causal structure) and derives therefrom the satisfaction of an energy condition (except in the trivial case where one imposes conditions directly on the Ricci or Einstein tensor, standing as a direct proxy for the stress-energy tensor by dint of the relation between them embodied by the Einstein field equation).[3] There are a plethora of results that show when various energy conditions may or must be violated both theoretically and according to observation, which I discuss in §3.2, but none that show nontrivially when one must hold. Indeed, this inability to prove them is an essential part of what seems to make them structure "at a deeper level" perhaps even than causality conditions (many of which can be derived from other fundamental assumptions), and so applicable across a *very* wide range of possible theories of spacetime.

In a similar vein, they occupy an odd methodological and theoretical niche quite generally. None is implied by any known general theory, though each can be formulated in the frameworks of a wide spectrum of different theories, and several can be shown to be inconsistent with a wide spectrum of theories (in the strong sense that one can derive their respective negations in the context of the theories). Indeed, they are among the very few physical propositions I know that can be used either to exclude as physically unreasonable individual solutions to the field equations of a particular theory (as for, *e.g.*, a wide class of FLRW spacetimes in general relativity that have strongly negative pressures[4]), or to exclude entire theories (such as the Hoyle Bondi steady-state theory of cosmology, as I discuss below in §3.2). Whether or not one should consider them as "part of" any given theory, therefore, seems a problematic question at best, and an ill-posed one at worst.

It is difficult to get a grip on their epistemic status as well. They seem in no sense to be laws, under any standard account in the literature, for none of them holds for all known "physically reasonable" types of matter, and each of them is in fact violated in what seem to be physically important circumstances. Neither do they appear to be empirical or inductive generalizations, for the same reason.[5] And yet we think that (at least) one of them—or something close to them—likely holds generically in

---

[3] The one possible exception to this claim I know of is the attempt by Wall [206] to derive the so-called averaged null energy condition (ANEC) from the Generalized Second Law of thermodynamics. While I find his arguments of great interest, I also find them problematic at best. See Curiel [48] for discussion.

[4] See Curiel [47] for discussion.

[5] It should be noted, however, that, to the best of my knowledge, there has never been direct experimental observation of a violation of any of the standard energy conditions I discuss in §2. We do, however, have extremely good *indirect* experimental and observational evidence for violations of several of them, as I will discuss in §3. See Curiel [47] for an extended discussion of evidence for their violation in cosmology, and Curiel [48] for one in the context of quantum field theory on curved spacetime. Even direct experimental verification of the Casimir effect does not yield direct measurement of negative energy densities, though the Casimir effect relies essentially on the existence of such; rather, the negative energy densities are inferred from measurement of the Casimir force itself [28].

the actual universe, at the level of classical (*i.e.*, non-quantum) physics at least, and even that one or more of them, appropriately reformulated, should hold generically at the quantum level as well.[6] Even more, as I have already indicated, there seem to be very good reasons for thinking that the sense in which they do obtain, whatever that may be, is grounded in structure at a level of our understanding even deeper than the Einstein field equation itself, which we surely do think of as a law, under any reasonable construal of the notion.

So what are they? The remainder of this paper consists of an attempt to come to grips with this question, by exploring their formulations, their consequences, their relations to other fundamental structures and principles, and their role in constraining the possible forms a viable theory of spacetime may take. Those who hope for a decisive answer to the question will leave disappointed. I feel I will have succeeded well enough if I am able only to survey the most important issues and questions, clarify and sharpen some of them, propose a few conjectures, and generally open the field up for other investigators to do more work in it.[7]

2 The Standard Energy Conditions

There are several different ways to formulate all the energy conditions standardly deployed in classical general relativity, both as a group and individually. I will focus here on three ways of formulating them as a group, what one may think of as the *geometric*, the *physical* and the *effective* ways, and will for a few of them discuss as well alternative individual formulations according to the geometric and physical ways, as they variously allow different insights into the character of the conditions.[8] The geometric and physical ways are easy to characterize: for the former, one writes down formal conditions expressed by use only of the value of a purely geometric tensor (such as the Ricci or Weyl tensor), perhaps as it is required to stand in relation to a fixed family of vectors or other tensors; for the latter, one writes down formal conditions expressed by use only of the value of the stress-energy tensor itself, perhaps as it is required to stand in relation to a fixed family of vectors or other tensors.[9] In every

---

[6]See Curiel [48] for discussion.

[7]This paper, in other words, has as its goal a more modest version of that of Earman's wonderful book *A Primer on Determinism*, to which the name of this paper is an homage.

[8]In this section, aside from a few idiosyncrasies, such as my classification of different types of formulation, I follow in part the exposition of [195, ch. 12] and in part that of [133, §2.5 and §2.8] for the formulations of the conditions themselves. See Curiel [47] for another formulation of them, based on the scale factor $a(t)$ in generic cosmological models, and discussion thereof.

[9]Another interesting way to study the properties and behavior of T_{ab} is by the Segré algebraic classification of symmetric rank-two covariant tensors. (See, *e.g.*, Hall [94].) It is beyond the scope of the current paper to discuss that.

case, the physical formulation is logically equivalent to the geometric formulation if the Einstein field equation is assumed to hold.[10]

The effective way requires a bit of groundwork to explain. According to a useful classification of stress-energy tensors given by [107, p. 89], a stress-energy tensor is said to be of type I if at every point there is a $1 + 3$ orthonormal frame with respect to which it is diagonal, *i.e.*, if its only nonzero components as computed in the given frame are on the diagonal in its matrix form. In this case, it is natural to interpret the timelike-timelike component as the ordinary (mass-)energy density ρ as represented in the given frame, and the three spacelike-spacelike components to be the three principal pressures $p_{\hat{\mu}}$ ($\hat{\mu} \in \{1, 2, 3\}$) as represented in the frame, to be understood by analogy with the case of a fluid or an elastic body. The effective formulation of an energy condition can then be stated as a quantitative relation among ρ and $p_{\hat{\mu}}$. Since all known "physically reasonable" classical fields (and indeed many unreasonable ones) have associated stress-energy tensors of type I, this is no serious restriction.[11] Thus, except for one special case to be discussed below, the effective formulation should be understood to be in all ways physically equivalent to the geometric and the physical formulations, under the assumption that the Einstein field equation holds, and matter is not too exotic. Under that assumption, the effective formulations become especially useful in cosmological investigations, since the matter fields in standard cosmological models, the FLRW spacetimes, can always be thought of as fluids.

It will be convenient to break the conditions up into two further classes, those (*pointilliste*) that constrain behavior at individual points and those (*impressionist*) that constrain average behavior over spacetime regions. I shall first list the definitions of all the former, then discuss the significance and interpretation of each as it will be useful to have them all in hand at once for the purposes of comparison, then do the same for the latter class.

2.1 Pointilliste Energy Conditions

null energy condition (NEC)

>   **geometric**   for any null vector k^a, $R_{mn}k^m k^n \geq 0$
>   **physical**   for any null vector k^a, $T_{mn}k^m k^n \geq 0$
>   **effective**   for each $\hat{\mu}$, $\rho + p_{\hat{\mu}} \geq 0$

---

[10] This equivalence between the physical and the geometrical formulations does not hold in general if *and only if* the Einstein field equation holds. The biconditional holds in general relativity (for minimally coupled fields, at least). In other spacetime theories with field equations similar to but distinct from the Einstein field equation, the biconditional will not in general hold. I will discuss this further in §5.

[11] The one possible exception to this claim is a null fluid, which has a stress-energy tensor of the form $T_{ab} = \rho k_a k_b + p_1 x_a x_b + p_2 y_a y_b$, where k^a is null and x^a and y^a are unit spacelike vectors orthogonal to k^a and to each other.

weak energy condition (WEC)

geometric   for any timelike vector ξ^a, $G_{mn}\xi^m\xi^n \geq 0$
physical   for any timelike vector ξ^a, $T_{mn}\xi^m\xi^n \geq 0$
effective $\rho \geq 0$, and for each $\hat{\mu}$, $\rho + p_{\hat{\mu}} \geq 0$

strong energy condition (SEC)

geometric   for any timelike vector ξ^a, $R_{mn}\xi^m\xi^n \geq 0$
physical   for any timelike vector ξ^a, $(T_{mn} - \frac{1}{2}Tg_{mn})\xi^m\xi^n \geq 0$
effective $\rho + \sum_{\hat{\mu}} p_{\hat{\mu}} \geq 0$, and for each $\hat{\mu}$, $\rho + p_{\hat{\mu}} \geq 0$

dominant energy condition (DEC)

geometric
1.   for any timelike vector ξ^a, $G_{mn}\xi^m\xi^n \geq 0$, and $G^a{}_n\xi^n$ is causal
2.   for any two co-oriented timelike vectors ξ^a and η^a, $G_{mn}\xi^m\eta^n \geq 0$

physical
1.   for any timelike vector ξ^a, $T_{mn}\xi^m\xi^n \geq 0$, and $T^a{}_n\xi^n$ is causal
2.   for any two co-oriented timelike vectors ξ^a and η^a, $T_{mn}\xi^m\eta^n \geq 0$

effective $\rho \geq 0$, and for each $\hat{\mu}$, $|p_{\hat{\mu}}| \leq \rho$

strengthened dominant energy condition (SDEC)

geometric
1.   for any timelike vector ξ^a, $G_{mn}\xi^m\xi^n \geq 0$, and, if $R_{ab} \neq 0$, then $G^a{}_n\xi^n$ is timelike
2.   either $G_{ab} = 0$, or, given any two co-oriented causal vectors ξ^a and η^a, $G_{mn}\xi^m\eta^n > 0$

physical
1.   for any timelike vector ξ^a, $T_{mn}\xi^m\xi^n \geq 0$, and, if $T_{ab} \neq 0$, then $T^a{}_n\xi^n$ is timelike
2.   either $T_{ab} = 0$, or, given any two co-oriented causal vectors ξ^a and η^a, $T_{mn}\xi^m\eta^n > 0$

effective $\rho \geq 0$, and for each $\hat{\mu}$, $|p_{\hat{\mu}}| \leq \rho$

(It is not a typo that the given effective forms of the DEC and the SDEC are identical; this is the one special case, mentioned above, in which the effective form of the energy condition diverges from the geometrical and physical forms. Of course, it is the case that when one restricts attention to stress-energy tensors of type I, then the geometrical and physical forms of the DEC and SDEC also coincide.) I first sketch the most more or less straightforward interpretations of the conditions, before discussing problems with those interpretations.

The idea of average radial acceleration (explained in detail in the technical appendix §2.5 below) offers one seemingly promising route toward an interpretation of the geometric and physical forms of the NEC. Roughly speaking, the average radial acceleration of a geodesic γ at a point p is the averaged magnitude of the acceleration of neighboring geodesics relative to γ in directions orthogonal to γ. If the average radial acceleration is negative, then this represents the fact that, again roughly speaking, neighboring geodesics tend to fall inwards towards γ at p. Thus, according to equation (2.4), the geometric form of the NEC requires that null geodesic congruences tend to be convergent in sufficiently small neighborhoods of every spacetime point (or at least not divergent). Assuming the Einstein field equation, the physical interpretation of negative average radial acceleration for causal geodesics is that, again roughly speaking, the "gravitational field" generated by the ambient stress-energy is "attractive". Thus, according to equation (2.5), the interpretation of the physical form is that particles following null geodesics will observe that "gravity" tends locally to be "attractive" (or at least not repulsive) when acting on nearby particles also following null geodesics. Another possible interpretation of the physical form of the NEC is that an observer traversing a null curve will measure the ambient (ordinary) energy density to be positive.

The interpretation of the effective form of the NEC is that the natural measure either of mass–energy or of pressure in any given spacelike direction can be negative as determined by an observer traversing a null curve, but not both, and, if either is negative, it must be less so than the other is positive. In so far as one may think of pressure as a momentum flux, therefore, and so equivalent relativistically to a mass–energy flow, the effective form requires that ordinary mass–energy density at any point cannot be negatively dominated by momentum fluxes in any given spacelike direction as determined by an observer traversing a null curve: one cannot indefinitely "mine" energy from a system by subjecting it to negative momentum flux.

The interpretation of the physical form of the WEC is straightforward: the (ordinary) total energy density of all matter fields, as measured in a natural way by any observer traversing a timelike curve, is never negative. The interpretation of the geometric form is not straightforward. Indeed, I know of no simple, intuitive picture that captures the geometrical significance of the condition.[12] The interpretation of the effective form is similar to that for the NEC. Ordinary mass–energy density must be nonnegative as experienced by any observer traversing a timelike curve, and the

---

[12] It has gone oddly unremarked in the physics and philosophy literatures, but is surely worth puzzling over, that the Einstein tensor itself, the fundamental constituent of the Einstein field equation, has no simple, natural geometrical interpretation, in the way, *e.g.*, that the Riemann tensor can naturally be thought of as a measure of geodesic deviation. Perhaps one could try to use the Bianchi identity to construct a geometric interpretation for G_{ab}, or the Lanczos tensor (see footnote 22), but it is not immediately obvious to me what such a thing would look like, if possible. One can give a geometrical interpretation of G_{ab} at a point by considering all unit timelike vectors at the point; the Einstein tensor can then be reconstructed by defining it to be the unique symmetric two-index covariant tensor at that point such that its double contraction with every unit timelike vector equals minus one-half the spatial scalar curvature of the spacelike hypersurface with vanishing extrinsic curvature orthogonal to the given vector. (See Malament [133, ch. 2, §7].) This may be only a matter of taste, but I find this interpretation obscure and Baroque, certainly not simple and natural, in large part because it relies on structure in a family of three-dimensional objects to fix the meaning of a four-dimensional object.

pressure in any given spacelike direction can never be so negative as to dominate that value.[13]

It is easy to see, by considerations of continuity, that the WEC implies the NEC. Tipler [189] proved two propositions that give some insight into the relation between the NEC and the WEC, and into the character of the WEC itself. He first showed that, in a natural sense, the WEC is the weakest local energy condition one can define. ("Local" here means something like: holding at a point, for all observers.) In particular, he proved the following: if $T_{mn}\xi^m\xi^n$ is finitely bounded from below for all timelike ξ^a, $i.e.$, if there exists a $b > 0$ such that $T_{mn}\xi^m\xi^n \geq -b$ for all timelike ξ^a, then WEC holds ($i.e.$, the infimum of all such b is 0). He next proved that one cannot do better by imposing further natural constraints on the condition: if $T_{mn}\xi^m\xi^n$ is finitely bounded from below for all unit timelike ξ^a, and T_{ab} is of type I, then the NEC holds. The effective form of the WEC, therefore, is in fact essentially equivalent to the NEC. Thus, though the WEC is not the weakest condition in a logical sense one can impose, it is the weakest in a loose, physical sense: one cannot do better by imposing further natural restrictions.

The interpretation of the geometric form of the SEC is similar to that of the NEC. According to equation (2.4), the geometric form of the SEC requires that timelike geodesic congruences tend to be convergent in sufficiently small neighborhoods of every spacetime point. This implies that congruences of null geodesics at that point are also convergent. Similarly, according to equation (2.5), the interpretation of the physical form is that observers following timelike geodesics will see that "gravity" tends locally to be "attractive" in its action on stuff following both timelike and null geodesics.[14] The effective form of the SEC has part of its interpretation the same as that of the WEC, $viz.$, ordinary mass–energy density at any point cannot be negatively dominated by momentum fluxes in any given spacelike direction as determined by an observer traversing a timelike curve. It also says, however, that ordinary mass–energy density cannot be negatively dominated by the sum of the individual pressures (momentum fluxes) at any point, as determined by an observer traversing a timelike curve. I know of no compelling elucidation of the physical content of that relation. The SEC does not imply the WEC, for the SEC can be satisfied even if the ordinary mass–density is negative. The SEC does, however, imply the NEC.

---

[13]Classically, some fluids such as water are known to exhibit negative pressures in some regimes as measured by observers traversing timelike curves ($e.g.$, us), but these negative pressures are never large enough to dominate the fluid's mass–energy. Indeed, when one considers how large the relativistic mass–energy of, say, 1 g of water is, and so correlatively how extraordinarily intense a momentum flux would have to be to achieve a mass–energy content comparable to that, one gets a good feel for just how "exotic" any stuff would be that violates the NEC.

[14]This explication of the physical form of the SEC clearly illustrates why it is problematic to try to think of general relativity as a theory of "gravity", in the sense of a force exerted on a body: for bodies traversing non-geodetic curves, that is, for bodies experiencing nontrivial acceleration, one has no natural way to judge whether "the force of gravity" is acting attractively or repulsively, not even when one fixes a standard of rest (a fiducial body traversing a timelike geodesic). $Pace$ particle physicists, general relativity simply cannot be comprehended as a theory describing a dynamical "force" at all.

As for the WEC, the interpretations of the geometrical forms of the DEC and the SDEC are not clear. The interpretations of their physical forms are apparent: every timelike observer will measure ordinary mass–energy density to be nonnegative, and will also measure total flux of energy–momentum to be causal, with the flow oriented in the same direction as the observer's proper time. The SDEC, as the name suggests, is slightly stronger in that it requires energy–momentum flux as measured by any timelike observer to be strictly timelike for nontrivial stress-energy distributions. The DEC (and *a fortiori* the SDEC) are, therefore, standardly taken to rule out "superluminal propagation of stress-energy". (See, *e.g.*, the exemplary remarks of Wald [203, p. 219].) As already noted, the effective forms of the DEC and SDEC are identical. Their interpretation, besides the now-familiar demand that locally measured energy density be nonnegative, is that pressures be strictly bounded both above and below by the energy density. This means that the effective fluid can be neither too "stiff" nor too "lax", but must lie in a middling Goldilocks regime.[15] The second given geometric and physical forms of the SDEC make it manifest that the SDEC is in fact logically stronger than the DEC. Of course, any T_{ab} that satisfied the DEC but violated the SDEC would have to be not of Hawking-Ellis type I, for it is only in that case that the two come apart. Clearly, the SDEC implies the DEC, which implies the WEC.

Before turning to examine the so-called impressionist energy conditions, I briefly discuss a few problems with the interpretations I have sketched of the pointilliste conditions. The interpretations of the geometrical and physical forms of the NEC based on average radial acceleration is undermined by the fact that convergence of null geodesics at a point does not in general imply convergence of all timelike geodesics at that point. This is why I hedged the proposed interpretations with slippery terms like 'tends to': even if the NEC is satisfied at a point, an observer traversing a timelike geodesic may still see "gravity acting repulsively" in a small neighborhood. The existence of a positive cosmological constant is a case in which NEC is satisfied, but, by the failure of the SEC, there is still divergence of timelike geodesics: "gravity acts repulsively" on matter following timelike geodesics, even though it "acts attractively" on stuff following null geodesics.[16] The other proposed interpretation of the physical form of the NEC—that observers traversing null curves will measure nonnegative energy density—suffers from the fact that it is difficult to see

---

[15] See Curiel [47] for a discussion of the consequences of allowing the effective fluid to be too lax, which is to say, allowing the barotropic index w to be less than -1, in the context of cosmology. $w := \frac{p}{\rho}$, and so is a useful measure of the "stiffness" of whatever (nearly) homogeneous, isotropic stuff fills spacetime in cosmological models.

[16] It should be kept in mind that the physical consequences of a "positive" versus a "negative" cosmological constant in this context depend on one's conventions for writing the Einstein field equation and on one's conventions for the metric signature. With the conventions I am using, a positive value of Λ itself leads to negative momentum flux in spacelike directions, and that is the condition that leads to accelerated expansion on the cosmological scale, as actually observed, and so the theoretical need for "dark energy".

what physical sense can be made of the idea of an observer traveling at the speed of light making (ordinary) energy measurements. One cannot try to ameliorate this problem by positing that the condition means only that a physical system traversing a null curve will "experience" only nonnegative energy densities in its couplings with other systems, irrespective of whether it is an observer making measurements: ordinary energy density is not an observer-independent quantity, and so it can mediate no physical interaction in any way with intrinsic physical significance. No physical system will "experience" ordinary energy density at all.[17]

The interpretation of the effective form of the NEC suffers the same difficulty: what physical content does it have to compare the magnitude of ordinary energy density and that of momentum flux in a given spacelike direction, as determined by an observer traversing a *null* curve? There is an even more serious problem here, though, which the effective form makes particularly clear, showing the limitations of the physical significance of the NEC. Assuming a well behaved barotropic equation of state for the effective fluid described by the stress-energy tensor, *i.e.*, a fixed relation $\rho(p)$ expressing ρ as an invertible function of the single isotropic pressure p, and assuming the medium is not too strongly dispersive, the speed of sound is $c_s^2 = \dfrac{\mathrm{d}p}{\mathrm{d}\rho}$. It should be clear that the NEC does *not* require that $c_s \leq 1$; in other words, stuff can satisfy the NEC while still permitting superluminal propagation of physically significant structure. It is thus unclear in the end what real physical significance the requirement that mass–energy density not be negatively dominated by momentum fluxes has.

The problems with the effective interpretation of the WEC are much the same as for the NEC: it is not clear what physical significance the given relations among energy density and pressure can have when they permit superluminal propagation of physical structure. The fact that the WEC requires energy density always to be positive may make one at first glance think that it will be violated in the ergosphere of a Kerr black hole, where, as is well known, ordinary systems can have in a natural sense negative energy [150, 154]. In fact, though, there is an equivocation on 'energy' here that points to a subtle and important point. The energy that can be negative near a Kerr black hole is the energy defined by the stationary Killing field of the spacetime, not the ordinary energy density as measured by any observer using tools at rest with respect to herself. (Because the stationary Killing field is spacelike in the ergosphere, no observer can have any of its orbits as worldine.) Now, as I remarked in footnote 17, ordinary

---

[17]We decompose T_{ab} into energy density, momentum flux and stress in our *representations* of our experiments, for various pragmatic and psychological reasons; the decomposition represents nothing of intrinsic physical significance about the world. This fact perhaps lies at the root of most if not all the difficulties and puzzles that plague the energy conditions, especially why they do not seem to be derivable from other fundamental principles. Of course, this fact also makes it even more puzzling that they should have such profound, physically significant consequences as they do. What is going on here?

energy density, not being an observer-independent quantity, is not a particularly natural concept in general relativity. The energy defined by a stationary Killing field, however, *is* observer-independent and so has *prima facie* physical significance, even more so given that it obeys both a local and an integral conservation law. Why is it not troubling that *this* quantity, a manifestly deep and important one, can be negative, whereas the negativity of the observer-dependent ordinary energy density throws us into fits? Why do we depend so strongly on conditions formulated using quantities that, under their standard physical interpretation, are not observer-independent, especially when proving results about quantities and structures such as event horizons that are observer-independent? I don't know. Perhaps the lesson here is that the geometric form of the energy conditions are the ones to be thought of as fundamental, in so far as they rely for their statement and interpretation only on invariant, geometrical structures and concepts. It would then be an interesting problem why in the context of some theories, such as general relativity, the physical interpretation of the conditions turns out to have questionable significance. Perhaps this is telling us to look for theories in which these important geometric conditions have physically significant interpretations. I will return to discuss this question in §5.

With regard to the SEC, because the convergence of all timelike geodesics at a point does imply the convergence of null geodesics there, the proposed interpretations of its geometric and physical forms, that "gravity tends to be attractive", are on firmer ground than for the NEC. There is still a problem, though, even here. Averaged radial acceleration is, after all, only an average, factitious quantity. That it be negative does not say that individual freely falling ordinary bodies cannot in fact accelerate away from each other for no apparent reason, only that, on average, they do not do so. Thus, the idea that average geodetic convergence should be thought of as a representation of the attractiveness of gravity is dicey at best. And, again, there is the issue that this condition says nothing at all about the "effect of gravity" on bodies accelerating under the action of other forces.

The DEC (and *a fortiori* the SDEC) are standardly taken to rule out "superluminal propagation of stress-energy". Once again, however, it is clear that the DEC does not preclude superluminal speeds of sound for fields, so it is not clear what work the prohibition on superluminal propagation of stress-energy is doing. Even if we put that point aside, though, there are other problems, as Earman [62] argues, claiming the DEC ought not be interpreted as prohibiting superluminal propagation of stress-energy. His argument goes in two steps. He first argues for the positive conclusion that the proper way to conceive of a prohibition on superluminal propagation is the existence of a well posed (in the sense of Hadamard) initial value formulation for all fields on spacetime. Then, based on Geroch [91], he shows that physical systems can have well posed initial value formulations even when the DEC is violated. Earman's arguments are buttressed by a recent argument due to Wong [212]. As Wong notes (along with Earman), the evidence almost always cited in support of the idea that DEC prohibits superluminal propagation of stress-energy is the theorem that states that, if a covariantly divergence-free T_{ab} is required to satisfy the DEC and it vanishes on a closed, achronal set, then it vanishes in the domain of dependence of that set

[102, 107].[18] Wong, I think rightly, points out that this theorem in fact shows only that DEC prohibits "the edge of a vacuum" (or vacuum fluctuations, in a quantum context) from propagating superluminally, not arbitrary stress-energy distributions. Given the nonlinearity of the Einstein field equation, I find it plausible that there may be problems in trying to naively generalize this result to arbitrary stress-energy tensors, whether they obey the DEC or not.

Comparing the strengths and weaknesses of the interpretations of the different forms of the conditions amongst themselves reveals some interesting questions. Consider the NEC: on the face of it, the geometric form has a relatively unproblematic interpretation, whereas the interpretations of the physical and effective forms are beset with more serious problems. The case is just the opposite for the WEC: the geometric form has no clear interpretation, whereas the physical form and at least part of the effective form (the positivity of energy density) are relatively unproblematic. The DEC occupies yet more treacherous ground, in so far as the geometric form has no clear interpretation, the physical interpretation (as Earman's and Wong's arguments show) is muddled at best, and the effective is only partially unproblematic. And yet these statements are, modulo the assumption of the Einstein field equation, logically equivalent. Ought unclarity of interpretation of one form push us to question the seeming clarity of interpretation of other forms? How can this happen, that the interpretation of one proposition can be problematic while the interpretation of a proposition logically equivalent is not (or, at least, is less so)? Can we lay all the blame on the assumption of the Einstein field equation? I don't think so, for, if we could, then surely the forms that had interpretive problems would all be of the same type, but that is not the case here. Sometimes it is the geometric that is less problematic, and other times it is the more problematic.

This is not the place to try to address these questions. I will remark only that this topic would provide very rich fodder for an investigation into the relations between pure geometry and the physical systems that geometry purports to represent in a given theory, what must be in place in order to extract physically significant information from the geometry of those systems, and what the difference is between having an interpretation of a piece of pure mathematics and having a physical interpretation of it in the context of a theory. I have the sense that it is often a tacit assumption in philosophical discussions of the meaning of theoretical terms that, if a mathematical structure has a clear physical interpretation in a theory, then it itself must have a clear mathematical interpretation already. These examples show that this need not be so. They also provide interesting case studies of how theoretically equivalent statements can seemingly have very different physical meanings.

I conclude this section with an observation of what is *not* here: there are no standard energy conditions based on the Weyl conformal tensor $C^a{}_{bcd}$ or on the Bel–Robinson

---

[18] A region of spacetime is achronal if no two of its points stand in timelike relation to each other. The domain of dependence $D(\Sigma)$ of a closed achronal set Σ is the collection of all points p in spacetime such that every inextendible causal curve passing through p intersects Σ.

tensor T_{abcd}.[19] I find this odd. Because there is no object in general relativity that one can reasonably interpret as the stress-energy tensor of the "gravitational field", the standard pointilliste energy conditions do not directly constrain the behavior of anything one may want to think of as gravitational stress-energy, and yet one may still want to try to do so.[20] The possible need for trying to do so becomes clear when one considers how strange, even pathological, purely vacuum spacetimes can be, such as Taub-NUT spacetime and some gravitational plane wave spacetimes.[21] Because the Weyl tensor is not directly constrained by the stress-energy tensor of matter, in the sense that it may be nonzero even when T_{ab} is zero, it is often thought to represent "purely gravitational" degrees of freedom.[22] The Bel–Robinson tensor, moreover, may usefully be thought of as a measure of a kind of "super-energy" associated with purely gravitational phenomena, and directly measures in a precise sense the intensity of gravitational radiation in infinitesimal regions. These two tensors, therefore, would seem perfect candidates to serve as the basis for conditions that would constrain the behavior of purely gravitational phenomena and, more particularly, of vacuum spacetimes. I think it would be of great interest to investigate whether there are natural conditions based on these two tensors that would constrain behavior in vacuum spacetimes so as to rule out such pathologies. I conjecture that there are indeed such conditions.[23] One potentially promising place to start a search for such

---

[19]For characterization and discussion of the Bel–Robinson tensor and its properties, see Penrose and Rindler [155], Senovilla [177, 178], Garecki [85] and García-Parrado Gómez-Lobo [84].

[20]There does not exist in general relativity a satisfactory definition for a "gravitational" stress-energy tensor, one that represents localized stress-energy of purely "gravitational" systems. (See Curiel [51].) One may want to think of this as a limitation on the possible physical content of the standard pointilliste energy conditions, as I discuss at the end of §2.1.

[21]See, e.g., Misner [137] and Ellis and Schmidt [71], respectively, and Curiel [45] for further discussion.

[22]Still, $C^a{}_{bcd}$ and T_{ab} are not entirely independent of each other. If we define the so-called Lanczos tensor

$$J_{abc} := \frac{1}{2}\nabla_{[b}R_{a]c} + \frac{1}{6}g_{c[a}\nabla_{b]}R$$
$$= 4\pi\nabla_{[b}T_{a]c} - \frac{1}{12}g_{c[b}\nabla_{a]}T \tag{2.1}$$

then the Bianchi identities may be rewritten

$$\nabla_n C^n{}_{abc} = J_{abc}$$

The similarity of this equation to the sourced Maxwell equation suggests regarding the Bianchi identities as field equations for the Weyl tensor, specifying how at a point it depends on the distribution of matter at nearby points. (This approach is especially useful in the analysis of gravitational radiation; see, for example, Newman and Penrose [142], Newman and Unti [143], and Hawking [99].) Thus, conditions imposed on the Weyl tensor might still be plausibly interpretable as energy conditions in spacetimes with nontrivial T_{ab}.

[23]It is well known that the Bel–Robinson tensor automatically satisfies the so-called "dominant super-energy condition", viz., $T_{mnrs}\xi^m\xi^n\xi^r\xi^s \geq 0$, for all causal vectors ξ^a in all spacetimes. Because of the complete universality of the condition, however, it cannot rule out pathologies.

conditions might be the Weyl Curvature Hypothesis of Penrose [152], and recent work attempting to formulate expressions for gravitational entropy based on these two tensors.[24]

2.2 Impressionist Energy Conditions

Before exhibiting the impressionist energy conditions, a little technical background is in order. If γ is a timelike curve, then it is natural to parameterize the line integral of a quantity along γ by proper time. If γ is a null curve, however, one does not have a natural parameterization of it available. In this case, it is convenient to use a generalized affine parameter.[25] The generalized affine parameter is especially useful in that it does not depend on the tetrad basis chosen in one crucial respect: whether or not the generalized affine parameter of the curve increases without bound.

In order to express the impressionist conditions in effective form, it will be convenient to define direction cosines for causal tangent vectors. Fix a $1 + 3$ orthonormal frame with respect to which the stress-energy tensor (assumed, recall, for the effective form, to be of Hawking-Ellis type I) is diagonal. Let k^{μ} be the components of the null vector k^a with respect to the fixed frame. Then define the normalization function v_n and the direction cosines $\cos \alpha_{\mu}$ so that $\cos \alpha_0 = 1$ and $k^{\mu} = v_n(k^a) \cos \alpha_{\mu}$. Let ξ^{μ} be the components of the timelike vector ξ^a with respect to the fixed frame. Then define the normalization function v_t, the real number β, and the direction cosines $\cos \alpha_{\mu}$ so that $\cos \alpha_0 = 1$, $\xi^0 = v_t(\xi^a) \cos \alpha_0$ and $\xi^{\hat{\mu}} = v_t(\xi^a)\beta \cos \alpha_{\hat{\mu}}$.

Although in principle one could define impressionist energy conditions based on spacetime regions of any dimension or topology, in practice, at least in the classical regime, they have all been defined using curves of various types. In my exposition of them here, I will give what is in effect only a template for the ones actually used to prove theorems, which often qualify the basic template in some way. I will explain or at least mention some of those qualifications in my discussion below in this section, and also in §3. All the impressionist energy conditions based on curves have this in common: the characteristic property that is postulated is required to hold on every curve in some fixed class Γ of curves on spacetime.

averaged null energy condition (ANEC)

geometric   for every γ in the fixed class of null curves Γ,

$$\int_{\gamma} R_{mn} k^m k^n \, d\theta \geq 0$$

where γ has tangent vector k^a and θ is a generalized affine parameter along γ
physical   for every γ in the fixed class of null curves Γ,

---

[24] See, e.g., Cotsakis and Klaoudatou [44] and Clifton [42].
[25] See, e.g., Schmidt [171] for a definition and discussion.

$$\int_\gamma T_{mn}k^m k^n \, d\theta \geq 0$$

where γ has tangent vector k^a and θ is a generalized affine parameter along γ

effective   for every γ in the fixed class of null curves Γ,

$$\int_\gamma \left(\rho + \sum_{\hat\mu} p_{\hat\mu} \cos^2 \alpha_{\hat\mu} \right) v_n^2(k^a) \, d\theta \geq 0$$

where γ has tangent vector k^a and θ is a generalized affine parameter along γ

averaged weak energy condition (AWEC)

geometric   for every γ in the fixed class of timelike curves Γ,

$$\int_\gamma G_{mn}\xi^m \xi^n \, ds \geq 0$$

where γ has tangent vector ξ^a and s is proper time

physical   for every γ in the fixed class of timelike curves Γ,

$$\int_\gamma T_{mn}\xi^m \xi^n \, ds \geq 0$$

where γ has tangent vector ξ^a and s is proper time

effective   for every γ in the fixed class of timelike curves Γ,

$$\int_\gamma \left(\rho + \beta^2 \sum_{\hat\mu} p_{\hat\mu} \cos^2 \alpha_{\hat\mu} \right) v_t^2(\xi^a) \, ds \geq 0$$

where γ has tangent vector ξ^a and s is proper time

averaged strong energy condition (ASEC)

geometric   for every γ in the fixed class of timelike curves Γ,

$$\int_\gamma R_{mn}\xi^m \xi^n \, ds \geq 0$$

where γ has tangent vector ξ^a and s is proper time

physical   for every γ in the fixed class of timelike curves Γ,

$$\int_\gamma \left(T_{mn} - \frac{1}{2}T g_{mn} \right) \xi^m \xi^n \, ds \geq 0$$

where γ has tangent vector ξ^a and s is proper time
effective   for every γ in the fixed class of timelike curves Γ,

$$\int_\gamma \left\{ \left(\rho + \beta^2 \sum_{\hat{\mu}} p_{\hat{\mu}} \cos^2 \alpha_{\hat{\mu}} \right) v_t^2(\xi^a) - \frac{1}{2} \xi^n \xi_n \left(\rho - \sum_{\hat{\mu}} p_{\hat{\mu}} \right) \right\} ds \geq 0$$

where γ has tangent vector ξ^a and s is proper time

Before discussing their respective interpretations, a few remarks are in order. No reasonable impressionist analogue of either of the pointilliste dominant conditions are known.[26] In practice, one generally requires that Γ consist of a suitably large family of inextendible geodesics of the appropriate type. For the ANEC, if Γ consists of null geodesics, then one can replace the generalized affine parameter with the ordinary affine parameter. In no case can one allow arbitrary parameterizations for null curves in the defining integral, as that would simply reduce the ANEC to the NEC. If one further requires for the ANEC that the curves in Γ be achronal, then the condition is often called the 'averaged achronal null energy condition' (AANEC). For the AWEC, if Γ contains enough timelike geodesics and the spacetime is well behaved, then there may be null geodesics that are limit curves of subfamilies of Γ; in this case, the relevant characteristic integral will be nonnegative for those null geodesics, and the AWEC with the fixed Γ can be said to imply the ANEC for the family of limiting null geodesics. Even in well-behaved spacetimes, however, there may be null geodesics that are not the limit of any family of timelike geodesics, so in general the AWEC does not imply the ANEC. The ASEC does not imply either the AWEC or the ANEC. Clearly, the NEC, WEC and SEC respectively imply the ANEC, AWEC, and ASEC.

I am sorry to say the discussion of the possible interpretations of, or even just motivations for, the standard impressionist energy conditions is a simple one to have: there are no compelling geometrical, physical or effective interpretations of these conditions, not even hand-waving, rough or approximate ones, and no compelling physical or philosophical motivations for them.

I should perhaps clarify what I mean in claiming that there are no compelling interpretations or motivations of these conditions. One can certainly describe in simple, clear, physical language the sorts of spacetimes in which they will be satisfied—geodesics experience more positive than negative energy, the regions in which the pointilliste conditions are violated are bounded in various ways, *etc.*—but it is difficult, at best, to understand these classes of spacetimes as being related in any but accidental ways. There is nothing principled or lawlike that makes these spacetimes similar or the same in any deep sense. It is not easy to imagine principled conditions one could impose on theories of matter or fields—say, a form for the Lagrangian, or manifestation of a symmetry, *etc.*—that would ensure the sort of behavior captured

---

[26]One could in flat spacetimes, and possibly in stationary spacetimes, circumvent the obvious problems with formulating a dominant-like impressionist energy condition, but, being confined to flat (and possibly stationary) spacetimes, such a condition would have little import or relevance.

by the averaged conditions. This somewhat vague qualm is substantiated by the ease with which violations of the averaged conditions can be found, in both the classical and the quantum cases, as I discuss in §3.2.

More to the point, there is at least one interesting way of making this vague qualm more precise, that at the same time shows clearly the artificiality of the impressionist conditions as compared to the pointilliste conditions: none of the quantities constrained by the impressionist conditions enter the equations of motion or the field equations of any known kinds of physical system, and, correlatively, no couplings between any known kinds of physical system are mediated by those quantities; the opposite is true for the pointilliste conditions, whose constrained quantities promiscuously appear in equations of motion, field equations and couplings for many if not most known kinds of physical system. Finally, the restriction to geodesics has no compelling physical or philosophical basis that I can see, but appears to be dictated by pragmatic considerations about the technical tractability of required calculations.

Still, there is more to say about them, even though none has a clear, principled interpretation or motivation. These conditions were all constructed by reverse engineering—an investigator looked for the weakest condition she could impose on the averaged behavior of some quantity depending on curvature or stress-energy in order to derive the result of interest to her. (Indeed, I think it is not going too far to say that many of them represent a case of outright gerrymandering by the relativity community.[27]) Other researchers were impressed by the weakness of the condition used to derive the important result, and so picked it up and used it themselves. And so the impressionist conditions have been passed down through the generations of relativists, hand to hand from teacher to student, powerful, talismanic runes to be brought out and invoked with precise ceremony on formal occasions, but whose inner significance is beyond our ken, though their very familiarity often obscures that fact.[28]

This is not to say the impressionist energy conditions have no foundational or physical interest at all. It is often important to find the weakest conditions one can to prove theorems whose conclusions have great weight or significance, such as the positive energy theorems or the singularity theorems, if only, for example, to get as clear as one can on what those conclusions really depend on. If one wants to try to extend or modify one's global theory while ensuring that certain results remain true, for example, it behooves one to find the weakest conditions from which one

---

[27] The only physicists I know of to express similar concerns are Visser and Barceló [6, 199]; indeed they seem to be of the opinion that it is difficult to think of all energy conditions, not just the impressionist ones, as anything more than pragmatically convenient tools whose formulation is driven by the technical needs in proving desired theorems.

[28]   und Das und Den,
         die man schon nicht mehr sah
         (so täglich waren sie und so gewöhnlich),
         auf einmal anzuschauen: sanft, versöhnlich
         und wie an einem Anfang und von nah
              — Rainer Maria Rilke, "Der Auszug des verlorenen Sohnes"

can derive those results. For we who are interested in the foundations of the theory in and of itself, however, these impressionist conditions have little to offer. Still, because they have been used to prove deep results of great interest in themselves, it is important to understand what sorts of system violate and what sorts satisfy the conditions (which I will discuss in §3).

Before moving on, it will be edifying to examine in a little detail two of the most important technical qualifications made to the templates I gave of the averaged conditions. Tipler [189], which if my history is not mistaken was the first use of an averaged condition to prove results of any depth, required the additional constraint that the characteristic integral of the averaged condition at issue can equal zero for any curve only if its integrand (*e.g.*, $T_{mn}\xi^m\xi^n$ for the physical AWEC) equals zero along the entire curve. As Borde [21] points out, this constraint raises problems for the physical plausibility, or at least possible scope, of the conditions.[29] To see the problem, let us for the sake of definiteness focus attention for the moment on the physical AWEC. Then Tipler's constraint rules out cases where the integral equals zero because the relevant curve passes endlessly in and out of regions of positive and negative energy density. This may not sound so bad at first, until one realizes it means that, for a spacetime to satisfy the constrained condition, every curve in the fixed class must eventually traverse only regions of nonnegative energy density, both to the past and the future: violations of the WEC are to be allowed only in bounded regions in the interior of spacetime, so to speak. There seems even less physical justification for demanding this than for the bare AWEC in the first place.

To try to address this problem, Borde proposed modifications to the averaged conditions. The technical details of his proposals, while ingenious, are not worth working through for my purposes, as they are complicated and shed little light on the issues I am discussing. The gist of his proposed modifications is this: rather than requiring that the salient integral equal zero only when its integrand equal zero everywhere along the curve, we require only that, if the integral equal zero, then the integrand must be suitably periodic along the entire curve, *i.e.*, roughly speaking, that the integrand visit a neighborhood of zero frequently and that the lengths of the intervals it spends visiting those neighborhoods not approach zero as one heads along the curve in either direction. This allows application of the averaged condition to situations in which the total integral may essentially be zero even though there are large and long violations of the relevant pointilliste condition, such as may occur for the SEC during inflationary periods of a spacetime. In this sense, Borde's modifications do seem an improvement on Tipler's original version. One cannot help but feel though, given the intricacy and physical opacity of the mathematical machinery required to formulate Borde's condition, that the problems of physical interpretation in the sense I sketched above—not having in hand a principled justification for the condition founded on general, fundamental principles, but rather only reverse engineering the weakest suitable condition one can manage to prove the results one

---

[29]Chicone and Ehrlich [38] also pointed out that there were lacunæ in Tipler's proofs, unrelated to Borde's problems, but that is by the by for our purposes, as they also showed how to fix the problems.

wants for the particular class of spacetimes one is interested in—become perhaps even more severe than before.

2.3 Appendix: A Failed Attempt to Derive the NEC and SEC

It is sometimes claimed (*e.g.*, Liu and Rebouças [130]) that one can derive the NEC and the SEC from the Raychaudhuri equation. Even though I think the argument fails, it is of interest to try to pinpoint exactly why it fails, as it sheds light on why it appears to be difficult to derive the energy conditions from other fundamental principles (the difficulty strongly suggested by the lack of convincing derivations). I will sketch the argument only for the SEC, as that for the NEC is essentially the same, with only a few inessential technical differences.

Raychaudhuri's equation expresses the rate of change of the scalar expansion of a congruence of geodesics, as one sweeps along the congruence, as a function of the expansion itself, of the congruence's shear and twist tensors, and of the Ricci tensor. For a congruence of timelike geodesics with tangent vector ξ^a, it takes the form

$$\xi^n \nabla_n \theta = -\frac{1}{3}\theta^2 - \sigma_{mn}\sigma^{mn} + \omega_{mn}\omega^{mn} - R_{mn}\xi^m \xi^n \qquad (2.2)$$

where θ is the expansion of the congruence, σ_{ab} its shear and ω_{ab} its twist.[30] If the total sum on the right-hand side is negative, then the expansion of the congruence is decreasing with proper time, *i.e.*, the geodesics in the congruence are everywhere converging on each other. The first term on the right-hand side is manifestly negative, as is the second, since σ_{ab} is spacelike in both indices, and so $\sigma_{mn}\sigma^{mn} \geq 0$. For a hypersurface orthogonal congruence, it follows directly from Frobenius's Theorem that $\omega_{ab} = 0$. Thus, if we assume that "gravity is everywhere attractive", and we interpret this to mean that congruences of timelike geodesics which have vanishing twist should always converge, then, in order to ensure that the total right-hand side of equation (2.2) is always negative, we require that $R_{mn}\xi^m\xi^n \geq 0$, which is just the geometrical form of the SEC.

It should be clear why I fail to find the argument compelling. In fact, all one can conclude from the demand that the righthand side of equation (2.2) be nonpositive (when $\omega_{ab} = 0$) is that

$$R_{mn}\xi^m \xi^n \geq -\frac{1}{3}\theta^2 - \sigma_{mn}\sigma^{mn} \qquad (2.3)$$

everywhere. Of course, this is not the SEC, but only a weaker form of the geometric formulation, one that sets a nonconstant lower bound on "how negative" mass–energy

---

[30]See, *e.g.*, Wald [203, ch. 9, §2] for a derivation and explanation of the Raychaudhuri equation for both timelike and null congruences. There is a generalization of the Raychaudhuri equation that treats congruences of accelerated curves, but nothing would be gained for our purposes by discussing it.

and momentum-energy flux can get (invoking the physical form of the condition).[31] When one considers that one can, in every spacetime, find at every point a congruence of timelike geodesics that has divergent expansion as one approaches that point, one realizes that the inequality (2.3) is vacuous, for the right-hand side of the inequality can be made as negative as one likes. (Proof: in any spacetime, at any point p, consider the family of timelike geodesics defined by the family of unit, past-directed, timelike vectors at p, parametrized by proper time so that each geodesic's parameter has the value 0 at p; there will be some real number ϵ such that the class of geodetic-segments defined by considering all geodesics in the family for proper time values in the open interval $(-\epsilon, 0)$ defines a proper congruence; that congruence will have divergent expansion along all its members as one approaches proper time 0, $i.e.$, the point p, as can be seen by the fact that any spacelike volume swept along the flow of the congruence toward p will converge to 0.)

The heart of the problem should now be clear. Geodesic congruences are a dime a dozen. You can't throw a rock in a relativistic spacetime without hitting a zillion of them, most of them having no intrinsic physical significance. Because the pointilliste energy conditions, moreover, constrain the behavior of curvature terms only at individual points, and that by reference to all timelike or null (or both) vectors at those points, one can always find geodesic congruences that are as badly behaved as one wants, in just about any way one wants to make that idea precise, with respect to how various measures of curvature evolve along the congruences. Nonetheless, geodesic congruences seem to be about the only structure one has naturally available to work with, if one wants to try to constrain the behavior of curvature as measured by the contractions of curvature tensors with causal vectors. So long as one wants to work with geodesic congruences, therefore, it seems one must find some way to restrict the class one allows as relevant to those that are "physically significant" in some important and clear way. I know of no way to try to address that problem in any generality. Of course, one could always try to work with structures other than geodesic congruences, but, again, I know of no other natural candidates to try to use to constrain the behavior of measures of curvature, given the typical form of the energy conditions.

Even if one could find natural, compelling ways to restrict attention to a privileged class of congruences in such a way as to resolve the technical problems I raised for this kind of argument, there would still be interpretative problems with this kind of argument. As I discussed at the end of §2.1 above, I do not find it convincing to interpret the fact that causal congruences are convergent as a representation of the idea that "gravity is attractive". Without that interpretation, however, one has little motivation for invoking Raychaudhuri's equation in the first place without ancillary physical justification.

---

[31] Because the lower bound is variable, the propositions of Tipler [189] I discussed in §2.1 do not allow one to infer that this weaker condition is in fact equivalent to the WEC.

2.4   Appendix: Very Recent Work

Recently, Abreu et al. [1] introduced a new classical energy condition

flux energy condition (FEC)

  geometric
  1.   for any timelike vector ξ^a, $G^a{}_n \xi^n$ is causal
  2.   for any timelike vector ξ^a, $G^m{}_r G_{ms} \xi^r \xi^s \geq 0$
  physical
  1.   for any timelike vector ξ^a, $T^a{}_n \xi^n$ is causal
  2.   for any timelike vector ξ^a, $T^m{}_r T_{ms} \xi^r \xi^s \geq 0$
  effective   for each $\hat{\mu}$, $\rho^2 \geq p_{\hat{\mu}}^2$

There is, as is to be expected, no simple interpretation of its geometric form. The simplest interpretation of its physical form is that the total flux of energy–momentum as measured by any timelike observer is always causal, albeit the temporal direction of the flux is not restricted. Because isotropic tachyonic gases always satisfy $\rho < \frac{1}{3}p$, with weaker bounds for anisotropic tachyonic material, the effective form may be interpreted as ruling out the possibility of tachyonic matter. Otherwise, I know of no compelling interpretation of it, as it allows energy density to be unboundedly negative, so long as the absolute value of pressure is not too great.

Abreu et al. [1] argue that the FEC gives better support to the claim that the cosmological equation-of-state parameter w (the so-called barotropic index—see footnote 15) must be ≤ 1, and so better substantiates arguments in favor of entropy bounds they give based on that assumption. Martín-Moruno and Visser [134, 135] investigated its properties and proposed a quantum analogue of it, which, they claim, works in several respects better than the standard quantum energy conditions.[32] The FEC, therefore, shows *prima facie* promise as being of real physical interest. It is, moreover, manifestly weaker than all the other standard energy conditions, as its characteristic nonlinearity (most easily seen in the second given articulations of its geometric and physical forms, and in its effective form) ensures that essentially no limit is placed on the possible negativity of the ordinary mass–energy of matter. If, therefore, it bears out its promise for leading to, or at least supporting, results of interest, it would be a great improvement on the standard energy conditions. Because, however, its properties and consequences are virtually unknown as compared to the standard conditions, I shall not discuss it further.

Even more recently, Martín-Moruno and Visser [135] proposed two more energy conditions, the determinant energy condition (DETEC) and the trace-of-square energy condition (TOSEC), and also proposed quantum analogues for them. Again, these energy conditions seem *prima facie* interesting, but even less work has been done on and with them than the FEC, so I shall not discuss them here either.

---

[32]See Curiel [48] for extended discussion of energy conditions in quantum field theory on curved spacetime.

2.5 Technical Appendix: Average Radial Acceleration

To characterize the idea of the average radial acceleration of a causal geodesic,[33] let ξ^a be a future-directed causal vector field whose integral curves γ are affinely parametrized geodesics. If γ is timelike, then assume ξ^a to be unit. Let λ^a be a vector field on γ such that at one point $\lambda^n \xi_n = 0$ and $\pounds_\xi \lambda^a = 0$. (Note that if ξ^a is null, then λ^a may be proportional to ξ^a; otherwise it must be spacelike.) Then automatically $\lambda^n \xi_n = 0$ at all points of γ. λ^a is usefully thought of as a "connecting field" that joins the image of γ to the image of another, "infinitesimally close" integral curve of ξ^a. Then $\xi^m \nabla_m (\xi^n \nabla_n \lambda^a)$ represents the acceleration of that neighboring geodesic relative to γ. According to the equation of geodesic deviation,

$$\xi^m \nabla_m (\xi^n \nabla_n \lambda^a) = R^a{}_{mnr} \xi^m \lambda^n \xi^r$$

Now, fix an orthonormal triad-field $\{\overset{\mu}{\lambda}{}^a\}_{\mu \in \{1, 2, 3\}}$ along γ such that each $\overset{\mu}{\lambda}{}^a$ forms a connecting (relative acceleration) field along γ. The magnitude of the radial component of the relative acceleration in the μth direction then is $- \overset{\mu}{\lambda}_r \xi^m \nabla_m (\xi^n \nabla_n \overset{\mu}{\lambda}{}^r)$. Fix a point $p \in \gamma$. The *average radial acceleration* A_r of γ at p is defined to be

$$A_r := -\frac{1}{k} \sum_\mu \overset{\mu}{\lambda}_r \xi^m \nabla_m (\xi^n \nabla_n \overset{\mu}{\lambda}{}^r)$$

where k is 3 if ξ^a is timelike and 2 if null. It is straightforward to verify that the average radial acceleration is independent of the choice of orthonormal triad, so it encodes a quantity of intrinsic geometric (and physical) significance accruing to ξ^a. A simple calculation using the equation of geodesic deviation then shows that

$$A_r = -\frac{1}{k} R_{mn} \xi^m \xi^n \tag{2.4}$$

If the Einstein field equation is assumed to hold, it follows that

$$A_r = -\frac{8\pi}{k} (T_{mn} - \frac{1}{2} T g_{mn}) \xi^m \xi^n \tag{2.5}$$

which reduces in the case of null vectors to

$$A_r = -4\pi T_{mn} \xi^m \xi^n \tag{2.6}$$

---

[33] I follow the exposition of Malament [133, §2.7], with a few emendations.

3 Consequences and Violations

To study the role of energy conditions in spacetime theories, I will look at results
that do not depend on the imposition of any field equations (*e.g.*, the Einstein field
equation) and yet directly constrain spacetime geometry. One often hears the claim
that such-and-such result (*e.g.*, various singularity theorems, various versions of the
geodesic postulate, the Zeroth Law of black hole mechanics, *etc.*) that assumes an
energy condition does require the Einstein field equation for its proof, but one must
be careful of such claims. It is almost always the case, in fact, that the Einstein field
equation is logically independent of the result (in the strong sense that one can assume
the negation of the Einstein field equation and still derive the result); the Einstein field
equation is used in such cases *only* to provide a physical interpretation of the assumed
energy condition; mathematically, one in general needs only the geometric form of
the condition, which is why I distinguish the geometric from the physical form.[34]
In this section, every consequence of the energy conditions I discuss is of this type:
it is logically independent of the Einstein field equation, and relies on the Einstein
field equation only for the physical interpretation of the assumed geometric energy
condition.[35] Many of the violations of the energy conditions I list here, however, do
rely on assuming the Einstein field equation for their derivation, in so far as they use
the Lagrangian formulation of the relevant forms of matter to derive the violation,
or in so far as they rely on the effective form of the energy conditions in conjunction
with, *e.g.*, the Friedmann equations to derive the violation.

I will begin with a list of the consequences of the energy conditions, *i.e.*, the results
each energy condition is used to derive, and then discuss the roles the conditions play
in the derivations of those results. I then list the classical cases in which each energy
condition is known to fail, then discuss how the known failures may or may not
undermine our confidence in the consequences.[36] In several of the references I give
in the list of consequences, no explicit mention is made of energy conditions, but, if
one works through their arguments, one will see that the relevant energy condition
is indeed being implicitly assumed. In other works I cite, an energy condition is
explicitly assumed, but in fact, according to the arguments of those works, either a
weaker one is sufficient or a stronger one is required; in such cases, I cite the result
under the sufficient or required condition. For almost none of the statements in the list

---

[34]There is perhaps room for debate over this claim, at least in a few cases. Some elements of the black
hole uniqueness theorems, *e.g.*, "use" the Einstein field equation to show that certain distinguished
spacelike hypersurfaces must be spatially conformally flat when the entire spacetime is assumed to
be vacuum; in such a case, the spatial conformal flatness follows from the vanishing of the Ricci
tensor, which follows from the vanishing of the stress-energy tensor by the Einstein field equation.
I would still argue in such cases that the Einstein field equation is not necessary for the proof of
the theorem—only $R_{ab} = 0$ is—and, again, the Einstein field equation is used only to provide the
necessary condition a physical interpretation.

[35]In cosmology, several of the most interesting results do require assumption of the Einstein field
equation. For this reason, and also because it is such a large and rich field on its own, I explore the
role and character of energy conditions in the context of cosmology at some length in Curiel [47].

[36]See Curiel [48] for examination of the cases of failure in the quantum regime.

of consequences is it the case that the energy condition alone is necessary or sufficient; it is rather that the energy condition is one assumption among others in the only known way (or ways) to prove the result. When I list the same proposition as a consequence of more than one energy condition (*e.g.*, "prohibition on spatial topology change" under both WEC and ANEC), it means that there are different proofs of the statement using different ancillary assumptions. When I qualify a spacetime as "spatially open" or "spatially closed", it should be understood that the spacetime is globally hyperbolic and the openness or closedness refers to the topology of spacelike Cauchy surfaces in a natural slicing of the spacetime.

3.1 Consequences

NEC

1. formation of singularities after gravitational collapse in spatially open space-times [148]
2. formation of singularities in asymptotically flat spacetimes with non-simply connected Cauchy surface [82, 128]
3. formation of an event horizon after gravitational collapse [148–150]
4. trapped and marginally trapped surfaces and apparent horizons must be inside asymptotically flat black holes [203]
5. Hawking's Area Theorem for asymptotically flat black holes (Second Law of black hole mechanics) [103]
6. the area of a generalized black hole always increases[37] (Second Law of generalized black hole mechanics) [111]
7. asymptotically predictable black holes cannot bifurcate[38] [203]
8. the domain of outer communication of a stationary, asymptotically flat, causally well behaved spacetime is simply connected[39] [40, 79, 81]
9. a stationary, asymptotically flat black hole has topology $\mathbb{S}^2$, if the domain of outer communication is globally hyperbolic and the closure of the black hole is compact[40] [40, 78]

---

[37] Hayward [111] defines a generalized notion of black hole, one applicable to spacetimes that are not asymptotically flat, by the use of what he calls "trapping horizons". In the same paper, he shows that generalized black holes obey laws analogous to the standard Laws of black hole mechanics.

[38] A spacetime is asymptotically predictable if it is asymptotically flat, and there is a partial Cauchy surface whose boundary is the event horizon, such that future null infinity is contained in its future domain of dependence.

[39] The domain of outer communication of an asymptotically flat spacetime is, roughly speaking, the exterior of the black hole region. See Chruściel et al. [41, §2.4] for a precise definition. This theorem is similar to, but stronger than, the original Topological Censorship Theorem of Friedman et al. [77]; see footnote 59. The theorem due to Galloway and Woolgar [81] in fact requires only the ANEC.

[40] This is also a constituent of the proof of the full No-Hair Theorem, but is important enough a result to warrant its own entry in the list; see footnote 41.

10. almost all the constituents of the black hole No Hair Theorem for asymptotically flat black holes[41] [12, 31, 119, 120, 136, 141, 164, 165, 185, 186, 201, 202]
11. generalized black holes are regions of "no escape" [110]
12. limits on energy extraction by gravitational radiation from colliding asymptotically flat black holes [103]
13. positivity of ADM mass[42] [4, 153]
14. the Generalized Second Law of Thermodynamics[43] [73]
15. Bousso's covariant universal entropy bound[44] [73]
16. the Shapiro "time-delay" is always a delay, never an advance[45] [200]
17. chronology implies causality[46] [107]
18. standard formulations of the classical Chronology Protection Conjecture[47] [105]

WEC

1. asymptotically flat spacetimes without naked singularities are asymptotically predictable [104]
2. asymptotically flat black holes cannot bifurcate [104]
3. the event horizon of a stationary black hole is a Killing horizon[48] [104, 107]

---

[41] The No Hair Theorem states that an asymptotically flat, stationary black hole is completely characterized by three parameters, *viz.*, its mass, angular momentum and electric charge. The proof of this theorem logically comprises many steps, each of interest in its own right, and historically stretched from the original papers of Israel [119, 120] to the final results of Mazur [136]. There are too many constituents of the proof to list each individually. A few remaining constituents require the DEC; see that list for details. Heusler [113] provides an excellent, relatively up-to-date overview of all the known results. There is an analogous No Hair Theorems for the generalized black holes of Hayward [111], but I will not discuss them.

[42] Earlier proofs relied on the DEC; see that list for details.

[43] The total entropy of the world, *i.e.*, the entropy of ordinary matter plus the entropy of a black hole as measured by its surface area, never decreases.

[44] Bousso [24, 25], clarifying and improving on earlier work by Bekenstein [13, 15–17], 't Hooft [184], Smolin [182, 183], Corley and Jacobson [43], and Fischler and Susskind [72], conjectured that in any spacetime satisfying the DEC the total entropy flux S_L through any null hypersurface L satisfying some natural geometrical conditions must be such that $S_L \leq A/4$, where A is a spatial area canonically associated with L. Flanagan et al. [73] managed to prove the bound using the weaker NEC.

[45] One can understand this result physically as a prohibition on a certain form of "hyper-fast" travel or communication. Roughly speaking, this is travel in spacetime in which the traveler is measured by external observers, in a natural way, to travel faster than the speed of light, even though the traveler's worldline is everywhere timelike. It is closely related, though not equivalent, to the idea of traversable wormholes.

[46] Chronology holds if there are no closed timelike curves; causality holds if there are no closed causal curves.

[47] This states, roughly, that the formation of closed timelike curves always requires either the presence of singularities or else pathological behavior "at infinity".

[48] A Killing horizon is a null hypersurface generated by the orbits of a non-degenerate null Killing field.

4. Third Law of black hole mechanics[49] [121]
5. limits on energy extraction by gravitational radiation from asymptotically flat colliding black holes [104]
6. formation of singularities after gravitational collapse in spatially open spacetimes [89, 189]
7. cosmological singularities in spatially open or flat spacetimes [89, 96]
8. cosmological singularities in globally hyperbolic spacetimes that are non-compactly regular near infinity[50] [83]
9. prohibition on spatial topology change [87, 187]
10. geodesic theorems for "point-particles" [64, 70]
11. mass limits for stability of hydrostatic spheres against gravitational collapse [19]
12. some standard forms of the Cosmic Censorship Hypothesis [123]

SEC

1. cosmological singularities in spatially closed spacetimes [86, 89, 98, 101, 106, 109]
2. cosmological singularities in spatially open spacetimes [89, 97, 100, 106, 109]
3. cosmological singularities in spacetimes with partial Cauchy surfaces [89, 97, 100, 101, 109]
4. formation of singularities after gravitational collapse in spatially closed spacetimes [89, 101, 109]
5. formation of singularities after gravitational collapse in spatially open spacetimes [89, 109]
6. Lorentzian splitting theorem[51] [80, 214]
7. a given globally hyperbolic extension of a spacetime is the maximal such extension [163]
8. existence and uniqueness of constant-mean-curvature foliations for spacetimes with compact Cauchy surfaces [10, 26, 161, 162]

DEC

1. formation of a closed trapped surface after gravitational collapse of arbitrary (*i.e.*, not necessarily close to spherical) matter distribution [174]

---

[49] No physical process can reduce the surface gravity of an asymptotically flat black hole to zero in a finite amount of time.

[50] Roughly speaking, a globally hyperbolic spacetime is noncompactly regular near infinity if it has a (partial) Cauchy surface that is the union of well behaved nested sets, each having compact boundary, that are themselves noncompact near infinity.

[51] I will give two versions of the theorem; see Galloway and Horta [80] for proofs of both. In order to state the first version of the theorem, define a *timelike line* to be an inextendible timelike geodesic that realizes the supremal Lorentzian distance between every two of its points [68]. Then the theorem, as first conjectured by Yau [214], is as follows: let $(\mathcal{M}, g_{ab})$ be a timelike geodesically complete spacetime satisfying the SEC; if it contains a timelike line, then it is isometric

2. a stationary, asymptotically flat black hole is topologically $\mathbb{S}^2$[52] [104]
3. a generalized black hole is topologically $\mathbb{S}^2$[53] [111]
4. constituents of the black hole No Hair Theorems for asymptotically flat black holes[54] [12, 32, 107]
5. Zeroth Law of black hole mechanics[55] [7]
6. Zeroth Law of generalized black hole mechanics[56] [111]
7. every past timelike geodesic in spatially open, nonrotating spacetimes with nonzero spatially averaged energy densities is incomplete[57] [179, 180]
8. positivity of ADM energy [172, 211]
9. positivity of Bondi energy [112, 115, 131, 173]
10. asymptotic energy-area inequality in the spherically symmetric case[58] [112]
11. if a covariantly divergence-free T_{ab} vanishes on a closed, achronal set, it vanishes in the domain of dependence of that set [102, 107]

(Footnote 51 continued)

to $(\mathbb{R} \times \Sigma, t_a t_b - h_{ab})$, where (Σ, h_{ab}) is a complete Riemannian manifold and t^a is a timelike vector field in $\mathcal{M}$. (In particular, $(\mathcal{M}, g_{ab})$ must be globally hyperbolic and static.)

In order to state the second, we need two more definitions. First, the *edge* of an achronal, closed set Σ is the set of points $p \in \Sigma$ such that every open neighborhood of p contains a point $q \in I^-(p)$, a point $r \in I^+(p)$ and a timelike curve from q to r that does not intersect Σ. Second, let Σ be a nonempty subset of spacetime; then a future inextendible causal curve is a *future Σ-ray* if it realizes the supremal Lorentzian distance between Σ and any of its points lying to the future of Σ [80]; *mutatis mutandis* for a *past Σ-ray*. (If γ is a Σ-ray, it necessarily intersects Σ.) The second version of the theorem is as follows: let $(\mathcal{M}, g_{ab})$ be a spacetime that contains a compact, acausal spacelike hypersurface Σ without edge and obeys the SEC; if it is timelike geodesically complete and contains a future Σ-ray γ and a past Σ-ray η such that $I^-(\gamma) \cap I^+(\eta) \neq \emptyset$, then it is isometric to $(\mathbb{R} \times \Sigma, t_a t_b - h_{ab})$, where (Σ, h_{ab}) is a compact Riemannian manifold and t^a is a timelike vector field in $\mathcal{M}$. (In particular, $(\mathcal{M}, g_{ab})$ must be globally hyperbolic and static.)

I discuss the physical meaning of the splitting theorems below.

[52] This is also a constituent of the proof of the full No Hair theorem, but is important enough a result to warrant its own entry in the list; see footnote 41. Hawking's original proof was not rigorous; in particular, it did not completely rule out a toroidal topology. See Gannon [83] for a rigorous proof of the theorem in electrovac spacetimes, and Galloway [78] and Chruściel and Wald [40] for a rigorous proof using the NEC for otherwise arbitrary stress-energy tensors but more stringent constraints on the global topology of the spacetime.

[53] See footnote 37.

[54] See footnote 41.

[55] The surface gravity is constant on the event horizon of a stationary, asymptotically flat black hole.

[56] The total trapping gravity of a generalized black hole is bounded from above, and achieves its maximal value if and only if the trapping gravity is constant on the trapping horizon, which happens when the horizon is stationary. (See footnote 37.)

[57] This theorem is particularly strong: it implies that any singularity-free spacetime satisfying the other conditions must have everywhere vanishing averaged spatial energies, making them highly non-generic.

[58] This inequality, first conjectured by Penrose [151], states that if a spacelike hypersurface in a spherically symmetric, asymptotically flat spacetime contains an outermost marginally trapped sphere of radius R (in coordinates respecting the spherical symmetry), then the ADM energy $\geq \frac{1}{2} R$. The DEC need hold only on the spacelike hypersurface, not in the whole spacetime.

12. standard statements of the initial value formulation of the Einstein field equation with nontrivial T_{ab} is well posed (in the sense of Hadamard) [107, 203]
13. natural definition of the center of mass, multipole moments and equations of motion for an extended body [55–58, 65, 67, 169, 170]
14. some standard forms of the Cosmic Censorship Hypothesis [92, 123, 152, 203]

SDEC

1. geodesic theorem for "arbitrarily small" bodies, neglecting self-gravitational effects [93, 133, 208]
2. geodesic theorem for "arbitrarily small" bodies, including self-gravitational effects [66]

ANEC

1. a stationary, asymptotically flat black hole is topologically $\mathbb{S}^2$ [122]
2. focusing theorems for congruences of causal geodesics [21]
3. formation of singularities after gravitational collapse in spatially open spacetimes [166, 176]
4. Topological Censorship Theorem[59] [77]
5. prohibition on traversable wormholes [140]
6. prohibition on spatial topology change [22]
7. positivity of ADM energy [156]

AWEC ∅
ASEC

1. cosmological singularities in spatially closed spacetimes[60] [176, 189]
2. cosmological singularities in spatially open spacetimes[61] [176, 189]

There is a striking absentee from the list of consequences: strictly speaking, the First Law of black hole mechanics (for asymptotically flat black holes)—conservation of mass–energy—does not require for its validity the assumption of

---

[59] The theorem states: fix an asymptotically flat, globally hyperbolic spacetime satisfying the ANEC; let γ be a causal curve with endpoints on past and future null infinity that lies in a simply connected neighborhood of null infinity; then every causal curve with endpoints on past and future null infinity is smoothly deformable to γ. Roughly speaking, this theorem says that no observer remaining outside a black hole can ever have enough time to probe the spatial topology of spacetime: isolated, nontrivial topological structure with positive energy will collapse into black holes too quickly for light to cross it. Loosely speaking, the region outside black holes is topologically trivial.

[60] Strictly speaking, Tipler's proof requires the ASEC with the additional constraint that its characteristic integral can equal 0 for any geodesic only if its integrand ($R_{mn}\xi^m\xi^n$) equals 0 along the entire geodesic. Senovilla's proof does not require these extra assumptions, though it does require the existence of a Cauchy surface with vanishing second fundamental form.

[61] Strictly speaking, Tipler's proof of this theorem requires the WEC as well as the ASEC, and also requires the same further constraint on the ASEC as described in footnote 60. Senovilla's proof also requires what is described in footnote 60.

any energy condition (unlike the other three Laws).[62] The issue is somewhat delicate in the details, however. The delicacy arises from the fact that all the most rigorous and the most physically compelling derivations of the Law I know [7, 205] assume that the surface gravity of the black hole is constant on the event horizon. This, of course, is the Zeroth Law of black hole mechanics, and all known proofs of the most general form of the Zeroth Law rely on the DEC. The qualification "most general" is required because there are weaker forms of the Zeroth Law that require no energy condition for their proof: any sufficiently regular Killing horizon must be bifurcate, and the appropriate generalization of surface gravity for a bifurcate Killing horizon must be constant on the entire horizon, without the need to impose any energy condition [113, 125, 159, 160, 204].[63] This is a weaker form of the Zeroth Law, in so far as it is not known whether the event horizons of all "physically reasonable" black holes are sufficiently regular in any of the senses required, though in fact the event horizons of all known exact black hole solutions are, and the condition of sufficient regularity has strong physical plausibility on its own, at least if one accepts any version of Cosmic Censorship—it almost necessarily follows that any non-sufficiently regular horizon will eventuate in a naked singularity.

Whether one considers the First Law a consequence of the DEC, therefore, depends on whether one thinks it suffices simply to assume the Zeroth Law in its most general form, whether one thinks one should include a derivation of the most general form of the Zeroth Law in a derivation of the First Law, or whether one thinks that the weaker form of the Zeroth Law, which requires no energy condition, suffices for the purposes of the First Law. The delicacy is exacerbated by the fact that (at least) two conceptually distinct formulations of the First Law appear in the literature, what (following Wald [204, ch. 6, §2]) I will call the physical-process version and the equilibrium version. The former fixes the relations among the changes in an initially stationary black hole's mass, surface gravity, area, angular velocity, angular momentum, electric potential and electric charge when the black hole is perturbed by throwing in an "infinitesimally small" bit of matter, after the black hole settles back down to stationarity. The latter considers the relation among all those quantities for two black holes in "infinitesimally close" stationary states, or, more precisely, for two "infinitesimally close" black hole spacetimes.

The roles the assumption of the Zeroth Law plays in the proofs of the two versions of the First Law differ significantly, moreover, so it is not clear one could give a single principled answer to the question of whether or not the First Law is a consequence

---

[62]Hayward [111] does give a proof of what he calls the First Law for generalized black holes (footnote 37), and that does explicitly require the NEC, but the physical interpretation of Hayward's result is vexed (as he himself admits), so I did not list it among the consequences of the NEC. The physical interpretation of that result would be an interesting problem to resolve, as it would likely shed light on the already vexed problem of understanding energy in general relativity.

[63]Roughly speaking, a Killing horizon is sufficiently regular in the relevant sense if: it is (locally) bifurcate; or the null geodesic congruence constituting it is geodesically complete; or the twist of the null geodesic congruence has vanishing exterior derivative; or the domain of exterior communication is static; or the domain of exterior communication is stationary, axisymmetric, and the 2-surfaces orthogonal to the two Killing fields are hypersurface orthogonal (and so integrable themselves).

of the DEC that covered both versions at once. For example, in the physical-process version, but not in the equilibrium version, one must assume that the black hole settles back down to a stationary state after one throws in the small bit of matter, and so, *a fortiori*, that the event horizon is not destroyed when one does so, resulting in a naked singularity. I know of no rigorous proofs of the stability of an event horizon under generic small perturbations. All the most compelling arguments in favor of a reasonably broad kind of stability I know, however, do assume constraints on the form of the matter causing the perturbation, constraints that usually look a lot like energy conditions.[64]

Why is there this problem with understanding the relation of the First Law to the energy conditions? The difficulty seems especially surprising in light of the fact that it is the only one of the Laws that constrains mass–energy! Is it, perhaps, that mere conservation doesn't care whether mass–energy is negative or positive?

As striking as the difficulty in that case is, however, I still find more striking the number, variety and depth of what are indubitably consequences that the energy conditions *do* have, especially without input from the Einstein field equation. The two most numerous types of theorems in the list of consequences are those pertaining to singularities and those to black holes (including horizons), respectively. Indeed, it was the epoch-making result of Penrose [148] showing that a singularity would inevitably result from gravitational collapse in an open universe that first demonstrated the power that the qualitative abstraction of energy conditions gives in proving far-reaching results of great physical importance. I will first discuss some interesting features of the singularity theorems and the role that energy conditions play in their proofs, then do the same for theorems about black holes, positive energy, geodesic theorems and entropy bounds.[65] In §3.2, I will then review the violations of the energy conditions and discuss whether they give us grounds for doubting the physical relevance of the positive consequences.

The weakest condition, the NEC, already has remarkably strong consequences. Among the singularity theorems it supports, to my mind the most astonishing is the one due to Gannon [82] and Lee [128]:[66] in any asymptotically flat spacetime with a non-simply connected Cauchy surface, a singularity is bound to form. Topological complexity by itself, with the only constraint on metrical structure being the mild one of the NEC, suffices for the formation of singularities (in the guise of the incompleteness of a causal geodesic). The theorem gives one no information about the singularity, whether it will be a timelike or null geodesic that is incomplete, or whether it will be associated with pathology in the curvature, or something that looks like collapse of a material body, or will be cosmological in character (such as a big bang or big crunch), but the simple fact that nontrivial topology plus the weak-

---

[64]See, *e.g.*, Press and Teukolsky [124, 158], Kay and Wald [124], Carter [33, 34], and Kokkotas and Schmidt [126].

[65]I will not discuss the role of energy conditions in ensuring that the initial value formulation of general relativity is well posed, as the relation between the two is complex and very little is known about it. Although I will make a few remarks on the subject in §5, it is work for a future project.

[66]Gannon and Lee discovered it independently, roughly simultaneously.

est energy condition, irrespective of dynamics, suffices for geodesic incompleteness already shows the profound power of these conditions. It is tempting to relate Gannon's and Lee's singularity theorem to Topological Censorship, especially in so far as the latter requires only the ANEC, which the NEC implies. If one assumes that the singularity predicted by the theorem will be hidden behind an event horizon, then the theorem gives some insight into why nontrivial spatial topological structure will always (quickly come to be) hidden inside a black hole. (See footnote 59.) It also suggests that, in some rough sense, nontrivial topological structure may have mass–energy associated with it (perhaps of an ADM-type). It would be of some interest to see whether that idea can be made precise; one possible approach would be to see whether one could attribute some physically reasonable, nonzero ADM-like mass to flat, topologically nontrivial spacetimes. If so, I think this would give insight into the vexed question of the meaning of "mass" and "energy" in general relativity. If such a definition were to be had, I conjecture that nontrivial topological structure could have either positive or negative mass–energy, depending on the form of the structure; otherwise, it would not seem necessary to assume an energy condition in order to derive the Topological Censorship Theorem.[67]

Another striking feature of the list is that the only important consequences of the SEC (and the ASEC) are singularity theorems,[68] and among them the most physically salient ones, whereas the DEC, contrarily, is used in only one type of singularity theorem (Senovilla [179, 180]), and that of a character completely different from the other singularity theorems. The singularity theorems following from the SEC are the most physically salient both because they tend to have the weakest ancillary assumptions, and because they apply to physically important situations, both for collapsing bodies and for cosmology. I have no compelling explanation for why the SEC should have no important consequences other than singularity theorems. Perhaps it has to do with the fact that the SEC has a relatively clear geometrical interpretation (convergence of timelike geodesics) that is manifestly relevant to the formation of singularities, whereas its physical and effective interpretations are obscure at best. If so, then one may want to consider the SEC a case of gerrymandering, the relativity community simply having posited the weakest formal condition it could find to prove the results it wants. This line of thought becomes especially attractive when one contemplates the many possible violations of the SEC and even more the strong preponderance of indirect observational evidence that the SEC has been widely violated on cosmological scales at many different epochs in the actual universe, and is likely being violated

---

[67] A good place to start might be the investigation of asymptotically flat spacetimes with nontrivial second Stiefel-Whitney class, as it is known that such spacetimes cannot support a global spinor structure [88, 90]. That shows already that there is something physically *outré* about those spacetimes.

[68] Although the proposition that a given globally hyperbolic extension of a spacetime is the maximal such extension depends for its only known proof on the assumption of the SEC, this is not really a counter-example to my claim: roughly speaking, the proof works by showing that the given globally hyperbolic extension cannot be extended (and so is maximal) because to do so would result "immediately" in singularities, contradicting the assumption of extendibility.

right now.[69] The result of Ansoldi [3], however, that black holes with singularity-free interiors necessarily violate the SEC, may push one towards the opposite view, in so far as it comes close to making the SEC both necessary and sufficient for the occurrence of certain types of singularities. (The construction of singularity-free FLRW spacetimes violating the SEC, in Bekenstein [14], buttresses this line of thought; I discuss this further below.)

I have no explanation for why the DEC should be used in almost no singularity theorems, except for the simple observation that the only real addition the DEC makes to the NEC and the WEC, that energy–momentum flux be causal, has no obvious connection to the convergence of geodesics. The one type of singularity theorem (Senovilla [179, 180]) it is used in, moreover, is the only one to make substantive, explicit assumptions (over and above the energy conditions themselves) about the distribution of stress-energy, in this case in the demand for nonzero averaged spatial energy density. Perhaps that is why the DEC comes into play in this theorem, though I have no real insight into how or why the DEC may bear on averaged spatial energy density and its relation to the convergence of geodesic congruences.

The Lorentzian splitting theorems may be thought of as rigidity theorems for singularity theorems invoking the SEC, for the splitting theorems show that, under certain other assumptions, there will be no singularities only when the spacetime is static and globally hyperbolic.[70] Static and globally hyperbolic spacetimes, however, are "of measure zero" in the space of all spacetimes, and so being free of singularities is, under the ancillary conditions, unstable under arbitrarily small perturbations.[71] Thus, they go some way towards proving the conjecture of Geroch [86] that essentially all spatially closed spacetimes either have singularities or do not satisfy the SEC.[72]

As a group, the singularity theorems are perhaps the most striking example of the importance of ascertaining the status and nature of the energy conditions, because all the assumptions used in proving essentially all of them have strong observational or theoretical support *except* the energy conditions, as Sciama [175] emphasized even before there were serious observational grounds for doubting any of the energy

---

[69]See §3.2 for discussion, and Curiel [47] for a more extensive and thorough analysis.

[70]See footnote 51 for a statement and explication of the splitting theorems. See Beem et al. [11, ch. 14] for a beautiful discussion of the rationale behind and intent of rigidity theorems, as well as an exposition of many of the most important ones.

[71]One should bear in mind that this argument is hand-waving at best. First, there is no known natural measure on the space of spacetimes; second, even if there were, being a measure on an infinite-dimensional space, it is possible that every open set (in some natural topology, of which there is also not one known) would have zero measure or infinite measure. (There is no Borel measure on an infinite-dimensional Fréchet manifold; thus measure and topology tend to come apart.) In that case, in a natural sense "arbitrarily small" perturbations of a static, globally hyperbolic spacetime could in fact yield another static, globally hyperbolic spacetime. This problem is not unique to this argument but plagues all hand-waving arguments invoking "measure zero" sets in the space of all spacetimes, which are a dime a dozen, especially in the cosmology literature. See Curiel [49] for detailed discussion of all these issues.

[72]If this conjecture were to be precisely formulated and proven, perhaps one could view it as providing something like an *a posteriori* partial physical interpretation of the SEC.

conditions. This raises the question of the necessity of the energy conditions for the singularity theorems. That some of the impressionist energy conditions can be used to prove essentially identical theorems already shows that satisfaction of the pointilliste conditions is not necessary for validity of at least some of the theorems. The original singularity theorem, the demonstration by Penrose [148] that singularities should form after gravitational collapse in spatially open universes, holds under the weaker assumption of the ANEC [166, 176]. Likewise, the existence of cosmological (*i.e.*, non-collapse) singularities in both spatially open and closed universes can be shown under the assumption of the ASEC [176, 189], without the full SEC. So far as I know, there is no proof that gravitational collapse will lead to singularities in the case of spatially closed spacetimes under the weaker assumption of an impressionist energy condition. I conjecture that there are such theorems; it would be of some interest to formulate and prove one or to construct a counter-example.

With the possible exception of the First Law of black hole mechanics (for asymptotically flat black holes), every fundamental result about black holes requires an energy condition for its proof, with the majority relying either on the NEC or the DEC. Roughly speaking, the results pertaining to black holes fall into three categories: those constraining the topological and Killing structure of horizons; those constraining the kinds of property black holes can possess; and those constraining the relations among the horizon and the properties. Almost all of the first category invokes the NEC for their proof. One can perhaps see why the NEC is relevant for the results about the topological and Killing structure of horizons associated with asymptotically flat black holes: such a black hole is defined as an event horizon, which is the boundary of the causal past of future null infinity, and the boundary of the causal past or future of any closed set is a null surface, *i.e.*, is generated by null geodesics and so may be thought of as a null geodesic congruence. The proofs of many of those results, moreover, tend to have the same structure: very broadly speaking, one assumes the result is not true and then derives a contradiction with the fact that null geodesic congruences, by dint of the NEC, must be convergent (or at least not divergent). This suggests that the NEC is necessary for these theorems, a suspicion strengthened by the facts that, first, there is no weaker energy condition that one could attempt to replace it with (except perhaps the FEC, if it turns out to be viable—see §2.4), and, second, no such results are known to follow from any of the impressionist energy conditions. Again, it would be of interest to see whether the impressionist energy conditions could be used to prove theorems about the topological and Killing structure of black hole horizons, or else to construct counter-examples to the results in spacetimes in which the impressionist but not the pointilliste conditions hold. The NEC is also used to prove many results about the kinds of properties required to characterize black holes (the constituents of the No Hair Theorems), *viz.*, that stationary black holes can be entirely characterized by three parameters, mass, angular momentum and electric charge. I have no physically compelling story to tell about why the NEC relates intimately to these kinds of result. Again, the lack of such results depending on impressionist conditions suggests that the pointilliste conditions are necessary, and, again, it is would be of some interest either to prove analogous results using the impressionist conditions or to find counter-examples.

Every consequence of the DEC pertaining to black holes is of the kind that constrains topological or Killing structure of the horizons. There is, however, no common thread to the role the DEC plays in the proofs of those various results analogous to the way that the NEC plays essentially the same role in the proofs of many of its consequences. It is thus difficult even to hazard a guess about the necessity of the DEC for these consequences. It would be of great interest to work through the various results to see whether counter-examples to them satisfying or violating the DEC could be found, or whether proofs using weaker energy conditions can be found. That there is no impressionist analogue to the DEC may suggest that the DEC is necessary for these results.

Roughly speaking, the idea of the Cosmic Censorship Hypothesis is that "naked singularities" should not be allowed to occur in nature, where, continuing in the same rough vein, a naked singularity is one that is visible from future null infinity. Now, the relation of the energy conditions to the status of the Cosmic Censorship Hypothesis is complicated, first and foremost, by the fact that there are a multitude of different formulations of the Hypothesis (thus calling into question the common practice of honoring the thing with the definite article and the capitalization of its name). Because the presence of naked singularities would seem to herald a spectacular breakdown in predictability and even determinism associated with dynamical evolution in general relativity (such as it is),[73] many attempts to make the Hypothesis precise focus on the initial value formulation of general relativity. The most common formulations invoke either the WEC or the DEC [123] as a constraint on the matter fields permissible for the initial value formulation. As initially plausible as are such attempts at formulating a precise version of the Hypothesis that would admit of rigorous proof, there are in fact cases where satisfaction of an energy condition actually seems to aid the development of a naked singularity after gravitational collapse, e.g., the WEC in the case of the self-similar collapse of a perfect fluid [123]. In such cases, one can show that the focusing effects the energy condition induces in geodesic congruences actually contribute directly to the lack of an event horizon. It is thus parlous to attempt to draw any concrete conclusions regarding the relation of the energy conditions to the Cosmic Censorship Hypothesis in our current state of knowledge.

With regard to results about positivity of global mass, because the NEC does not require the convergence of timelike geodesics (as I discussed in §2.1), and so does not entail that "gravity be attractive" for bodies traversing such curves, it is particularly striking that Penrose [153] and Ashtekar and Penrose [4] were able to prove positivity of ADM mass using only it, and that Penrose et al. [156] were able to prove it using the even weaker ANEC, and not the significantly stronger DEC, as all other known proofs require. All known proofs of the positivity of the Bondi mass do require the DEC, which is perhaps not surprising, in light of the fact that the Bondi energy essentially tracks mass–energy radiated away along null curves to future null infinity. If the DEC were to fail, then it seems plausible that the Bondi energy could become negative, if negative mass–energy radiated to null infinity. It

---

[73] See, e.g., Earman [61] for a thorough discussion, and Curiel [45] for arguments arriving at somewhat contrary conclusions.

would be of some interest to try to find a spacetime model with negative Bondi mass in which the DEC is not violated. Perhaps matter fields with "superluminal acoustic modes" that still satisfied the DEC (§2.1) might provide such examples.

The most precise, rigorous and strongest geodesic theorems [66, 93] both assume the SDEC.[74] Under the assumptions used to prove the theorem of Geroch and Jang [93], Malament [132] showed that the SDEC is necessary for the body to follow a geodesic, and not just any timelike curve. Weatherall [208] strengthened the result by showing that the SDEC is necessary for the geodesic to be timelike, not spacelike. He showed as well that the SDEC is not strong enough to ensure that the curve not be null: there is a spacetime with a null geodesic satisfying all the conditions of the Geroch-Jang Theorem. It is perhaps important that the example Weatherall [208] produces to show that a null curve can satisfy all of the theorem's conditions rely on a stress-energy tensor not of Hawking-Ellis type I. Since stress-energy tensors not of type I are generally considered "unphysical", it would be of interest to determine whether there are counter-examples to the Geroch and Jang [93] and Ehlers and Geroch [66] theorems that rely on stress-energy tensors of type I. Because of the character of the proofs of the theorems and of the counter-examples that Weatherall [208] produces, I conjecture that there are no such counter-examples, and thus that null curves satisfying the conditions of the theorem require nonstandard stress-energy tensors.[75]

Whether or not my conjecture is correct, I think the necessity of the strongest energy condition for the validity of the theorems poses a problem for many attempts to analyze and clarify the conceptual foundations of general relativity. Many attempts to provide interpretations of the formalism of general relativity, for instance, place fundamental weight on the so-called Geodesic Principle, that "small bodies", when acted on by no external forces, traverse timelike geodesics. The "fact" that the Geodesic Principle is a consequence of the Einstein field equation is often cited as justification for the validity of the Principle (*e.g.*, Brown [27]). The work of Malament [132] and Weatherall [208], however, show that, at best, such approaches to the foundations of general relativity must be more subtle where the Geodesic Principle is concerned, and, at worst, that the Principle may in fact not be suitable at all for playing a fundamental role in giving an interpretation of the theory.[76]

With regard to entropy bounds such as that of Bousso [24, 25], if in fact the NEC or DEC were necessary for their validity, this could spell serious trouble for

---

[74]The statement of the theorems in each of those papers in fact uses the DEC, but an examination of the proof shows that they both actually use the SDEC, in both cases in order to ensure that a constructed scalar quantity that can be thought of as the mass of an "arbitrarily small" body is strictly greater than zero.

[75]I have not had the opportunity to work through the arguments of Dixon [55–58], Ehlers and Rudolph [67] and Schattner [169, 170] to determine whether their results on the definability of the center of mass of an extended body and the formulation of equations of motion for that center of mass in fact rely on the SDEC rather than, as they explicitly assumed, the DEC. Because of the intimate connection of these relations with the geodesic theorems, this would be of some interest to determine.

[76]See Weatherall [209, this volume] for extended discussion of these issues.

many programs in quantum gravity, or at least for the ways that research in such programs are currently being carried out, in so far as many programs place enormous motivational, argumentative and interpretational weight on such entropy bounds, and we already know that essentially all energy conditions are promiscuously violated when quantum effects are taken into account.[77]

3.2 Violations

In some of the cases of violations I list, the circumstance or condition possibly leading to a violation of the germane energy condition (*e.g.*, for some subset of possible values for relevant parameters); in other cases, it necessarily does so. I will indicate which is which. When I list the same type of system as violating different energy conditions (*e.g.*, "big bang" singularities for both NEC and SEC), it means that different instances of that type of system (having different parameters) violate the different conditions.

NEC

1. conformally coupled massless and massive scalar fields **[possibly]** [6, 199]
2. generically non-minimally coupled massless and massive scalar fields **[possibly]** [6, 59, 74, 199]
3. "big bang" and "big crunch" singularities[78] **[possibly]** [35, 37]
4. "big rip" singularities[79] **[necessarily]** [35, 37]
5. sudden future singularities[80] **[possibly]** [8, 9, 35, 37]
6. naked singularities **[possibly]** [5, 123, 152]
7. closed timelike curves **[possibly]** [195]
8. Tolman wormholes and Einstein–Rosen bridges **[necessarily]** [5]
9. any fluid with a barotropic index $w < -1$[81] (such as those postulated in so-called phantom cosmologies) **[necessarily]** [53, 195]

---

[77]See Curiel [48] for more detailed discussion of all these issues.

[78]A big bang or a big crunch is a singularity in a standard cosmological model where the expansion factor $a(t) \to 0$ in a finite period of time to the past or future, respectively. See, *e.g.*, Weinberg [210] or Wald [203]. In the specific context of FLRW spacetimes, this condition implies that a singularity is "strong" in the sense of Tipler [188].

[79]A big rip is a singularity in a standard cosmological model where the expansion factor $a(t) \to \infty$ in a finite period of time. If, as is currently believed, the universe is expanding at an accelerated rate, and it continues to do so, it is possible that such a big rip will occur. See, *e.g.*, Caldwell [29], Caldwell et al. [30] and Chimento and Lazkoz [39].

[80]These are singularities in standard cosmological models in which the pressure of the effective fluid or some higher derivative of the expansion factor $a(t)$ diverges, even though the energy density and curvature remain well behaved. They are very strange, not least because they do not necessarily lead to curve incompleteness of any kind. See Curiel [50] for further discussion.

[81]See footnote 15.

10.   "hyper-fast" travel[82] **[possibly]** [200]

WEC

1.   naked singularities **[possibly]** [76]
2.   closed timelike curves **[possibly]** [195]
3.   physically traversable wormholes **[necessarily]** [139, 193, 194]
4.   cosmological steady-state theories of Bondi and Gold [20] and Hoyle [116][83] **[necessarily]**
5.   classical Dirac fields **[possibly]** [203]
6.   a negative cosmological constant (*e.g.*, anti-de Sitter Space) **[necessarily]** [107, 195]
7.   future eternal inflationary cosmologies **[possibly]** [23]
8.   "hyper-fast" travel[84] **[necessarily]** [2, 127, 144]

SEC

1.   "big bang" and "big crunch" singularities[85] **[possibly]** [35, 37]
2.   sudden future singularities[86] **[possibly]** [8, 9, 35, 37]
3.   cosmological "bounces"[87] **[necessarily]** [35, 37]
4.   just before or just after a cosmological "inflexion"[88] **[possibly]** [35, 37]
5.   spatially closed, expanding, singularity-free spacetimes **[necessarily]** [176]
6.   cosmological inflation **[necessarily]** [195]
7.   a positive cosmological constant, as in de Sitter spacetime, and the "dark energy" postulated to drive the observed accelerated expansion of the universe **[necessarily]** [29, 30, 54, 107]
8.   asymptotically flat black holes with regular (nonsingular) interiors **[necessarily]** [3]
9.   closed timelike curves **[possibly]** [195]
10.   physically traversable wormholes **[necessarily]** [114, 138]
11.   minimally coupled massless and massive scalar fields **[possibly]** [6, 199]
12.   massive Klein-Gordon fields **[possibly]** [195]
13.   typical gauge theories with spontaneously broken symmetries **[possibly]** [189]
14.   conformal scalar fields coupled with dust **[possibly]** [14]

---

[82] See footnote 45.

[83] See also Pirani [157], Hoyle and Narlikar [117], and Hawking and Ellis [107, §4.3, pp. 90–91; §5.2, p. 126].

[84] See footnote 45.

[85] See footnote 78.

[86] See footnote 80.

[87] A bounce, in the context of a standard cosmological model, is a local minimum of the expansion factor $a(t)$. See, *e.g.*, Bekenstein [14] and Molina-Paris and Visser [138].

[88] An inflexion, in the context of a standard cosmological model, is a saddle point of the expansion factor $a(t)$. See, *e.g.*, Sahni et al. [167] and Sahni and Shtanov [168].

15.   "hyper-fast" travel[89] **[necessarily]** [2, 127, 144]

DEC

1.   "big bang" and "big crunch" singularities[90] **[possibly]** [35, 37]
2.   sudden future singularities[91] **[possibly]** [8, 9, 35, 37]
3.   classical Dirac fields **[necessarily]** [155]

ANEC

1.   massless conformally coupled scalar fields[92] **[possibly]** [6, 199]
2.   massless and massive non-minimally coupled scalar fields **[possibly]** [59, 74]
3.   closed timelike curves **[possibly]** [195]
4.   traversable wormholes **[possibly]** [140]

AWEC

1.   cosmological steady-state theories of Bondi and Gold [20] and Hoyle [116] **[necessarily]** (my calculation)
2.   a negative cosmological constant (*e.g.*, anti-de Sitter Space) **[necessarily]** (my calculation)
3.   classical Dirac fields **[possibly]** (my calculation)
4.   closed timelike curves **[possibly]** [195]
5.   physically traversable wormholes **[possibly]** (my calculation)
6.   "hyper-fast" travel[93] **[possibly]** (my calculation)

ASEC

1.   a positive cosmological constant, as in de Sitter spacetime, and the "dark energy" postulated to drive the observed accelerated expansion of the universe **[necessarily]** (my calculation)
2.   cosmological inflation **[possibly]** (my calculation)
3.   massive Klein-Gordon fields **[possibly]** (my calculation)
4.   typical gauge theories with spontaneously broken symmetries **[possibly]** (my calculation)
5.   conformal scalar fields coupled with dust **[possibly]** (my calculation)

---

[89] See footnote 45.

[90] See footnote 78.

[91] See footnote 80.

[92] Urban and Olum [192] also show that AANEC can be violated by conformally coupled scalar fields in conformally flat spacetimes, such as the standard FLRW cosmological models.

[93] See footnote 45.

The most compelling empirical evidence for violations of energy conditions comes from cosmology. For instance, strongly substantiated cosmographic arguments comparing best estimates for the age of the oldest stars to the epoch of galaxy formation show that the SEC must have been violated in the relatively recent cosmological past (redshift $z < 7$) [196–198]. Visser's arguments, particularly as presented in the 1997 papers, are an especially striking example of the power of the energy conditions: years before there was any hard observational evidence for the acceleration of the current expansion of the universe, and so hard, direct support for the existence of a positive cosmological constant, Visser predicted on purely theoretical grounds that the most likely culprit for violation of SEC in the recent cosmological past must be a positive cosmological constant. In fact, if the current consensus that the expansion of the universe is accelerating is correct, and so some form of "dark energy" exists, then we know that the SEC is currently being violated on cosmological scales, entirely independently of any assumptions about the nature of the fields entering into the stress-energy tensor or cosmological constant [6, 35–37, 198, 199]. Finally, if any model of inflationary cosmology is correct, then we know that the SEC was necessarily violated at least during the inflationary period and possibly, depending on the particulars of the model, the ASEC as well. One glimmer of hope among the gloom, however, is that the presence of a positive cosmological constant does not yield violations of the NEC, so no matter how exotic so-called dark energy is, and whatever fundamental mechanism may underlie it, at the classical level at least it will still satisfy that condition.

Far and away the simplest theoretical mechanisms presently known for yielding violations of energy conditions, and in many ways the most plausible, come from models including scalar fields. Indeed, using classical scalar fields alone, without even having to resort to quantum weirdness, it is relatively easy to engineer violations of even the weakest conditions, the NEC and the ANEC, as the list of violations shows. We do not yet have indubitable evidence for the existence of a fundamental scalar field in nature. (The recently discovered Higgs field is without question phenomenologically a scalar field, but the jury is still out on whether or not it is a composite, bound state of underlying non-scalar entities.) The importance of scalar fields in fundamental theoretical physics, however, is indubitable.[94] For many theoretical and pragmatic reasons, the so-called inflaton field that drives cosmological inflation is most commonly modeled as a classical scalar field, and cosmological inflation necessarily violates SEC and, depending on particulars of the model, possibly ASEC. Many meson fields in the Standard Model (pions, kaons and many other mesons, including their "charmed", "truth" and "beauty" correlates), moreover, are modeled to an extraordinarily high degree of accuracy as scalar fields, even though we believe they in fact consist of bound states of (non-scalar) quark–antiquark pairs. It is also widely believed that the so-called "strong CP problem", the fact that no CP-violation in strong nuclear interactions has ever been observed, is best solved

---

[94] It would be an interesting project to try to determine why theoretical physicists are firmly wedded to scalar fields as fundamental constituents of reality in the face of an almost complete lack of evidence for them, and whether their reasons for the marriage are really sound.

by the postulation of a scalar field called the axion [147], though to the best of my knowledge it is not known whether any classical models of the axion violate any of the energy conditions (any more than those of other quantum fields do, at any rate).

Now, violations of the NEC are disturbing for at least two important reasons. First and perhaps foremost, they imply violations of all other pointilliste energy conditions. Second, they already would seem to allow not only violations of the ordinary Second Law of thermodynamics [52, 75], but of the Generalized Second Law as well: send lots of negative energy (with positive entropy) through the event horizon into a black hole, and *voilà!*—the area of the black hole shrinks, even though arbitrary amounts of entropy have disappeared from outside the event horizon. Perhaps the most troubling violation of the NEC from the above list is the case of a conformally coupled scalar field, given the naturalness of "conformal coupling" for scalar fields in quantum field theory [6, 199], which is why in the list of violations I singled it out from the class of generically non-minimally coupled scalar fields.

The particular example of a massive conformal scalar field coupled with dust given by Bekenstein [14] in an example of how to construct a nonsingular FLRW model, exploiting the fact that the system can be made to violate the SEC, has interesting possible physical significance, which is why I singled it out in the list of systems for which energy conditions can fail: the pions that mediate the strong nuclear force can to a very high degree of approximation be represented by just such scalar fields. Thus, Bekenstein argues, nuclear matter in the very early, dense stages of the actual universe may not have satisfied the SEC, which may suggest that the initial singularity in standard Big Bang models may be avoidable. This may give reason to doubt the stability of at least some of the singularity theorems in regimes where the energy conditions fail. Because the SEC would have been necessarily violated during an epoch of inflationary expansion, moreover, and because inflationary theories have such strong support among many cosmologists, such doubts should perhaps cause further concern for advocates of an initial Big Bang singularity. In light of the fact that the strongest theorems for big bang singularities rely on the SEC, and that the Lorentzian Splitting Theorems (in conjunction with the results of Senovilla [176] to the effect that spatially closed, expanding, singularity-free spacetimes necessarily violate the SEC) come close to showing that the SEC is necessary for those theorems, then I think it becomes quite reasonable to question the current confidence in the so-called Standard Model of cosmology, which rests on the idea that the universe "started with" a big bang. That, moreover, both a cosmological "bounce" and a Tolman wormhole (perhaps the two most natural possible replacements for an initial big bang singularity) require violation *only* of the SEC [114, 138], not any of the other energy conditions, only exacerbates the problem.

Tipler [189], in a line of argument intended to mitigate such doubts, has pointed out an amusing poignancy in the role that homogeneity (high symmetry) plays in Bekenstein's construction of nonsingular FLRW spacetimes that violate the SEC. It follows from a theorem Tipler proves that, if a black hole (marginally trapped surface) develops in one of Bekenstein's spacetimes, then, because they do satisfy the WEC, a singularity would necessarily develop. Of course, a marginally trapped surface would form only if there were deviations from homogeneity. We would expect, however, on physical grounds, that even slight deviations from homogeneity could lead to the development of marginally trapped surfaces. Thus, it is only the strict symmetry of the Bekenstein models that precludes singularities. This, of course, turns the standard (mistaken) pre-Penrose [148] argument on its head: that the singularities of the FLRW, Schwarzschild, and Oppenheimer and Snyder [145] spacetimes were simply an artifact of their unrealistic perfect symmetry. In the case of Bekenstein's spacetimes, it is only their unrealistic perfect symmetries that *precludes* singularities. Theorem 1 of Tipler [189], moreover, gives him even stronger grounds for thinking that violations of SEC will not necessarily block formation of singularities, at least for closed universes, so long as the period and extent of the failure is limited with respect to its satisfaction in the rest of spacetime, *i.e.*, so long as the ASEC holds.

The theorems predicting big bang and big crunch singularities face one more problem peculiar to them alone: all such theorems invoke energy conditions of various kinds, mostly the SEC, and yet one can show that, depending on the characteristics of a given big bang or big crunch singularity, the presence of the singularity itself implies a violation of the relevant energy condition. Roughly speaking, whether a big bang or big crunch implies a violation of a given energy condition depends on how "violent" the singularity is, which idea can be made precise by analysis of the nature of the matter fields present (*e.g.*, the value of the barotropic index of the ambient homogeneous cosmological fluid), or by the behavior of geodesic congruences in the immediate neighborhood of the singularity (*e.g.*, whether such singularities are strong in the sense of Tipler [189], and, if so, how quickly they squeeze spatial volumes to zero). What is one to say in such cases? Clearly, the known theorems do not apply to such singularities, but also clearly the exact spacetimes in which such singularities occur have been shown to exist. The only safe conclusion seems to be that, at least in the case of these kinds of singularity, violations of salient energy conditions need not preclude their existence. But then one must question the importance of the theorems themselves, especially in light of the growing body of observational evidence that, if there is a big bang or big crunch, it may well be of a type that violates energy conditions.

What about the remainder of the singularity theorems? Should any of the violations drive us to doubt their validity or physical relevancy? In order to try to answer this question with some generality, it will be useful to draw two distinctions, the first between types of violations, and the second between types of theorems.[95] First, roughly speaking, the violations fall into one of two classes, being associated either with a type of physical system (*e.g.*, conformally coupled scalar field, classical Dirac field), or with a type of "event" (very loosely construed, *e.g.*, traversable wormhole, closed timelike curve, or big rip singularity). Generally speaking, for the latter, the regions where the energy conditions are violated can be "localized" to a neighborhood of the "event". The scare-quotes are to remind us of the fact that some such events—*e.g.*, many types of singularities—are not localizable in any reasonable sense

---

[95] I do not think the classifications I sketch here are of relevance beyond the context of such discussions as this. I certainly do not think they capture anything of fundamental significance about the nature of violations of energy conditions or about singularities.

of the term.[96] The qualification "generally speaking" hedges against cases such as the traversable wormholes of Visser [194], for which travelers moving through the wormholes never experience a violation of any energy condition. Generally speaking, for violations of the former class (*viz.*, associated with a type of physical system), one cannot "localize" the regions of violation in any way, unless one can localize the system itself, or at least those spacetime regions in which the system is known to violate the energy conditions and one can also determine that the system violates them nowhere else.

As for the singularity theorems, they also fall roughly into two classes, which for lack of better terms I will refer to as pinpointing and not. Roughly speaking, pinpointing theorems, as the name suggests, in certain ways allow one to say where in spacetime the singularities occur, and so in a sense one can "localize" the singularities.[97] Such theorems demonstrate the existence of singularities associated with closed, trapped surfaces (for singularities contained in asymptotically flat black holes: Penrose [148], Hawking and Ellis [107]), or with trapping surfaces (for singularities contained in generalized black holes: Hayward [110, 111]), or with the "boundaries" of spacetime (such as big bang and big crunch singularities), or they place the defining incomplete, inextendible geodesic entirely in a compact subset of the spacetime (*e.g.*, Hawking and Ellis [107, pp. 290–292]). Singularity theorems that are not pinpointing, such as those of Gannon [82, 83], merely demonstrate the existence of incomplete, inextendible geodesics without giving one any information about "where the incompleteness of the geodesic" is in spacetime.

Now, the impact of possible violations will differ from theorem to theorem depending on whether the theorem at issue pinpoints or not, and on whether the violation can be localized in an appropriate sense to that region of spacetime in which the theorem locates the predicted singularity. For theorems that do not pinpoint, I think there is no principled reason to believe that any salient violations may or may not vitiate the theorem. For theorems that do pinpoint, there may be hope of showing that at least some salient violations may or likely will not vitiate the theorems, but one must work through them on a case by case basis to make the determination. If one has some reason to believe, for example, that a given type of salient violation can be segregated entirely from the region of spacetime in which a closed, trapped surface forms and evolves (because, *e.g.*, of the type of collapsing matter that eventuates in the trapped surface), then one also has some reason to believe that any theorem that both invokes the violated condition and places the singularity in such a closed, trapped surface may still hold despite the violation. It would take us too long to go through all the singularity theorems and all the types of violations to determine which violations can and cannot be relevantly segregated from the regions where the predicted singularities form or reside. I leave this as an exercise for the reader.

---

[96] See Curiel [45] for discussion.

[97] Again, see Curiel [45] for discussion of why the scare-quotes are called for.

Similar considerations about pinpointing, type of violation, and the possibility of segregation come into play when trying to determine whether a given violation should give us reason to doubt the soundness of any other type of given consequence of an energy condition. I see no way to draw clean, general conclusions.

In sum, it seems difficult to escape the conclusion that we are faced with the horns of an important dilemma: either we must learn to live with the "exotic" physics that violations of energy conditions lead to (wormholes, closed timelike curves, sudden future singularities, spatial topology change, naked singularities, *et al.*), and so become much more skeptical of the plethora of seemingly important results that rely on the conditions; or else we must reconstruct fundamental physical theory root and branch, *e.g.*, by prohibiting the use of essentially all scalar fields, in order to rule out the possibility of such violations. I personally find it more realistic, if not more palatable, to grasp the first horn. An investigation of the consequences of this conclusion for projects that purport to provide fundamental explication and interpretation of the conceptual and physical structure of general relativity is beyond the scope of this paper, but is, I think, urgently called for.

3.3 Appendix: The Principle of Equivalence

There is an interesting, though not obvious, possible connection between the principle of equivalence (in at least some of its guises) and energy conditions. (See Wallace [207, this volume] for discussion of the principle of equivalence.) Postulating the lack of a preferred flat affine connection is, to my mind, one of the most promising ways of trying to formulate the principle of equivalence in a way that one can make somewhat precise [190, 191], even if one cannot show that such a principle must be true in the context of the theory. Could one derive an energy condition, or the violation of one, from the existence of a preferred flat affine structure? One way to determine such a privileged flat affine connection would be by use of the existence of a distinguished family of particles possessing what, for lack of a better term, I will call "anti-inertial charge", which would couple with the "active gravitational mass" of ordinary matter in such a way as to result in the anti-inertial systems traversing curves whose images form the projective structure of a flat affine connection. For a force that picks out such a connection, one can assign to it a stress-energy tensor by solving the equation of geodesic deviation using it as a force that exactly cancels out the curvature terms due to the ordinary affine connection, and deriving an expression for an "effective" stress-energy tensor associated with the force.

One possible mechanism for producing anti-inertial charge is strongly suggested by the arguments of Bondi [18] showing that active and passive gravitational mass are not necessarily equal in general relativity, at least when negative mass is allowed.[98]

---

[98]I put aside the problem that "mass" is, in general, not a well defined concept in general relativity. If one likes, one can consider the following discussion to be restricted to test particles in static spacetimes.

In particular, negative masses uniformly repel all other mass, irrespective of the sign of the other masses, and likewise positive masses uniformly attract all other masses, and so, most strikingly, a system consisting of one positive and one negative mass will spontaneously accelerate, even when no forces other than "gravitational" are present. a clear violation of the weak equivalence principle, that, roughly speaking, all small enough freely falling bodies traverse the same worldlines, *viz.*, geodesics. (Arguably, the inequality of passive and active gravitational mass already constitutes a violation of the weak principle of equivalence, at least in one of its guises.) In this case, negative mass plays the role of an anti-inertial charge. In the case that Bondi describes, therefore, the projective structure of the flat affine connection could possibly be determined by the acceleration curves of systems having equal parts positive and negative active gravitational mass.

This line of thought suggests the following.

Conjecture 1 *If one were able to demonstrate the existence of a privileged flat affine connection, by the existence of a family of particles with anti-inertial charge, then one or more of the standard pointilliste energy conditions would be generically violated.*

3.4 Coda: The Trace Energy Condition

The history of what may be called the Trace Energy Condition (TEC) should give one pause before rejecting possible violations of the standard energy conditions on the grounds that the circumstances or types of matter involved in the violations seem to us today "too exotic". The TEC states that the trace of the stress-energy tensor can never be negative ($T = T^n{}_n \geq 0$—or, depending on one's metrical conventions, that it can never be positive). In its effective formulation, therefore, the condition requires that $p \leq \frac{1}{3}\rho$ in a medium with isotropic pressure. Before 1961, it seemed to have been more or less universally believed in the general relativity community that this condition would always be satisfied, even under the most extreme physical conditions. It is, for instance, assumed without argument, or even remark, in the seminal papers of Oppenheimer and Volkoff [146] and Harrison et al. [95] on possible equations of state for neutron stars. It was not seriously questioned until the work of Zel'dovich in the early 1960s, in which he showed that a natural solution for a quantum field theory relevant to modeling the matter in neutron stars leads to macroscopic equations of state of the form $p = \rho$.[99] In fact, it is widely believed today that matter at densities above 10 times that of atomic nuclei, as we expect to find in the interior of neutron stars, behaves in exactly that manner [181, ch. 8].[100]

---

[99]See Zel'dovich and Novikov [215] (especially p. 197) for a discussion.

[100]This coda was inspired by the discussion in Morris and Thorne [139].

4 Temporal Reversibility

For the purposes of the discussion in §5, and because it is of some interest in its own right, I will briefly discuss the relation of the energy conditions to the idea of temporal reversibility.

A spacetime is temporally orientable if one can consistently designate one lobe of the null cone at every point as the "future" lobe. A temporal orientation then is logically equivalent to the existence of a continuous timelike vector field ξ^a; by convention, the future lobe of the null cone at each point is that into which ξ^a points, and a causal vector is itself future-directed if it points into or lies tangent to the future lobe. To reverse the temporal orientation is to take $-\xi^a$ to point everywhere in the "future" direction. If T_{ab} is the stress-energy tensor in the original spacetime, then we want the time-reversed spacetime to have the stress-energy tensor T'_{ab} such that: the four-momentum of any particle as determined relative to any observer will be reversed in the time-reversed case; and the energy density of any particle as determined relative to any observer will stay the same. Formally

1. $T'_{an}(-\xi^n) = -T_{an}\xi^n$
2. $T'_{mn}(-\xi^m)(-\xi^n) = T_{mn}\xi^m\xi^n$

Clearly, then, $T'_{ab} = T_{ab}$. So, in sum, I claim the rule for constructing the time reverse of a (temporally orientable) relativistic spacetime is to leave everything the same except for the sense of parameterization of timelike (and null) curves, which should be reversed. (No problem arises with parameterization of spacelike curves: there is no natural or preferred sense for their parameterization in the first place.)

This makes physical sense. The best way to see this is to ask what should happen to the metric under time reversal. I claim the answer is: nothing at all. The metric stays the same. Temporal orientation is not a metrical concept. It is a concept at the level of differential topology and conformal structure. The temporal orientation is determined by how one parameterizes temporal curves (which in turn, of course, depends on whether one can do so in a way that consistently singles out a choice of "future lobe of null cone" at every point of the manifold in the first place). It also makes geometrical sense. If one fixes a $1 + 3$ tetrad $\{\overset{\mu}{\xi}{}^a\}_{\mu \in \{0, 1, 2, 3\}}$ (not necessarily orthonormal) such that the metric at a point can be expressed as $\sum_\mu \alpha_\mu \overset{\mu}{\xi}_a \overset{\mu}{\xi}_b$, for some real coefficients α_μ, then reversing the sign of $\overset{0}{\xi}{}^a$ clearly does not change the metric.[101] (One can always find such a tetrad at a single point, though it may not be extendible to a tetrad-field with the same property.)

It is a simple matter to verify that a spacetime satisfies any one of the standard energy conditions listed in §2 if and only if the time reverse of the spacetime does as

---

[101] Another way to see this is to note that the only reasonable choice for "changing the metric" under time reversal would be to multiply it by -1; that however, does not change the Einstein tensor, and so *a fortiori* cannot change the stress-energy tensor.

well. (The same holds as well for all the more recently proposed energy conditions discussed in §2.4.) On the face of it, this is somewhat surprising. A white hole, for instance, is the time reverse of a black hole, and surely that should violate some energy condition. But in fact, no, it shouldn't, as a perusal of the relevant Penrose diagram will show: a white hole will violate an energy condition if and only if its time-reversed black hole does so.

5 Constraints on the Character of Spacetime Theories

General relativity assumes the existence of a single object, the stress-energy tensor T_{ab} that encodes, for all fields of matter, all properties relevant to determining the relationship of the matter to the geometrical structure of spacetime. This relationship is governed by the Einstein field equation,

$$G_{ab} = 8\pi T_{ab}$$

This equation, conjoined with the definition of a spacetime model ($\mathcal{M}$, g_{ab}), constitutes the entirety of general relativity as a formal theory.

In order to do physics, however, we must give physical significance to the formal terms in the Einstein field equation, and this is where the idea of stress-energy enters. As its name suggests, the stress-energy tensor encodes for matter only information about what we normally think of as its energy, momentum and stress content. General relativity, then, assumes that what we normally think of as stress-energy content completely determines the relation of spacetime structure to matter—no other property of matter "couples" with spacetime structure at all, except in so far as it may have a part in determining the stress-energy of the matter. It is exactly this feature of general relativity that affords the energy conditions their power. Nonetheless, we fully expect, or at least fervently hope, that general relativity will one day give way to a deeper theory of gravity, one that will attend to the presumably quantum nature of phenomena in regions of extreme curvature.[102] It thus makes sense to explore alternative theories of spacetime even in the strictly classical regime, if only to get ideas about how to try to modify general relativity in the search for that deeper theory. Surely not everything is up for grabs, though. Even in the attempt to formulate alternative theories in the spirit of free exploration, some core structure or set of structures must be retained in order for the explorations to take place in the province of "spacetime theories". What is that core? Is there a single one?

In particular, for our purposes, the most important question is: what must be true about the relation of stress-energy to the local and global structures of spacetime, in a

---

[102]I will not discuss the relation of energy conditions to any programs in quantum gravity, as I do not feel any of them are mature enough as proposals for a physical theory to support serious analysis of this sort. See Curiel [46] for why I hold this view. See Wüthrich [213, this volume], among others, for arguments to the contrary.

candidate spacetime theory, for one to be able to formulate energy conditions and use them to derive results? What, we are thus led to ask, must a spacetime theory itself be like in order for it to be able to exploit the possibility that deep and extensive features of global structure depend only on purely qualitative properties of stress-energy? Any field equations it imposes must be "loose" enough to respect this fact. In particular, no global feature of the geometry, as constrained by a theory's field equations, should depend on anything but purely qualitative properties of stress-energy; *a fortiori*, no global feature of the geometry should depend on the species of matter present, so long as that species manifests a relevant qualitative property. It is otherwise difficult to see how generic, purely qualitative conditions could determine specific, concrete features of spacetime geometry.

A useful way to begin to try to address these questions, and at the same time to begin to figure out the place of energy conditions in relation to potentially viable alternative spacetime theories, is to ask oneself, following a line of questioning introduced early in Geroch and Horowitz [92], what one can envisage needing to hold onto in future developments of physical theory, come what may. Not the Einstein field equation itself, most likely. Very likely causal structure of some sort. What else?

What follows is my attempt at such a list of structures, roughly ordered by "fundamentality"—where I mean by this only something like: what we would or should be willing to give up before what else, what we have more and less confidence will survive in future theories (not anything having to do with recent debates in the metaphysics literature). Such an ordering should respect, at a minimum, the fact that one needs in place already some structure in order to be able to define other structure—one could not countenance giving up the former before the latter.[103]

In constructing the list, I have been guided by the tenet that any physically reasonable spacetime theory should "look enough like" general relativity so as to make all the elements of the list sensible in its context. Not all the elements in the list, however, should be understood to be restricted to the form they take in standard accounts of general relativity. For instance, "causal structure" need not mean Lorentzian light cone structure; it may signify, for example, only some relation among events required by some feature of ambient matter fields, such as respecting the characteristic cones of matter obeying symmetric, quasilinear, hyperbolic equations of motion, whether or not those cones conform to the standard Lorentzian metric of spacetime.[104] Any such list, moreover, will ineluctably be shaped in part by the biases, prejudices and aesthetic and practical predilections of the one constructing it, so the following attempt should be taken with a healthy dose of salt.[105]

---

[103]For a similar list, albeit constructed for a somewhat different purpose, and with a very different ordering than mine, see Isham [118, p. 10].

[104]See, *e.g.*, Geroch [91] and Earman [62].

[105]One could sharpen this list by distinguishing between local and global varieties of structure, *e.g.*, by allowing for the possibility that it makes sense to determine a local causal structure without necessarily requiring the existence of a global one. (In such a case, presumably something like transitivity of causal connectability would fail.) While I think such distinctions could have interest for some projects, they are too *recherché* for my purposes here.

1. event structure: primitive set of "events" constituting the fundamental building blocks of spacetime[106]
2. causal structure: primitive relation of "causal connectability" among events (not necessarily distinguishing between null and timelike connectability)
3. topology: spacetime dimension; notion of continuous curves and fields (maps to and from event structure); relative notions of "proximity" among events; global notions of "connectedness" and "hole-freeness" on event structure
4. projective structure, conformal structure, temporal orientability: notion of a set of events forming a "straight line", and so physically a distinguished family of curves (but not yet a distinction between accelerated and non-accelerated motion); distinction among spacelike, null and timelike curves; preferred orientation for parameterization of causal curves; null geodesics (but not timelike or spacelike); asymptotic flatness; singularities (incomplete, inextendible causal curves); horizons (event, apparent, particle, *etc.*, and so asymptotically flat black holes)
5. differential structure: notion of smooth (or at least finitely differentiable) curves and fields; and so of tangent vectors, tensors, Lie derivatives and exterior derivatives; and so of field equations and equations of motion; spinor structure
6. affine structure: notion of accelerated versus non-accelerated motion, and so timelike geodesics; spacelike geodesics, and so characterization of "rigid bodies"; "hyperlocal" conservation laws (covariant divergence), at least for quantities "represented by" contravariant indices on tensors; comparison (ratios) of lengths of curve-segments, and so integrals along curves
7. metric structure: principled distinction between Ricci and Weyl curvature ("matter" versus "vacuum"); "hyperlocal" conservation laws (covariant divergence) for any quantity; volume element, and so integrals, and so integral conservation laws (in the presence of symmetries) for spacetime regions of any dimension; variational principles; convergence and divergence of geodesic congruences (Raychaudhuri equation), and so trapping surfaces (generalized black holes)
8. Einstein field equation: fixed relation between properties of ponderable matter and spacetime geometry; initial value formulation and dynamics

Now, granting the interest of the list for the sake of argument, where, if at all, should one place energy conditions on it? No matter what else is the case, so long as definitional dependence (what one needs in place already to define or characterize structure of a particular sort) is one criterion used in ordering such a list, it seems that energy conditions, in their standard forms, must be not so fundamental as differential structure: one needs differential structure in order to write down any tensor, and so *a fortiori* to write down a stress-energy tensor. Because all the standard energy conditions (and pretty much all the nonstandard ones), rely on the distinction between causal and noncausal vectors in general, and often on the distinction between null and timelike, it seems likely that energy conditions will be less fundamental than conformal structure as well. Energy conditions, however, do not seem to require a notion of temporal orientability, as the discussion of §4 strongly suggests, and, except

---

[106]This does not presuppose that an "event" is a purely local entity, in any relevant sense of "local".

for the impressionist conditions, neither do they require a projective structure. They also seem to be largely independent of topological structure (except in so far as it is required to define differential structure). The impressionist energy conditions do require an affine structure (for the definition of a line-integral along a geodesic), but since they have much murkier physical significance and far fewer important applications than the pointilliste ones, I would almost certainly prefer to forego them before foregoing an affine structure.

Now, if one accepts my ordering, or anything close to it, energy conditions do not seem to fit anywhere neatly in it. So what can we conclude? One possibility is that energy conditions are not clearly a part of any broad conception of what a spacetime theory is, and thus, perhaps, are not themselves of fundamental importance in the study of the foundations of spacetime theories. Alternatively, one could choose to take the fact that energy conditions seem to fit nowhere neatly in the list as a reason to change my groupings of structure into levels or to change my proposed order of levels. All of these considerations are complicated by the fact that the geometrical and physical forms of the energy conditions are equivalent if one assumes the Einstein field equation, but, if one does not, as in most if not all alternative theories of spacetime, all bets are off. In such cases, should one hold on to the geometrical formulations, to try to ensure that one will still be able to derive the consequences listed in §3.1? Or should one hold on to the physical formulations, so as to be able to investigate possible violations of the sort listed in §3.2?

One reason to think they should form part of any broad conception of what constitutes a spacetime theory, irrespective of which formulation one wants to hold on to, rests on the remark of Geroch and Horowitz [92] I quoted on page 49, that without energy conditions the Einstein field equation "has no content." The conditions one needs to impose to make the initial value problem of general relativity merely consistent—the so-called Gauss-Codazzi constraints—look very much like conditions on the allowed forms of types of matter. So does the fact that the standard proofs showing existence and uniqueness of solutions to the initial value problem of general relativity require matter fields that yield quasilinear, hyperbolic equations of motion satisfying the DEC throughout all of spacetime (Hawking and Ellis [107, ch. 7, §7, pp. 254–5]; Wald [203, ch. 10, especially pp. 250 and 266–7]). This fact seems to place a constraint on spacetime theories—only theories that require nontrivial input about the nature of matter in order for the distribution of matter to constrain the geometry of spacetime ought to be counted as physically reasonable, at least if we want to try to hold on to the idea that a viable spacetime theory ought to support a cogent notion of dynamical evolution, and thus (at a minimum) ought to admit a well set initial value formulation.[107]

---

[107]One ought to keep in mind, of course, that we already know the DEC is not necessary for a well defined initial value formulation, as the arguments of Geroch [91] show. What is at issue here is whether the DEC is necessary for the initial value formulation to be well set in the sense of Hadamard—whether, that is, we can show not only existence and uniqueness of solutions, but also stability under small perturbations. There is some evidence that solutions to the initial value problem for some particular types of matter fields violating the NEC will be unstable [59, 200], but the arguments are murky and often hand-waving.

One can try to make this idea precise, and at the same time to capture the kernel of Geroch and Horowitz's remark, in the following way. First, note that globally hyperbolic spacetimes represent in a natural way possible solutions to the initial value problem of general relativity as it is normally posed.[108] Now, it is a trivial matter to find globally hyperbolic spacetimes that violate any energy condition. Proof: pick your favorite globally hyperbolic spacetime and some open set in it; from the formulæ in Wald [203, Appendix D], it follows that one can always find a conformal transformation of the metric that is the identity outside the open set and nontrivial inside such that at some point in the set the transformed stress-energy tensor will yield whatever one wants on contraction with a timelike or null vector; since conformal transformations preserve causal structure, the transformed spacetime is still globally hyperbolic.

Now, this fact poses a serious problem for any attempt to formulate a notion of dynamical evolution that would support any minimal notion of predictability or determinism. Fix a Cauchy surface in the original spacetime to the past of the open set one conformally jiggered in the proof I sketched. Take that Cauchy surface as initial data for the initial value problem of general relativity. Which spacetime will the Cauchy development off that Cauchy surface (the solution to the initial value problem with that initial data) yield? The original one? One of the conformally jiggered ones? Another one entirely? If one cannot give principled reasons for why exactly one of those spacetimes and no other is the natural result of dynamical evolution off the Cauchy surface according to the Einstein field equation, then one has captured one sense in which the Einstein field equation may "have no content."[109] The fact that the only known proof of the theorem that a given globally hyperbolic extension of a spacetime is the maximal such extension requires the WEC [163], in conjunction with the fact that the assumptions of standard proofs of the well posedness of the initial value formulation for general relativity imply the DEC throughout the entirety of the derived spacetime [203, ch. 11], suggest that it may be the energy conditions that intervene to ensure a cogent notion of dynamical evolution that supports some minimal notion of predictability or determinism.[110]

---

[108]But see, e.g., Ringström [163] for a discussion of the formidable subtleties and complexities involved in trying to make even this seemingly simple idea precise.

[109]This is *not* the infamous Hole Argument [63], nor is it in any way related to it, as conformal transformations are not in general associated with diffeomorphisms.

[110]One may want to object that, inside the region where the conformal transformation is nontrivial, one has actually changed the stress-energy tensor in such a way that what type of matter now is there is not the same as it was before, and so must obey different field equations; thus, the requirement that the same field equations apply throughout the evolution suffices to guarantee uniqueness. I, however, find the notion of "same field equations" to be, in our current state of understanding, hopelessly ambiguous. It is a highly nontrivial matter to ascertain whether or not some matter field obeys the "same field equations" in different spacetimes, or even in different regions of the same spacetime. Say that the field couples to the scalar curvature, but it so happens that in the spacetime at issue the scalar curvature vanishes everywhere. After the conformal transformation, the scalar curvature may no longer vanish in the region where the conformal transformation is nontrivial, and so the field equations will look as though they have "changed form" when passing from one region to the other.

Holding on to everything in my list except for the Einstein field equation, so long as whatever field equations do hold depend only on something like the stress-energy tensor that does not depend on idiosyncratic features of particular kinds of matter, I strongly suspect that one will likely face the same problem. Thus, once again, we seem pushed toward the view that energy conditions play some fundamental role or other in any reasonably broad conception of spacetime theories or at least any such conception that would include a cogent notion of dynamical evolution.

If one does think energy conditions belong as a part of any reasonably broad conception of what constitutes spacetime theory, one tempting way to try to capture the sense in which they may hold at a level of structure deeper than the Einstein Field Equation invokes the thermodynamical character of stress-energy: all stress-energy is fungible, is interchangeable, in the strong sense that the form it takes (electromagnetic, viscoëlastic, thermal, *etc.*), and so *a fortiori* any property or quality it may have idiosyncratic to that form, is irrelevant to its gravitational effects, both locally and globally. This is not a conclusion that follows by logical consequence from the observation that qualitative energy conditions suffice to prove theorems of great depth and strength about global structure. It is only one that is strongly suggested by what thermodynamics tells us about the nature of energy. I will not be able to discuss this idea further in this paper, however, as it would take us too far afield.[111]

The inability to derive the energy conditions from other propositions of a fundamental character constitutes an essential part of what pushes one to conceive of them as structure "at a deeper level" than many other elements on the list, perhaps even deeper than causality conditions (many of which can be derived from other fundamental assumptions), and so applicable across a *very* wide range of possible theories of spacetime. If, in the end, one does hold the view that they ought to be thought of as a fundamental part of a reasonably broad conception of what constitutes a spacetime theory, then perhaps, as I suggested in §2.1, the final lesson here is that the geometric form of the energy conditions are the ones to be thought of as fundamental, in so far as they rely for their statement and interpretation only on invariant, geometrical structures and concepts, whose consequences will hold irrespective of the exact field equation assumed by the given spacetime theory. If that is so, then perhaps one

---

[111] In one of the first papers in which he tried to provide a fundamental derivation of the field equation bearing his name, Einstein [69, pp. 148–9] explicitly used a similar line of thought to motivate the idea that all gravitationally relevant mass-energetic quantities associated with matter of any kind is exhaustively captured by the stress-energy tensor:

> The special theory of relativity has led to the conclusion that inert mass is nothing more or less than energy, which finds its complete mathematical expression in a symmetrical tensor of second rank, the energy tensor. Thus in the general theory of relativity we must introduce a corresponding energy tensor of matter $T^{\alpha}{}_{\sigma}$ .... It must be admitted that this introduction of the energy tensor of matter is not justified by the relativity postulate alone. For this reason we have here deduced it from the requirement that the energy of the gravitational field shall act gravitatively in the same way as any other kind of energy.

potentially fruitful way to use the (poorly named?) energy conditions as a constraint on the construction of spacetime theories is to search for theories in which these important geometric conditions have unproblematic, physically significant interpretations.

References

1. Abreu, G., C. Barceló, and M. Visser (2011). Entropy bounds in terms of the w parameter. *Journal of High Energy Physics 2011*(12), 092. doi:10.1007/JHEP12(2011)092. Preprint: arXiv:1109.2710v3 [gr-qc].
2. Alcubierre, M. (1994). The warp drive: Hyper-fast travel within general relativity. *Classical and Quantum Gravity 11*(5), L73–L77. doi:10.1088/0264-9381/11/5/001. Preprint: arXiv:gr-qc/0009013.
3. Ansoldi, S. (2007). Spherical black holes with regular center: A review of existing models including a recent realization with Gaussian sources. Delivered at the conference "Dynamics and Thermodynamics of Blackholes and Naked Singularities", Department of Mathematics of the Politecnico, Milan, 10–12 May 2007. Preprint: arXiv:0802.0330 [gr-qc].
4. Ashtekar, A. and R. Penrose (1990, October). Mass positivity from focussing and the structure of i^o. *Twistor Newsletter 31*, 1–5. Freely available at http://people.maths.ox.ac.uk/lmason/Tn/31/TN31-02.pdf.
5. Barceló, C. and M. Visser (1999, November). Traversable wormholes from massless conformally coupled scalar fields. *Physics Letters B 466*(2-4), 127–134. doi:http://dx.doi.org/10.1016/S0370-2693(99)01117-X. Preprint: arXiv:gr-qc/9908029.
6. Barceló, C. and M. Visser (2002). Twilight for the energy conditions? *International Journal of Modern Physics D 11*(10), 1553–1560. doi:10.1142/S0218271802002888. Preprint: arXiv:gr-qc/0205066.
7. Bardeen, J., B. Carter, and S. Hawking (1973). The four laws of black hole mechanics. *Communications in Mathematical Physics 31*(2), 161–170. doi:10.1007/BF01645742.
8. Barrow, J. (2004a). More general sudden singularities. *Classical and Quantum Gravity 21*(11), 5619–5622. doi:10.1088/0264-9381/21/23/020.
9. Barrow, J. (2004b). Sudden future singularities. *Classical and Quantum Gravity 21*(11), L79–L82. doi:10.1088/0264-9381/21/11/L03.
10. Bartnik, R. (1988). Remarks on cosmological spacetimes and constant mean curvature surfaces. *Communications in Mathematical Physics 117*(4), 615–624. Freely available at http://projecteuclid.org/euclid.cmp/1104161820.
11. Beem, J., P. Ehrlich, and K. Easley (1996). *Global Lorentzian Geometry* (Second ed.). New York: Marcel Dekker.
12. Bekenstein, J. (1972). Nonexistence of baryon number for static black holes. *Physical Review 5*(6), 1239–1246. doi:10.1103/PhysRevD.5.1239.
13. Bekenstein, J. (1973). Black holes and entropy. *Physical Review D 7*, 2333–2346. doi:10.1103/PhysRevD.7.2333.
14. Bekenstein, J. (1975). Nonsingular general-relativistic cosmologies. *Physical Review D 11*(8), 2072–2075. doi:10.1103/PhysRevD.11.2072.
15. Bekenstein, J. (1981). Universal upper bound on the entropy-to-energy ratio for bounded systems. *Physical Review D 23*, 287–298. doi:10.1103/PhysRevD.23.287.
16. Bekenstein, J. (1994a). Do we understand black hole entropy? In R. J. Ruffini, R. and G. MacKeiser (Eds.), *Proceedings of the 7th Marcel Grossmann Meeting on General Relativity*, pp. 39–58. Singapore: World Scientific Press. Preprint: arXiv:gr-qc/9409015v2.
17. Bekenstein, J. (1994b). Entropy bounds and black hole remnants. *Physical Review D 49*(4), 1912–1921. doi:10.1103/PhysRevD.49.1912.

18. Bondi, H. (1957, July). Negative mass in general relativity. *Reviews of Modern Physics 29*, 423–428. doi:10.1103/RevModPhys.29.423.
19. Bondi, H. (1964). Massive spheres in general relativity. *Proceedings of the Royal Society of London. Series A. Mathematical and Physical Sciences 282*, 303–317. doi:10.1098/rspa. 1964.0234.
20. Bondi, H. and T. Gold (1948). The steady-state theory of the expanding universe. *Monthly Notices of the Royal Astronomical Society 108*, 252–270.
21. Borde, A. (1987). Geodesic focusing, energy conditions and singularities. *Classical and Quantum Gravity 4*(2), 343–356. doi:10.1088/0264-9381/4/2/015.
22. Borde, A. (1994). Topology change in classical general relativity. arXiv:gr-qc/9406053.
23. Borde, A. and A. Vilenkin (1997). Violations of the weak energy condition in inflating spacetimes. *Physical Review D 56*(2), 717–723. doi:10.1103/PhysRevD.56.717. Preprint: arXiv:gr-qc/9702019.
24. Bousso, R. (1999a). A covariant entropy conjecture. *Journal of High Energy Physics 1999*(07), 004. doi:10.1088/1126-6708/1999/07/004. Preprint: arXiv:hep-th/9905177v3.
25. Bousso, R. (1999b). Holography in general spacetimes. *Journal of High Energy Physics 1999*(06), 028. doi:10.1088/1126-6708/1999/06/028. Preprint: arXiv:hep-th/9906022v2.
26. Brill, D. and F. Flaherty (1976). Isolated maximal surfaces in spacetime. *Communications in Mathematical Physics 50*(2), 157–165. Freely available at http://projecteuclid.org/euclid. cmp/1103900190.
27. Brown, H. (2005). *Physical Relativity: Space-time Structure from a Dynamical Perspective*. Oxford: Oxford University Press.
28. Brown, L. and G. Maclay (1969). Vacuum stress between conducting plates: An image solution. *Physical Review 184*(5), 1272–1279. doi:10.1103/PhysRev.184.1272.
29. Caldwell, R. (2002, 03 October). A phantom menace? Cosmological consequences of a dark energy component with super-negative equation of state. *Physics Letters B 545*, 23–29. doi:10. 1016/S0370-2693(02)02589-3. Preprint: arXiv:astro-ph/9908168v2.
30. Caldwell, R., M. Kamionkowski, and N. Weinberg (2003). Phantom energy: Dark energy with $w < -1$ causes a cosmic doomsday. *Physical Review Letters 91*, 071301. doi:10.1103/ PhysRevLett.91.071301. Preprint: arXiv:astro-ph/0302506.
31. Carter, B. (1971). Axisymmetric black hole has only two degrees of freedom. *Physical Review Letters 26*, 331–333. doi:10.1103/PhysRevLett.26.331.
32. Carter, B. (1973). Black hole equilibrium states. In B. DeWitt and C. DeWitt (Eds.), *Black Holes*, pp. 56–214. New York: Gordon and Breach.
33. Carter, B. (1997). Has the black hole equilibrium problem been solved? Invited contribution to the Eighth Marcel Grossman Meeting, Jerusalem, 1997. Preprint: arXiv:gr-qc/9712038.
34. Carter, B. (1999). The black hole equilibrium problem. In B. Iyer and B. Bhawal (Eds.), *Black Holes, Gravitational Radiation and the Universe*, pp. 1–16. Dordrecht: Kluwer Academic Publishers.
35. Cattoën, C. and M. Visser (2005). Necessary and sufficient conditions for big bangs, bounces, crunches, rips, sudden singularities and extremality events. *Classical and Quantum Gravity 22*, 4913–4930. doi:10.1088/0264-9381/22/23/001.
36. Cattoën, C. and M. Visser (2007). Cosmological milestones and energy conditions. *Journal of Physics Conference Series 68*, 012011. Paper given at 12th Conference on Recent Developments in Gravity. doi:10.1088/1742-6596/68/1/012011. Preprint: arXiv:gr-qc/0609064.
37. Cattoën, C. and M. Visser (2008). Cosmodynamics: Energy conditions, Hubble bounds, density bounds, time and distance bounds. *Classical and Quantum Gravity 25*(16), 165013. doi:10.1088/0264-9381/25/16/165013. Preprint: arXiv:0712.1619 [gr-qc].
38. Chicone, C. and P. Ehrlich (1980). Line integration of Ricci curvature and conjugate points in Lorentzian and Riemannian manifolds. *Manuscripta Mathematica 31*, 297–316. doi:10. 1007/BF01303279.
39. Chimento, L. and R. Lazkoz (2004, October). On big rip singularities. *Modern Physics Letters A 19*(33), 2479–2485. doi:10.1142/S0217732304015646. Preprint: arXiv:gr-qc/0405020.

40. Chruśchiel, P. and R. Wald (1994). On the topology of stationary black holes. *Classical and Quantum Gravity 11*(12), L147–L152. doi:10.1088/0264-9381/11/12/001. Preprint: arXiv:gr-qc/9410004.

41. Chruściel, P., J. L. Costa, and M. Heusler (2012). Stationary black holes: Uniqueness and beyond. *Living Reviews in Relativity 15*, 7. doi:10.12942/lrr-2012-7. URL (accessed online 11 Apr 2014): http://www.livingreviews.org/lrr-2012-7.

42. Clifton, T., G. Ellis, and R. Tavakol (2013). A gravitational entropy proposal. *Classical and Quantum Gravity 30*(12), 125009. doi:10.1088/0264-9381/30/12/125009. Preprint: arXiv:1303.5612v2 [gr-qc].

43. Corley, S. and T. Jacobson (1996). Focusing and the holographic hypothesis. *Physical Review D 53*(12), R6720–R6724. doi:10.1103/PhysRevD.53.R6720. Preprint: arXiv:gr-qc/9602043.

44. Cotsakis, S. and I. Klaoudatou (2007). Cosmological singularities and Bel-Robinson energy. *Journal of Geometry and Physics 57*(4), 1303–1312. doi:10.1016/j.geomphys.2006.10.007. Preprint: arXiv:gr-qc/0604029v2.

45. Curiel, E. (1999). The analysis of singular spacetimes. *Philosophy of Science 66*, S119–S145. Supplement. Proceedings of the 1998 Biennial Meetings of the Philosophy of Science Association. Part I: Contributed Papers. Stable URL: http://www.jstor.org/stable/188766. A more recent, corrected, revised and extended version of the published paper is available at: http://strangebeautiful.com/phil-phys.html.

46. Curiel, E. (2001). A plea for modesty: Against the current excesses in quantum gravity. *Philosophy of Science 68*(3), S424–S441. doi:10.1086/392926.

47. Curiel, E. (2014a). Energy conditions and methodology in cosmology. Unpublished manuscript.

48. Curiel, E. (2014b). Energy conditions and quantum field-theory on curved spacetime. Unpublished manuscript.

49. Curiel, E. (2015). Measure, topology and probabilistic reasoning in cosmology. Unpublished manuscript. Preprint: arXiv:1509.01878 [gr-qc].

50. Curiel, E. (2016a). The analysis of singular spacetimes revisited. Unpublished manuscript.

51. Curiel, E. (2016b). On geometric objects, the non-existence of a gravitational stress-energy tensor, and the uniqueness of the Einstein field equation. Forthcoming in *Studies in History and Philosophy of Modern Physics*. Preprint: arXiv:0908.3322 gr-qc. A more recently updated version is available at http://strangebeautiful.com/papers/curiel-nonexist-grav-seten-uniq-efe.pdf.

52. Davies, P. and A. Ottewill (2002, May). Detection of negative energy: 4-dimensional examples. *Physical Review D 65*, 104014. doi:10.1103/PhysRevD.65.104014. Preprint: arXiv:gr-qc/0203003.

53. Dąbrowski, M. and T. Denkiewicz (2009). Barotropic index w-singularities in cosmology. *Physical Review D 79*(6), 063521. doi:10.1103/PhysRevD.79.063521. Preprint: arXiv:0902.3107v3 [gr-qc].

54. Dąbrowski, M., T. Stachowiak, and M. Szydłowski (2003). Phantom cosmologies. *Physical Review D 68*(10), 103519. doi:10.1103/PhysRevD.68.103519.

55. Dixon, W. (1970a, 27 January). Dynamics of extended bodies in general relativity. I. Momentum and angular momentum. *Proceedings of the Royal Society of London. Series A. Mathematical and Physical Sciences 314*, 499–527. doi:10.1098/rspa.1970.0020.

56. Dixon, W. (1970b, 10 November). Dynamics of extended bodies in general relativity. II. Moments of the charge-current vector. *Proceedings of the Royal Society of London. Series A. Mathematical and Physical Sciences 319*(1539), 509–547. doi:10.1098/rspa.1970.0191.

57. Dixon, W. (1973). The definition of multipole moments for extended bodies. *General Relativity and Gravitation 4*(3), 199–209. doi:10.1007/BF02412488.

58. Dixon, W. (1974, 24 August). Dynamics of extended bodies in general relativity. III. Equations of motion. *Philosophical Transactions of the Royal Society A 277*(1264), 59–119. doi:10.1098/rsta.1974.0046.

59. Dubovsky, S., T. Grégoire, A. Nicolis, and R. Rattazzi (2006). Null energy condition and superluminal propagation. *Journal of High Energy Physics 2006*(03), 025. doi:10.1088/1126-6708/2006/03/025. Preprint: arXiv:hep-th/0512260v2.

60. Earman, J. (1986). *A Primer on Determinism.* Dordrecht: D. Reidel Publishing Co.
61. Earman, J. (1995). *Bangs, Crunches, Whimpers and Shrieks: Singularities and Acausalities in Relativistic Spacetimes.* Oxford: Oxford University Press.
62. Earman, J. (2013). No superluminal propagation for classical relativistic and relativistic quantum fields. Unpublished manuscript. http://philsci-archive.pitt.edu/10945/.
63. Earman, J. and J. Norton (1987, December). What price spacetime substantivalism? The hole story. *Philosophy of Science 38*(4), 515–525. doi:10.1093/bjps/38.4.515.
64. Eddington, A. (1923). *Mathematical Theory of Relativity* (Second ed.). Cambridge: Cambridge University Press.
65. Ehlers, J. (1987). Folklore in relativity and what is really known. In M. MacCallum (Ed.), *General Relativity and Gravitation*, pp. 61–71. Cambridge: Cambridge University Press. Proceedings of the 11th International Conference on General Relativity and Gravitation, Stockholm, July 6–12, 1986.
66. Ehlers, J. and R. Geroch (2004, January). Equation of motion of small bodies in relativity. *Annals of Physics 309*(1), 232–236. doi:10.1016/j.aop.2003.08.020. Preprint: arXiv:gr-qc/0309074.
67. Ehlers, J. and E. Rudolph (1977). Dynamics of extended bodies in general relativity: Center-of-mass description and quasi-rigidity. *General Relativity and Gravitation 8*(3), 197–217. doi:10.1007/BF00763547.
68. Ehrlich, P. and G. Galloway (1990). Timelike lines. *Classical and Quantum Gravity 7*(3), 297–307. doi:10.1088/0264-9381/7/3/006.
69. Einstein, A. (1916). The foundation of the general theory of relativity. In *The Principle of Relativity*, pp. 109–164. New York: Dover Press. Published originally as "Die Grundlage der allgemeinen Relativitätstheorie", *Annalen der Physik* 49(1916).
70. Einstein, A., L. Infeld, and B. Hoffmann (1938, January). The gravitational equations and the problem of motion. *Annals of Mathematics 39, Second Series*(1), 65–100.
71. Ellis, G. and B. Schmidt (1977). Singular space-times. *General Relativity and Gravitation 8*(11), 915–953. doi:10.1007/BF00759240.
72. Fischler, W. and L. Susskind (1998). Holography and cosmology. arXiv:hep-th/9806039v2.
73. Flanagan, E., D. Marolf, and R. Wald (2000). Proof of classical versions of the Bousso entropy bound and of the Generalized Second Law. *Physical Review D 62*(8), 084035. doi:10.1103/PhysRevD.62.084035. Preprint: arXiv:hep-th/9908070v4.
74. Flanagan, E. and R. Wald (1996). Does backreaction enforce the null energy condition in semiclassical gravity? *Physical Review D 54*(10), 6233–6283. doi:10.1103/PhysRevD.54.6233. Preprint: arXiv:gr-qc/9602052v2.
75. Ford, L. (1978, 12 December). Quantum coherence effects and the Second condition Law of thermodynamics. *Proceedings of the Royal Society of London. Series A. Mathematical and Physical Sciences 364*(1717), 227–236. doi:10.1098/rspa.1978.0197.
76. Ford, L. and T. Roman (1992). "Cosmic flashing" in four dimensions. *Physical Review D 46*(4), 1328–1339. doi:10.1103/PhysRevD.46.1328.
77. Friedman, J., K. Schleich, and D. Witt (1983, June). Topological censorship. *Physical Review Letters 71*, 1486–1489. doi:10.1103/PhysRevLett.71.1486. Preprint: arXiv:gr-qc/9305017. Erratum: *Physics Review Letters*, 75(1995):1872.
78. Galloway, G. (1993). On the topology of black holes. *Communications in Mathematical Physics 151*(1), 53–66. doi:10.1007/BF02096748.
79. Galloway, G. (1995). On the topology of the domain of outer communication. *Classical and Quantum Gravity 12*(10), L99–L101. doi:10.1088/0264-9381/12/10/002.
80. Galloway, G. and A. Horta (1996, May). Regularity of Lorentzian Busemann functions. *Transactions of the American Mathematical Society 348*(5), 2063–2084.
81. Galloway, G. and E. Woolgar (1997). The cosmic censor forbids naked topology. *Classical and Quantum Gravity 14*(1), L1–L7. doi:10.1088/0264-9381/14/1/001. Preprint: arXiv:gr-qc/9609007v2.
82. Gannon, D. (1975). Singularities in nonsimply connected space-times. *Journal of Mathematical Physics 16*(12), 2364–2367. doi:10.1063/1.522498.

83. Gannon, D. (1976). On the topology of spacelike hypersurfaces, singularities, and black holes. *General Relativity and Gravitation 7*(2), 219–232. doi:10.1007/BF00763437.
84. García-Parrado Gómez-Lobo, A. (2008). Dynamical laws of supergravity in general relativity. *Classical and Quantum Gravity 25*(1), 015006. doi:10.1088/0264-9381/25/1/015006. Preprint: arXiv:0707.1475 [gr-qc].
85. Garecki, J. (2001). Some remarks on the Bel-Robinson tensor. *Annalen der Physik 10*(11-12), 911–919. doi:10.1002/1521-3889(200111)10:11/12<911::AID-ANDP911>3.0.CO;2-M. Preprint: arXiv:gr-qc/0003006.
86. Geroch, R. (1966). Singularities in closed universes. *Physical Review Letters 17*, 445–447. doi:10.1103/PhysRevLett.17.445.
87. Geroch, R. (1967). Topology in general relativity. *Journal of Mathematical Physics 8*(4), 782–786. doi:10.1063/1.1705276.
88. Geroch, R. (1969). Spinor structure of space-times in general relativity I. *Journal of Mathematical Physics 9*, 1739–1744.
89. Geroch, R. (1970a). Singularities. In M. Carmeli, S. Fickler, and L. Witten (Eds.), *Relativity*, pp. 259–291. New York: Plenum Press.
90. Geroch, R. (1970b). Spinor structure of space-times in general relativity II. *Journal of Mathematical Physics 11*, 343–8.
91. Geroch, R. (2010). Faster than light? arXiv:1005.1614 [gr-qc].
92. Geroch, R. and G. Horowitz (1979). Global structure of spacetimes. See [108], Chapter 5, pp. 212–293.
93. Geroch, R. and P. Jang (1975). The motion of a body in general relativity. *Journal of Mathematical Physics 16*(1), 65–67. doi:10.1063/1.522416.
94. Hall, G. (1976). The classification of the Ricci tensor in general relativity theory. *Journal of Physics A: Mathematical and General 9*(4), 541–545. doi:10.1088/0305-4470/9/4/010.
95. Harrison, B., M. Wakano, and J. Wheeler (1957). Report. In *La Structure et l'Evolution de l'Univers*, Onzieme Conseil de Physique Solvay. Brussels: Editions Stoops.
96. Hawking, S. (1965). Occurrence of singularities in open universes. *Physical Review Letters 15*, 689–690. doi:10.1103/PhysRevLett.15.689.
97. Hawking, S. (1966a). The occurrence of singularities in cosmology. *Proceedings of the Royal Society of London. Series A. Mathematical and Physical Sciences 294*, 511–521. doi:10.1098/rspa.1966.0221.
98. Hawking, S. (1966b). The occurrence of singularities in cosmology. II. *Proceedings of the Royal Society of London. Series A. Mathematical and Physical Sciences 295*, 490–493. doi:10.1098/rspa.1966.0255.
99. Hawking, S. (1966c). Perturbations of an expanding universe. *Astrophysics Journal 145*, 544–554. doi:10.1086/148793.
100. Hawking, S. (1966d). Singularities in the universe. *Physical Review Letters 17*, 444–445. doi:10.1103/PhysRevLett.17.444.
101. Hawking, S. (1967). The occurrence of singularities in cosmology. III: causality and singularities. *Proceedings of the Royal Society of London. Series A. Mathematical and Physical Sciences 300*, 187–210. doi:10.1098/rspa.1967.0164.
102. Hawking, S. (1970). The conservation of matter in general relativity. *Communications in Mathematical Physics 18*(4), 301–306. doi:10.1007/BF01649448.
103. Hawking, S. (1971). Gravitational radiation from colliding black holes. *Physical Review Letters 26*(21), 1344–1346. doi:10.1103/PhysRevLett.26.1344.
104. Hawking, S. (1972). Black holes in general relativity. *Communications in Mathematical Physics 25*(2), 152–166. doi:10.1007/BF01877517.
105. Hawking, S. (1986). Chronology protection conjecture. *Physical Review D 46*(2), 603–611. doi:10.1103/PhysRevD.46.603.
106. Hawking, S. and G. Ellis (1969). The cosmic black body radiation and the existence of singularities in our universe. *Astrophysical Journal 152*, 25–36. doi:10.1086/149520.
107. Hawking, S. and G. Ellis (1973). *The Large Scale Structure of Space-Time*. Cambridge: Cambridge University Press.

108. Hawking, S. and W. Israel (Eds.) (1979). *General Relativity: An Einstein Centenary Survey.* Cambridge: Cambridge University Press.

109. Hawking, S. and R. Penrose (1970). The singularities of gravitational collapse and cosmology. *Philosophical Transactions of the Royal Society (London) A 314*, 529–548. doi:10.1098/rspa. 1970.0021.

110. Hayward, S. (1994a). Confinement by black holes. arXiv:gr-qc/9405055v.

111. Hayward, S. (1994b). General laws of black hole dynamics. *Physical Review D 49*(12), 6467–6474. doi:10.1103/PhysRevD.49.6467. Preprint: arXiv:gr-qc/9303006v3.

112. Hayward, S. (1996). Gravitational energy in spherical symmetry. *Physical Review D 53*(4), 1938–1949. doi:10.1103/PhysRevD.53.1938.

113. Heusler, M. (1996). *Black Hole Uniqueness Theorems.* Number 6 in Cambridge Lecture Notes in Physics. Cambridge: Cambridge University Press.

114. Hochberg, D., C. Molina-Paris, and M. Visser (1999). Tolman wormholes violate the strong energy condition. *Physical Review D 59*(4), 044011. doi:10.1103/PhysRevD.59.044011. Preprint: arXiv:gr-qc/9810029.

115. Horowitz, G. and M. Perry (1982). Gravitational energy cannot become negative. *Physical Review Letters 48*(6), 371–374. doi:10.1103/PhysRevLett.48.371.

116. Hoyle, F. (1948). A new model for the expanding universe. *Monthly Notices of the Royal Astronomical Society 108*, 372–382.

117. Hoyle, F. and J. Narlikar (1964). A new theory of gravitation. *Proceedings of the Royal Society of London. Series A. Mathematical and Physical Sciences 282*(1389), 191–207. doi:10.1098/ rspa.1964.0227.

118. Isham, C. (1994). Prima facie questions in quantum gravity. In J. Ehlers and H. Friedrich (Eds.), *Canonical Gravity: From Classical to Quantum,* Number 434 in Lecture Notes in Physics, Berlin, pp. 1–21. Springer. doi:10.1007/3-540-58339-4_13. Preprint: arXiv:gr-qc/9310031.

119. Israel, W. (1967). Event horizons in static vacuum space-times. *Physical Review 164*(5), 1776–1779. doi:10.1103/PhysRev.164.1776.

120. Israel, W. (1968). Event horizons in static electrovac space-times. *Communications in Mathematical Physics 8*(3), 245–260. doi:10.1007/BF01645859.

121. Israel, W. (1986). Third Law of black hole mechanics: A formulation of a proof. *Physical Review Letters 57*(4), 397–399. doi:10.1103/PhysRevLett.57.397.

122. Jacobson, T. and S. Venkataramani (1995). Topology of event horizons and topological censorship. *Classical and Quantum Gravity 12*(4), 1055–1062. doi:10.1088/0264-9381/12/4/ 012. Preprint: arXiv:gr-qc/9410023v2.

123. Joshi, P. (2003). Cosmic censorship: A current perspective. *Modern Physics Letters A 17*(15), 1067–1079. doi:10.1142/S0217732302007570. Preprint: arXiv:gr-qc/0206087.

124. Kay, B. and R. Wald (1987). Linear stability of Schwarzschild under perturbations which are non-vanishing on the bifurcation 2-sphere. *Classical and Quantum Gravity 4*(4), 893–898. doi:10.1088/0264-9381/4/4/022.

125. Kay, B. and R. Wald (1991, August). Theorems on the uniqueness and thermal properties of stationary, nonsingular, quasifree states on spacetimes with a bifurcate Killing horizon. *Physics Reports 207*(2), 49–136. doi:10.1016/0370-1573(91)90015-E.

126. Kokkotas, K. D. and B. Schmidt (1999). Quasi-normal modes of stars and black holes. *Living Reviews in Relativity 2*, 2. doi:10.12942/lrr-1999-2. URL (accessed online 11 Apr 2014): http://www.livingreviews.org/lrr-1999-2.

127. Krasnikov, S. (1998). Hyperfast travel in general relativity. *Physical Review D 57*(8), 4760–4766. doi:10.1103/PhysRevD.57.4760.

128. Lee, C. (1976). A restriction on the topology of Cauchy surfaces in general relativity. *Communications in Mathematical Physics 51*(2), 157–162. doi:10.1007/BF01609346.

129. Lehmkuhl, D., G. Schiemann, and E. Scholz (Eds.) (2016). *Towards a Theory of Spacetime Theories.* Einstein Studies. Berlin: Springer.

130. Liu, D. and M. Rebouças (2012). Energy conditions bounds on $f(t)$ gravity. *Physical Review D 86*(8), 083515. doi:10.1103/PhysRevD.86.083515. Preprint: arXiv:1207.1503v3 [astro-ph.CO].

131. Ludvigsen, M. and J. Vickers (1982). A simple proof of the positivity of the Bondi mass. *Journal of Physics A 15*(2), L67–L70. doi:10.1088/0305-4470/15/2/003.
132. Malament, D. (2012a). A remark about the "geodesic principle" in general relativity. In M. Frappier, D. Brown, and R. DiSalle (Eds.), *Analysis and Interpretation in the Exact Sciences: Essays in Honor of William Demopoulos*, Number 78 in The Western Ontario Series in Philosophy of Science, Chapter 14, pp. 245–252. Berlin: Springer Verlag.
133. Malament, D. (2012b). *Topics in the Foundations of General Relativity and Newtonian Gravitational Theory*. Chicago: University of Chicago Press. Uncorrected final proofs for the book are available for download at http://strangebeautiful.com/other-texts/malament-founds-gr-ngt.pdf.
134. Martín-Moruno, P. and M. Visser (2013a). Classical and quantum flux energy conditions for quantum vacuum states. *Physical Review D 88*, 061701(R). doi:10.1103/PhysRevD.88.061701. Preprint: arXiv:1305.1993 [gr-qc].
135. Martín-Moruno, P. and M. Visser (2013b). Semiclassical energy conditions for quantum vacuum states. *Journal of High Energy Physics 2013*(09), 1–36. doi:10.1007/JHEP09(2013)050. Preprint: arXiv:1306.2076.
136. Mazur, P. (1982). Proof of uniqueness of the Kerr-Newman black hole solution. *Journal of Physics A 15*(10), 3173–3180. doi:10.1088/0305-4470/15/10/021.
137. Misner, C. (1967). Taub-NUT space as a counterexample to almost anything. In J. Ehlers (Ed.), *Relativity Theory and Astrophysics: 1. Relativity and Cosmology*, Number 8 in Lectures in Applied Mathematics, pp. 160–169. Providence, RI: American Mathematical Society. Proceedings of the Fourth Summer Seminar on Applied Mathematics, Cornell University, Jul. 26–Aug. 20, 1965.
138. Molina-Paris, C. and M. Visser (1999, 27 May). Minimal conditions for the creation of a Friedmann-Robertson-Walker universe from a 'bounce'. *Physics Letters B 455*, 90–95. doi:10.1016/S0370-2693(99)00469-4. Preprint: arXiv:gr-qc/9810023.
139. Morris, M. and K. Thorne (1988). Wormholes in space-time and their use for interstellar travel: A tool for teaching general relativity. *American Journal of Physics 56*, 395–412. doi:10.1119/1.15620.
140. Morris, M., K. Thorne, and U. Yurtsever (1988). Wormholes, time machines, and the weak energy condition. *Physical Review Letters 61*(13), 1446–1449. doi:10.1103/PhysRevLett.61.1446.
141. Müller zum Hagen, H., D. Robinson, and H. Seifert (1973). Black holes in static vacuum space-times. *General Relativity and Gravitation 4*(1), 53–78. doi:10.1007/BF00769760.
142. Newman, E. and R. Penrose (1962). An approach to gravitational radiation by a method of spin coefficients. *Journal of Mathematical Physics 3*(3), 566–578. doi:10.1063/1.1724257.
143. Newman, E. and T. Unti (1962). Behaviour of asymptotically flat empty spaces. *Journal of Mathematical Physics 3*(5), 891–901. doi:10.1063/1.1724303.
144. Olum, K. (1998). Superluminal travel requires negative energies. *Physical Review Letters 81*(17), 3567–3570. doi:10.1103/PhysRevLett.81.3567.
145. Oppenheimer, J. and H. Snyder (1939). On continued gravitational contraction. *Physical Review 56*(5), 455–459. doi:10.1103/PhysRev.56.455.
146. Oppenheimer, J. and G. Volkoff (1939). On massive neutron cores. *Physical Review 55*, 374–381. doi:10.1103/PhysRev.55.374.
147. Peccei, R. and H. Quinn (1977, June). CP conservation in the presence of pseudoparticles. *Physical Review Letters 38*(25), 1440–1443. doi:10.1103/PhysRevLett.38.1440.
148. Penrose, R. (1965). Gravitational collapse and space-time singularities. *Physical Review Letters 14*(3), 57–59. doi:10.1103/PhysRevLett.14.57.
149. Penrose, R. (1968). Structure of spacetime. In C. DeWitt and J. Wheeler (Eds.), *Battelle Rencontres*, Chapter VII, pp. 121–235. New York: W. A. Benjamin.
150. Penrose, R. (1969). Gravitational collapse: The role of general relativity. *Revista del Nuovo Cimento Numero Speziale 1*, 257–276. Reprinted in *General Relativity and Gravitation 34*(2002,7):1141–1165, doi:10.1023/A:1016534604511.

151. Penrose, R. (1973, December). Naked singularities. *Annals of the New York Academy of Sciences 224*, 125–134. Proceedings of the Sixth Texas Symposium on Relativistic Astrophysics. doi:10.1111/j.1749-6632.1973.tb41447.x.

152. Penrose, R. (1979). Singularities and time-asymmetry. See [108], pp. 581–638.

153. Penrose, R. (1990, June). Light rays near i^o: A new mass positivity theorem. *Twistor Newsletter 30*, 1–5. Freely available at http://people.maths.ox.ac.uk/lmason/Tn/30/TN30-02.pdf.

154. Penrose, R. and R. Floyd (1971, 08 February). Extraction of rotational energy from a black hole. *Nature 229*, 177–179. doi:10.1038/physci229177a0.

155. Penrose, R. and W. Rindler (1984). *Spinors and Spacetime: Two-Spinor Calculus and Relativistic Fields*, Volume 1. Cambridge: Cambridge University Press.

156. Penrose, R., R. Sorkin, and E. Woolgar (1993). A positive mass theorem based on the focusing and retardation of null geodesics. arXiv:gr-qc/9301015.

157. Pirani, F. (1955, 22 March). On the energy-momentum tensor and the creation of matter in relativistic cosmology. *Proceedings of the Royal Society of London. Series A. Mathematical and Physical Sciences 228*(1175), 255–262. doi:10.1098/rspa.1955.0061.

158. Press, W. and S. Teukolsky (1973). Perturbations of a rotating black hole. II. Dynamical stability of the Kerr metric. *Astrophysical Journal 185*, 649–674. doi:10.1086/152445.

159. Rácz, I. and R. Wald (1992). Extensions of spacetimes with Killing horizons. *Classical and Quantum Gravity 12*(12), 2643–2656. doi:10.1088/0264-9381/9/12/008.

160. Rácz, I. and R. Wald (1996). Global extensions of spacetimes describing asymptotic final states of black holes. *Classical and Quantum Gravity 13*(3), 539–552. doi:10.1088/0264-9381/13/3/017.

161. Rendall, A. (1996a). Constant mean curvature foliations in cosmological spacetimes. *Helvetica Physica Acta 69*, 490–500. Preprint: arXiv:gr-qc/9606049.

162. Rendall, A. (1996b). Existence and non-existence results for global constant mean curvature foliations. Preprint: arXiv:gr-qc/9608045.

163. Ringström, H. (2009). *The Cauchy Problem in General Relativity*. ESI Lectures in Mathematics and Physics. Zürich: European Mathematical Society Publishing House.

164. Robinson, D. (1975). Uniqueness of the Kerr black hole. *Physical Review Letters 34*(14), 905–906. doi:10.1103/PhysRevLett.34.905.

165. Robinson, D. (1977). A simple proof of the generalization of Israel's theorem. *General Relativity and Gravitation 5*(8), 695–698. doi:10.1007/BF00756322.

166. Roman, T. (1988). On the "averaged weak energy condition" and Penrose's singularity theorem. *Physical Review D 37*(2), 546–548. doi:10.1103/PhysRevD.37.546.

167. Sahni, V., H. Feldman, and A. Stebbins (1992). A loitering universe. *Astrophysical Journal 385*, 1–8. doi:10.1086/170910.

168. Sahni, V. and Y. Shtanov (2005). Did the universe loiter at high redshifts? *Physics Review D 71*, 084018. doi:10.1103/PhysRevD.71.084018. Preprint: arXiv:astro-ph/0410221.

169. Schattner, R. (1979a). The center of mass in general relativity. *General Relativity and Gravitation 10*(5), 377–393. doi:10.1007/BF00760221.

170. Schattner, R. (1979b). The uniqueness of the center of mass in general relativity. *General Relativity and Gravitation 10*(5), 395–399. doi:10.1007/BF00760222.

171. Schmidt, B. (1971). A new definition of singular points in general relativity. *General Relativity and Gravitation 1*, 269–280.

172. Schoen, R. and S.-T. Yau (1981). Proof of the positive mass theorem. II. *Communications in Mathematical Physics 79*(2), 231–260. doi:10.1007/BF01942062.

173. Schoen, R. and S.-T. Yau (1982). Proof that the Bondi mass is positive. *Physical Review Letters 48*(6), 369–371. doi:10.1103/PhysRevLett.48.369.

174. Schoen, R. and S.-T. Yau (1983). The existence of a black hole due to condensation of matter. *Communications in Mathematical Physics 90*(4), 575–579. doi:10.1007/BF01216187.

175. Sciama, D. (1976). Black holes and their thermodynamics. *Vistas in Astronomy 19, Part 4*, 385. doi:10.1016/0083-6656(76)90052-0.

176. Senovilla, J. (1998). Singularity theorems and their consequences. *General Relativity and Gravitation 30*(5), 701–848. doi:10.1023/A:1018801101244. *N.b.*: in the article itself (as, *e.g.*, downloaded from the *General Relativity and Gravitation* website), the paper is indicated as appearing in volume 29, 1997; as the author of the paper has assured me, this is erroneous.

177. Senovilla, J. (2000). Super-energy tensors. *Classical and Quantum Gravity 17*(14), 2799–2842. doi:10.1088/0264-9381/17/14/313. Preprint: arXiv:gr-qc/9906087.

178. Senovilla, J. (2002). Superenergy tensors and their applications. Invited lecture presented at the 1st Conference on Lorentzian Geometry, "Benalmadena 2001". Preprint arXiv:math-ph/0202029.

179. Senovilla, J. (2007). A singularity theorem based on spatial averages. *Pramana 69*(1), 31–48. doi:10.1007/s12043-007-0109-2. Preprint: arXiv:gr-qc/0610127v3.

180. Senovilla, J. (2008, January). A new type of singularity theorem. In A. Oscoz, E. Mediavilla, and M. Serra-Ricart (Eds.), *Spanish Relativity Meeting—Encuentros Relativistas Españoles, ERE2007: Relativistic Astrophysics and Cosmology*, Number 30 in EAS Publications, pp. 101–106. doi:10.1051/eas:0830009. Preprint: arXiv:0712.1428 [gr-qc].

181. Shapiro, S. and S. Teukolsky (1983). *Black Holes, White Dwarfs and Neutron Stars*. New York: Wiley Interscience.

182. Smolin, L. (1995). The Bekenstein bound, topological quantum field theory and pluralistic quantum field theory. arXiv:gr-qc/9508064.

183. Susskind, L. (1995). The world as a hologram. *Journal of Mathematical Physics 36*(11), 6377–6396. doi:10.1063/1.531249. Preprint: arXiv:hep-th/9409089v2.

184. 't Hooft, G. (1988). On the quantization of space and time. In M. Markov, V. Berezin, and V. Frolov (Eds.), *Quantum Gravity*, pp. 551–567. Singapore: World Scientific Press.

185. Teitelboim, C. (1972a). Nonmeasurability of the lepton number of a black hole. *Lettere al Nuovo Cimento 3*(10), 397–400. doi:10.1007/BF02826050.

186. Teitelboim, C. (1972b). Nonmeasurability of the quantum numbers of a black hole. *Physical Review D 5*(12), 2941–2954. doi:10.1103/PhysRevD.5.2941.

187. Tipler, F. (1977a, September). Singularities and causality violations. *Annals of Physics 108*(1), 1–36. doi:10.1016/0003-4916(77)90348-7.

188. Tipler, F. (1977b, November). Singularities in conformally flat spacetimes. *Physics Letters A 64*(1), 8–10. doi:10.1016/0375-9601(77)90508-4.

189. Tipler, F. (1978). Energy conditions and spacetime singularities. *Physical Review D 17*(10), 2521–2528. doi:10.1103/PhysRevD.17.2521.

190. Trautman, A. (1965). Foundations and current problems of general relativity. In S. Deser. and K. Ford (Eds.), *Lectures on General Relativity*, Number 1 in Brandeis Summer Institute in Theoretical Physics, pp. 1–248. Englewood Cliffs, NJ: Prentice-Hall, Inc.

191. Trautman, A. (1966). General relativity. *Soviet Physics Uspekhi 9*(3), 319–339.

192. Urban, D. and K. Olum (2010). Averaged null energy condition violation in a conformally flat spacetime. *Physical Review D 81*(2), 024039. doi:10.1103/PhysRevD.81.024039.

193. Visser, M. (1989a, December). Traversable wormholes from surgically modified Schwarzschild spacetimes. *Nuclear Physics B 311*(1), 203–212. doi:10.1016/0550-3213(89)90100-4. Preprint: arXiv:0809.0927 [gr-qc].

194. Visser, M. (1989b, May). Traversable wormholes: Some simple examples. *Physical Review D 39*, 3182–3184(R). doi:10.1103/PhysRevD.39.3182.

195. Visser, M. (1996). *Lorentzian Wormholes: From Einstein to Hawking*. Woodbury, NY: American Institute of Physics Press.

196. Visser, M. (1997a, 04 April). Energy conditions in the epoch of galaxy formation. *Science 276*(5309), 88–90. doi:10.1126/science.276.5309.88. Preprint: arXiv:gr-qc/9710010v2. Originally presented at the Eighth Marcel Grossmann Conference on General Relativity, Jerusalem, Israel, June 1997.

197. Visser, M. (1997b). General relativistic energy conditions: The Hubble expansion in the epoch of galaxy formation. *Physical Review D 56*, 7578–7587. doi:10.1103/PhysRevD.56.7578. Preprint: arXiv:gr-qc/9705070.

198. Visser, M. (2005). Cosmography: Cosmology without the Einstein equation. *General Relativity and Gravitation 37*(9), 1541–1548. doi:10.1007/s10714-005-0134-8. Based on a talk delivered at ACRGR4, the 4th Australasian Conference on General Relativity and Gravitation, Monash University, Melbourne, January 2004. Preprint: arXiv:gr-qc/0411131.
199. Visser, M. and C. Barceló (2000). Energy conditions and their cosmological implications. Plenary talk presented at Cosmo99, Trieste, Sept/Oct 1999. Preprint: arXiv:gr-qc/0001099.
200. Visser, M., B. Bassett, and S. Liberati (2000, June). Superluminal censorship. *Nuclear Physics B – Proceedings Supplements 88*(1-3), 267–270. doi:10.1016/S0920-5632(00)00782-9. Preprint: arXiv:gr-qc/9810026.
201. Wald, R. (1972). Electromagnetic fields and massive bodies. *Physical Review D 6*(6), 1476–1479. doi:10.1103/PhysRevD.6.1476.
202. Wald, R. (1973). On perturbations of a Kerr black hole. *Journal of Mathematical Physics 14*(10), 1453–1461. doi:10.1063/1.1666203.
203. Wald, R. (1984). *General Relativity*. Chicago: University of Chicago Press.
204. Wald, R. (1994). *Quantum Field Theory in Curved Spacetime and Black Hole Thermodynamics*. Chicago: University of Chicago Press.
205. Wald, R. and S. Gao (2001). "Physical process version" of the First Law and the Generalized Second Law for charged and rotating black holes. *Physical Review D 64*(8), 084020. doi:10.1103/PhysRevD.64.084020.
206. Wall, A. (2010). Proving the achronal averaged null energy condition from the Generalized Second Law. *Physical Review D 81*(2), 024038. doi:10.1103/PhysRevD.81.024038. Preprint: arXiv:0910.5751 [gr-qc].
207. Wallace, D. (2016). The relativity and equivalence principles for self-gravitating systems. See [129], Chapter 8.
208. Weatherall, J. (2012). A brief remark on energy conditions and the Geroch-Jang theorem. *Foundations of Physics 42*, 209–214. doi:10.1007/s10701-011-9538-y.
209. Weatherall, J. (2016). Inertial motion, explanation, and the foundations of classical spacetime theories. See [129], Chapter 2. Preprint: arXiv:1206.2980 [physics.hist-ph].
210. Weinberg, S. (1972). *Gravitation and Cosmology: Principles and Applications of the General Theory of Relativity*. New York: Wiley and Sons Press.
211. Witten, E. (1981). A new proof of the positive energy theorem. *Communications in Mathematical Physics 65*(3), 381–402. doi:10.1007/BF01208277.
212. Wong, W. (2011). Regular hyperbolicity, dominant energy condition and causality for Lagrangian theories of maps. *Classical and Quantum Gravity 28*, 215008. doi:10.1088/0264-9381/28/21/215008.
213. Wüthrich, C. (2016). Raiders of the lost spacetime. See [129], Chapter 11.
214. Yau, S.-T. (1982). Problem section. In S.-T. Yau (Ed.), *Seminar on Differential Geometry*, Volume 102, pp. 669–706. Princeton, NJ: Princeton University Press.
215. Zel'dovich, Y. and I. Novikov (1971). *Stars and Relativity*. Chicago: University of Chicago Press. *Relativistic Astrophysics*, Vol. 1.

Background Independence, Diffeomorphism Invariance and the Meaning of Coordinates

Oliver Pooley

Abstract Diffeomorphism invariance is sometimes taken to be a criterion of background independence. This claim is commonly accompanied by a second that the genuine physical magnitudes (the "observables") of background-independent theories and those of background-dependent (non-diffeomorphism-invariant) theories are essentially different in nature. I argue against both claims. Background-dependent theories can be formulated in a diffeomorphism-invariant manner. This suggests that the nature of the physical magnitudes of relevantly analogous theories (one background free, the other background dependent) is essentially the same. The temptation to think otherwise stems from a misunderstanding of the meaning of spacetime coordinates in background-dependent theories.

1 What is so Special about General Relativity?

According to a familiar and plausible view, the core of Einstein's general theory of relativity (GR) is what was, in 1915, a radically new way of understanding gravitation. In pre-relativistic theories, whether Newtonian or specially relativistic, the structure of spacetime is taken to be fixed, varying neither in time nor from solution to solution. Gravitational phenomena are assumed to be the result of the action of gravitational forces, diverting gravitating bodies from the natural motions defined by this fixed spacetime structure. According to GR, in contrast, freely falling bodies are force free; their trajectories are natural motions. Gravity is understood in terms of a mutable spacetime structure. Bodies act gravitationally on one another by affecting the curvature of spacetime. "*Space acts on matter, telling it how to move. In turn, matter reacts back on space, telling it how to curve*" [41, 5]. Note that the first of the claims in the quotation is as true in pre-relativistic theories as it is in GR, at least according to the substantivalist view, which takes spacetime structure in such a theory to be an independent element of reality. The novelty of GR lies in the second

O. Pooley (✉)
Oriel College Oxford, Oxford OX1 4EW, UK
e-mail: oliver.pooley@philosophy.ox.ac.uk

© Springer Science+Business Media, LLC 2017
D. Lehmkuhl et al. (eds.), *Towards a Theory of Spacetime Theories*,
Einstein Studies 13, DOI 10.1007/978-1-4939-3210-8_4

claim: spacetime curvature varies, in time (and space) and across models, and the material content of spacetime affects how it does so.

This sketch of the basic character of GR has two separable elements. One is the interpretation of the metric field, g_{ab}, as intrinsically geometrical: gravitational phenomena are to be understood in terms of the curvature *of spacetime*. The second is the stress on the dynamical nature of the metric field: the fact that it has its own degrees of freedom and, in particular, that their evolution is affected by matter. While I believe that both of these are genuine (and novel) features of GR, my focus in this paper is on the second. Those who reject the emphasis on geometry are likely to claim that the second element by itself encapsulates the true conceptual revolution ushered in by GR. Non-dynamical fields, such as the spacetime structures of pre-relativistic physics, are now standardly labelled *background fields* (although which of their features qualifies them for this status is a subtle business, to be explored in what follows). On the view being considered, the essential novelty of GR is that such background structures have been excised from physics; GR is the prototypical *background-independent* theory[1] (as it happens, a prototype yet to be improved upon).

Although this paper is about this notion of background independence, the question of the geometrical status of the metric field cannot be avoided entirely. In arguing against the interpretation of GR as fundamentally about spacetime geometry, Anderson writes

> What was not clear in the beginning but by now has been recognised is that one does not need the "geometrical" hypotheses of the theory, namely, the identification of a metric with the gravitational field, the assumption of geodesic motion, and the assumption that "ideal" clocks measure proper time as determined by this metric. Indeed, we know that both of these latter assumptions follow as approximate results directly from the field equations of the theory without further assumptions. [3, 528]

There is at least the suggestion here that GR differs from pre-relativistic theories not only in lacking non-dynamical, background structures but also in terms of how one of its structures, the "gravitational field", acquires geometrical meaning: the appropriate behaviour of test bodies and clocks can be derived, approximately, in the theory. Does this feature of GR really distinguish it from special relativity (SR)?

Consider, in particular, a clock's property of measuring the proper time along its trajectory. In a footnote, Anderson goes on to explain that "the behaviour of model clocks and what time they measure can be deduced from the equations of

---

[1] In what follows I focus specifically on the notion of background independence that is connected to the idea that background structures are non-dynamical fields. In doing so, I am ignoring several other (not always closely related) definitions of background independence, including those given by Gryb [35] (which arises more naturally in the context of Barbour's 3-space approach to dynamics) and by Rozali [55] (which arises naturally in string theory). A more serious omission is the lack of discussion of the definition given by Belot [9], which is motivated by ideas closely related to the themes of this paper. I hope to explore these connections on another occasion.

sources of the gravitational and electromagnetic fields which in turn follow from the field equations" [3, 529]. But the generally relativistic "equations of sources of the gravitational and electromagnetic fields" are, on the assumption of minimal coupling, exactly the same as the equations of motion of an analogue specially relativistic theory.[2] It follows that whatever explanatory modelling one can perform in GR, by appeal to such equations, to show that some particular material system acts as a good clock and discloses proper time, is equally an explanation of the behaviour of the same type of clock in the context of SR. Put differently, it is as true in SR as it is in GR that the "geometrical" hypothesis linking the behaviour of ideal clocks to the (in this context) non-dynamical background "metric" field is in principle dispensable.[3]

2 Einstein on General Covariance

The previous section's positive characterisation of GR's essential difference from its predecessors goes hand-in-hand with a negative claim: GR does not differ from its predecessors in virtue of being a *generally covariant* theory. In particular, the general covariance of GR does not embody a "general principle of relativity" (asserting, for example, the physical equivalence of observers in arbitrary states of relative motion). In contrast, the restricted, Lorentz covariance of standard formulations of specially relativistic physics *does* embody the (standard) relativity principle. In Michael Friedman's words, "the principle of general covariance has no physical content whatever: it specifies no particular physical theory; rather it merely expresses our commitment to a certain style of formulating physical theories" [32, 55].

Notoriously, of course, Einstein thought otherwise, at least initially.[4] The restricted relativity principle of SR and Galilean-covariant Newtonian theories is the claim that the members of a special class of frames of reference, each in uniform translatory

---

[2]That it is only in the GR context that material fields merit the label "sources of the gravitational field" is, of course, irrelevant.

[3]In this context, it is interesting to consider Fletcher's proof that the clock hypothesis holds up to arbitrary accuracy for sufficiently small light clocks [31]. As is explicit in Fletcher's paper, his result is as applicable to accelerating clocks in SR as it is to arbitrarily moving clocks in GR. Fletcher's proof assumes only that light travels on null geodesics; it does not make any assumptions about the fundamental physics, or even (specific) assumptions about the deformation of the spatial dimensions of the clock. All of this is consistent with one of the morals of the "dynamical approach to special relativity", defended in Brown [11] and Brown and Pooley [13], that it is no more of a brute fact in SR than in GR that real rods and clocks, which are more or less complex solutions of the laws governing their constituents, map out geometrical properties in the way that they do. What Fletcher's proof illustrates is that some interesting results are nonetheless obtainable from minimalist, high-level physical assumptions. (Note that, in contrast to the position taken in Brown and Pooley [13], I am here assuming that the structure encoded by the flat metric field of special relativity corresponds to a primitive element of reality, as was entertained in Brown and Pooley [13, 82, fn 22].)

[4]The evolution of Einstein's views is covered in detail by Norton[43, §3]. In this section, I largely follow Norton's narrative.

motion relative to the others, are physically equivalent. In such theories, although no empirical meaning can be given to the idea of absolute rest, there is a fundamental distinction between accelerated and unaccelerated motion. Einstein thought this was problematic, and offered a thought experiment to indicate why.

Consider two fluid bodies, separated by a vast distance, rotating relative to one another about the line joining their centres. Such relative motion is in principle observable, and so far our description of the set-up is symmetric with respect to the two bodies. Now, however, imagine that one body is perfectly spherical while the other is oblate. A theory satisfying only the restricted principle of relativity is compatible with this kind of situation. In such a theory, the second body might be flattened along the line joining the two bodies only because that body is rotating, not just with respect to other observable bodies, but with respect to the theory's privileged, non-accelerating frames of reference. Einstein deemed this an inadequate explanation. He claimed that appeal to the body's motion with respect to the invisible inertial frames was an appeal to a "merely factitious cause". In Einstein's view, a truly satisfactory explanation should cite "*observable facts of experience*" [24, 113]. A theory which in turn explains the (local) inertial frames in terms of the configuration of (observable) distant masses—that is, a theory satisfying (a version of) *Mach's Principle*—would meet such a requirement.

In his quest for a relativistic theory of gravity, Einstein did not attempt to implement (this version of) Mach's principle directly. Instead he believed that the *equivalence principle* (as he understood it) was the key to extend the relativity principle to cover frames uniformly accelerating with respect to the inertial frames. In standard SR, force-free bodies that move uniformly in an inertial frame F are equally accelerated by inertial "pseudo forces" relative to a frame F' that is uniformly accelerating relative to F. According to Einstein's equivalence principle, the physics of frame F' is strictly identical to that of a "real" inertial frame in which there is a uniform gravitational field. In other words, the same laws of physics hold in two frames that accelerate with respect to each other. According to one frame, there is a gravitational field; according to the other, there is not. The laws that hold with respect to both frames, therefore, must cover gravitational physics. Einstein took it to follow that there is no fact of the matter about whether a body is moving uniformly or whether it is accelerating under the influence of gravitation. The existence of a gravitational field becomes frame-relative, in a manner allegedly analogous to the frame-relativity of particular electric and magnetic fields in special relativity.[5]

The equivalence principle, then, led Einstein to believe both that relativistic laws covering gravitational phenomena would extend the relativity principle and that the gravitational field would depend, in a frame-relative manner, on the metric field, g_{ab}. A theory implementing a general principle of relativity would affirm the physical equivalence of frames of reference in arbitrary relative motion. Einstein took the physical equivalence of two frames to be captured by the fact that the equations

---

[5]For a recent, sympathetic discussion of this aspect of Einstein's understanding of the equivalence principle, see Janssen [37].

expressing the laws of physics take the same form with respect to each of them.[6] But general covariance is the property that a theory possesses if its equations retain their form under smooth but otherwise arbitrary coordinate transformation. Einstein noted that such coordinate transformations strictly include "those which correspond to all relative motions of three-dimensional systems of co-ordinates" [24, 117]. He therefore maintained that any generally covariant theory satisfies a general postulate of relativity.[7]

Einstein soon modified his view. Essentially the view expressed by Friedman in the quotation given above—that any theory can be given a generally covariant formulation—was put to Einstein by Kretschmann [39].[8] In his response, Einstein conceded the basic point [25]. He identified three principles as at the heart of GR: (a) the (general) principle of relativity; (b) the equivalence principle; and (c) Mach's principle. The relativity principle, at least as characterised in his reply to Kretschmann, was no longer conceived of in terms of the physical equivalence of frames of reference in various types of relative motion. Instead it had simply become the claim that the laws of nature are statements only about spatiotemporal coincidences, from which it was alleged to be an immediate corollary that such laws "find their natural expression" in generally covariant equations. Mach's principle was also given a GR-specific rendition: the claim was that the metric was completely determined by the masses of bodies.

In another couple of years, as a result of findings by de Sitter and Klein, Einstein was also forced to accept that his theory did not vindicate Mach's ideas about the origin of inertia. His official objection to the spacetime structures of Newtonian and specially relativistic theories changed accordingly, in order to fit this new reality.[9] Einstein conceded that taking Newtonian physics at face value involves taking Newton's Absolute Space to be "some kind of physical reality" [28, 15]. That it has to be conceived of as something real is, he says, "a fact that physicists have only come to understand in recent years" [28, 16]. It is absolute, however, not merely in the substantivalist sense that it exists absolutely. Now Einstein placed emphasis on the fact that it is not *influenced* "either by the configuration of matter, or by anything else" [28, 15]. This violation of the *action–reaction principle*, rather than its status as an unobservable causal agent, came to be seen as what is objectionable about pre-relativistic spacetime. In Einstein's words, "it is contrary to the mode of thinking

---

[6]Recall Einstein's 1905 statement of the restricted principle of relativity: "The laws by which the states of physical systems undergo change are not affected, whether these changes of state be referred to the one or the other of two systems of co-ordinates in uniform translatory motion" [23, 41].

[7]"Es ist klar, daß eine Physik, welche diesem Postulat [i.e. general covariance] genügt, dem allgemeinen Relativitätspostulat gerecht wird" [24, 776].

[8]Kretschmann's position is more subtle than the headline lesson that is standardly taken from it. In particular, he relied on a key premise, closely analogous to the central premise of Einstein's 'point-coincidence' response to his own hole argument that the factual content of a theory is exhausted by spatiotemporal coincidences between the objects and processes it posits; see Norton [43, §5.1]. The assumption that the basic objects of a theory must be well defined in the sense of differential geometry has come to play a similar role in modern renditions of Kretschmann's claim.

[9]For more on the evolution of this aspect of Einstein's thinking, see Brown and Lehmkuhl [12].

in science to conceive of a thing (the space-time continuum) which acts itself, but which cannot be acted upon" [27, 62].[10] It is clear that, while GR fails to fulfil the Machian goal of providing a reductive account of the local inertial frames, it does not suffer from this newly identified (alleged) defect of pre-relativistic theories. The metric structure of GR conditions the evolution of the material content of spacetime, but it is also, in turn, affected by that content.

This potted review of Einstein's early pronouncements is intended to show that he was one of the original advocates of the view outlined in Section 1, namely, that GR differs from its predecessors, not through lacking the kind of spacetime structures that such theories have, but by no longer treating that structure as a non-dynamical background. It also shows that, despite being responsible for the idea that the general covariance of GR has physical significance as the expression of the theory's generalisation of the relativity principle, Einstein himself quickly retreated from this idea. He continued (mistakenly) to espouse the idea that GR generalised the principle of relativity, via the equivalence principle, but GR's general covariance was no longer taken to be a sufficient condition of its doing so. Instead the implication in the opposite direction was stressed. General covariance was taken to be a *necessary* condition of implementing a general relativity principle: there can be no special coordinate systems adapted to preferred states of motion in a theory in which there are no preferred states of motion!

In the immediate wake of Kretschmann's criticism, one of Einstein's most revealing statements concerning the status of general covariance comes in his response to a paper by Ernst Reichenbächer. There, Einstein contrasts a theory that includes an acceleration standard with one that does not

> if acceleration has absolute meaning, then the nonaccelerated coordinate systems are preferred by nature, i.e., the laws then must—when referred to them—be different (and simpler) than the ones referred to accelerated coordinate systems. Then it makes no sense to complicate the formulation of the laws by pressing them into a generally covariant form.
>
> Vice versa, if the laws of nature are such that they do not attain a preferred form through the choice of coordinate systems of a special state of motion, then one cannot relinquish the condition of general covariance as a means of research. [26, 205]

From a modern perspective, several things are notable about this passage. First, GR qualifies as a theory whose laws do not attain a "preferred form through the choice of coordinate systems of a special state of motion", not because (as Einstein believed) acceleration does not have an absolute meaning in the theory, but because the structure that defines absolute acceleration is no longer homogeneous; in general, it is not possible to define, over a neighbourhood of a point in spacetime, a coordinate system whose lines of constant spatial coordinate are both non-accelerating absolutely

---

[10] Similarly, Anderson writes that violation of what he calls a general principle of reciprocity "seems to be fundamentally unreasonable and unsatisfactory" [1, 192]. As far as I know, neither he nor Einstein explain why, exactly, such violation is supposed to be objectionable. At the very least, given Newton's open-eyed advocacy of absolute space, it seems peculiar to describe it as "contrary to the mode of scientific thinking."

and not accelerating with respect to each other. GR lacks a non-generally covariant formulation,[11] but not for the reason Einstein suggests.

Second, while the equations expressing a theory's laws might be *simpler* in a coordinate system adapted to the theory's standard of acceleration, it does not follow that these equations, and the equations that hold with respect to accelerated coordinate systems, express different laws. In fact, it is much more natural to see the formally different equations as but different coordinate-dependent expressions of the same relations holding between coordinate-independent entities. As Anderson says of entities that occur explicitly in a generally covariant formulation of some laws but which were not apparent in the non-(generally)-covariant equations: "these elements were there in the first place, although their existence was masked by the fact that they had been assigned particular values. That is, the $g^{\mu\nu}$ [of a generally covariant formulation of a special relativity] are present in [the Lorentz-covariant form of] special relativity with the fixed preassigned values of the Minkowski metric" [1, 192].[12]

Finally, while calculation might not be aided by complicating the formulation of the laws by expressing them generally covariantly, conceptual clarity can be. Real structures that are only implicit in the non-covariant formalism are laid bare in the generally covariant formalism, and their status can then be subjected to scrutiny.

In fact, Einstein himself says something quite consonant with these observations earlier in the same paper

> the coordinate system is only a *means of description* and in itself has nothing to do with the *objects to be described*. Only a law of nature in a generally covariant form can do complete justice in this situation, because in any other way of describing, statements about the means of description are jumbled with statements about the object to be described. [26, 203]

Einstein's idea seems to be that coordinates should not have a function beyond the mere labelling of physical entities, the qualitative character of which is to be fully described by other means. But this is a basis, not for an argument in favour of laws that can only be expressed generally covariantly (seemingly Einstein's intention), but for an argument for the generally covariant formulation of laws in general, whatever they be. Ironically, it is an argument that is most relevant to pre-relativistic theories, not GR, because only in this context can one choose to encode physically meaningful quantities (spacetime intervals) via special choices of coordinate system, and thereby 'jumble up' the mode of description with that described.

---

[11] Even this can be disputed. Fock, for example, argued that harmonic coordinates, defined via the condition $(g^{\mu\nu}\sqrt{-g})_{,\mu} = 0$, have a preferred status in GR, analogous to that of Lorentz charts in special relativity.

[12] The same view of the meaning of the preferred coordinates of the non-covariant form of Newtonian gravitation theory is clearly articulated by Trautman [65, 418]. It was thoroughly assimilated in the philosophical literature; see, e.g., Friedman [32, 54–55]. The perspective is explored further in Sections 4 and 10, where I argue that its relevance for discussions of alleged differences between the observables of GR and pre-relativistic theories has not been fully appreciated.

3 Dissent from Quantum Gravity

Let me sum up the picture presented so far. General covariance *per se* has no physical content: the essence of Kretschmann's objection to Einstein is that any sensible theory can be formulated in a generally covariant manner. It follows that GR does not differ from SR in virtue of having a generally covariant formulation. However, GR does differ from SR in *lacking* a *non*-covariant formulation. Some authors have made this fact the basis for claiming that GR, but not SR, satisfies a "principle of general covariance". For example, Bergmann writes "The hypothesis that the geometry of physical space is represented best by a formalism which is covariant with respect to general coordinate transformations, and that a restriction to a less general group of transformations would not simplify that formalism, is called *the principle of general covariance*" [10, 159].

In SR, the existence of a non-covariant formulation is connected with the failure of a general principle of relativity. The privileged coordinate systems of SR, in which the equations expressing the laws simplify, encode (*inter alia*) a standard of non-accelerated motion. There can be no preferred coordinate systems (of such a type) in a theory that implements a general principle of relativity. This might suggest that GR's lack of a non-covariant formulation is connected to the generalisation of a relativity principle, but (*pace* Einstein) it stems from no such thing. Rather, the lack of preferred coordinates is due to the fact that the spacetime structures of a generic solution, including those structures common to SR and GR that define absolute acceleration (in essentially the same way in both theories), lack symmetries and so cannot be encoded in special coordinates.

Finally, this lack of symmetry is entailed by, *but does not entail*, the fundamental distinguishing feature of GR, namely, that the structure encoded by the metric of GR is, unlike that of SR, dynamical. A fully dynamical field, free to vary from solution to solution, will generically lack symmetries. So a background independent theory, in which all fields are dynamical, will lack a non-covariant formulation (of the relevant kind). The converse, however, is not true. In principle we can define a theory involving a background metric with no isometries, and such a theory will only have a generally covariant formulation.[13]

Something like this collection of commitments, though not uncontroversial, represents a mainstream view, at least amongst more recent textbooks in the tradition of Synge [63] and Misner et al. [41]. Unfortunately, there is a fly in the ointment, for it apparently conflicts with a dominant view amongst many in the quantum gravity community, in particular, the founding fathers of loop quantum gravity. Workers in

---

[13] Smolin demurs: "if one believes that the geometry of space is going to have an absolute character, fixed in advance, by some a priori principles, you are going to be led to posit a homogeneous geometry. For what, other than particular states of matter, would be responsible for inhomogeneities in the geometry of space?" [58, 201]. But why does a background geometry need to be fixed by "a priori principles"? Its being what it is could simply be brute fact, inhomogeneities notwithstanding.

this field often endorse the idea that GR's background independence, understood as the absence of 'fixed', non-dynamical spacetime structure, is its defining feature. But they go on to link this property to the theory's general covariance, or, to use the more favoured label, its *diffeomorphism invariance*. For example, Lee Smolin claims that "both philosophically and mathematically, it is diffeomorphism invariance that distinguishes general relativity from other field theories" [57, 234]. And Carlo Rovelli, who has perhaps written the most on the link between background independence and diffeomorphism invariance, says of the background independence of classical GR that "technically, it is realised by the gauge invariance of the action under (active) diffeomorphisms" [53, 10], and (perhaps in less careful moments) he treats the two as synonymous [33, 279].

On the face of it, these claims conflict with the Kretschmann view. They appear to assert that a formal property of GR, its "(active) diffeomorphism invariance", has physical content in virtue of realising, or expressing, a physical property of the theory, namely, its background independence. Since specially relativistic theories are not background independent (as we have been understanding this term), it should follow that they cannot be formulated in a diffeomorphism invariant manner. At the very least, if one follows Kretschmann in supposing that any theory can be formulated in a generally covariant manner, then (active) diffeomorphism invariance, as understood by Rovelli *et al.*, cannot be the same as general covariance as understood in the Kretschmann tradition. And, indeed, the same authors routinely draw distinctions of this kind.

Much of the rest of this paper is concerned to see how far one can push back against the Rovelli–Smolin line, in the spirit of Kretschmann and Friedman. What the exercise reveals is that the connection between diffeomorphism invariance and background independence is messier, and less illuminating, than recent discussions originating in the quantum gravity literature might suggest. It also sheds light on a different but closely related topic. In the same discussions, the diffeomorphism invariance and/or background independence of GR is frequently taken to have profound implications for the nature of the theory's observables. It is important that a merely technical sense of "observable" is not all that is at issue. The claim often appears to be that GR and pre-relativistic theories differ in terms of the kind of thing that is observable in a non-technical sense. In other words, it is alleged that the theories differ over the fundamental nature of the physical magnitudes that they postulate.[14] This, I believe, is a mistake, as I hope some of the distinctions to be reviewed below help to show.

The first task is to clarify what might be meant by "diffeomorphism invariance" as distinct from "general covariance". I then revisit the notion of a background field, as characterised informally above, for finer grained distinctions should be drawn here too.

---

[14] Amongst philosophers, Earman [20] and Rickles [49] are proponents of variants of this view.

4   General Covariance Versus Diffeomorphism Invariance

Several authors have drawn what they presumably take to be the crucial, bipartite distinction between types of general covariance and diffeomorphism invariance. Norton, for example, distinguishes "active" and "passive" general covariance [42, 1226, 1230]. Rovelli distinguishes "active diff invariance" from "passive diff invariance" [52, 122]. Earman distinguishes merely "formal" from "substantive" general covariance [20, 21]. Ohanian and Ruffini distinguish "general covariance" from "general invariance" [44, 276–9]. Finally, Giulini distinguishes "covariance under diffeomorphisms" from "invariance under diffeomorphisms" [34, 108]. As this cornucopia of terminology indicates, several different distinctions are in play, and linked to further ancillary notions (for example, that between "active" and "passive" transformations) in myriad ways. In the face of this morass, my strategy will be to articulate as clearly as I can what I take to be the most useful distinction, before relating it to several of the ideas just listed.

In differentiating distinct notions of general covariance and diffeomorphism invariance, it will be useful to consider various concrete formulations of theories that exemplify the properties in question. Further, when contrasting specially and generally relativistic theories, it is a good policy to eliminate unnecessary and potentially misleading differences by choosing theories that are as similar as possible. My running example, for both the specially and generally relativistic cases, will be theories of a relativistic massless real scalar field, Φ.

In the context of SR, such a field obeys the Klein–Gordon equation, but there are at least three "versions" of this equation to consider:

$$\frac{\partial^2 \Phi}{\partial x^2} + \frac{\partial^2 \Phi}{\partial y^2} + \frac{\partial^2 \Phi}{\partial z^2} - \frac{\partial^2 \Phi}{\partial t^2} = 0, \tag{1}$$

$$\eta^{\mu\nu} \Phi_{;\nu\mu} = 0, \tag{2}$$

$$\eta^{ab} \nabla_a \nabla_b \Phi = 0. \tag{3}$$

These equations are most plausibly understood as (elements of) different formulations of one and the same theory, not as characterising different theories. This requires that the equations are understood as but different ways of picking out the very same set of models (and thereby the very same set of physical possibilities). On the picture that allows this, one also gains a better understanding of the content of each equation.

What is that picture? Start with equation (3). The roman indices occurring in the equation are "abstract indices", indicating the type of geometric object involved. This equation, therefore, is not to be interpreted (as the other two are) as relating the coordinate components of various objects. Rather, it is a direct description of (the relations holding between) certain geometric object fields defined on a differentiable manifold. Its models are triples of the form $\langle M, \eta_{ab}, \Phi \rangle$: differential manifolds equipped with a (flat) Lorentzian metric field η_{ab} and a single scalar field Φ. (I am taking the torsion-free, metric-compatible derivative operator, ∇, to be defined in terms of the metric field; it is not another primitive object, over and above η_{ab} and Φ.)

Equations (1) and (2) are to be understood as ways of characterising the very same models, but now given under certain types of coordinate description. In particular, in the case of equation (1), one is choosing coordinates that are specially adapted to symmetries of one of the fields of the model, namely, the flat Minkowski metric. Such coordinates are singled out via the "coordinate condition" $\eta_{\mu\nu} = \text{diag}(-1, 1, 1, 1)$. In the case of equation (2), one is allowing any coordinate system adapted to the differential structure of the manifold, M.

We are now in a position to draw the crucial distinction between general covariance (as it has been implicitly understood in the previous sections) and diffeomorphism invariance for, on one natural way of further filling in the details, although it is generally covariant, *the theory just given fails to be diffeomorphism invariant*.

First, general covariance. We define this as follows:

General Covariance. A formulation of a theory is *generally covariant* iff the equations expressing its laws are written in a form that holds with respect to all members of a set of coordinate systems that are related by smooth but otherwise arbitrary transformations.

It is clear that such a formulation is possible for our theory. It is what is achieved in the passage from the traditional form of the equation (1), to equation (2). General covariance in this sense is sometimes taken to be equivalent to the claim that the laws have a *coordinate-free formulation* (32, 54; 34, 108). This takes us to equation (3): if the laws relate geometric objects of types that are intrinsically characterisable, without recourse to how their components transformations under changes of coordinates, then one should be able, with the introduction of the right notation, to describe the relationships between them directly, rather than in terms of relationships that hold between the objects' coordinate components.

In order to address the question of the theory's diffeomorphism invariance, one needs to be more explicit than we have so far been about how one should understand equation (3). In particular, what, exactly, is the referent of the 'η_{ab}' that occurs in this equation? Here is one very natural way to set things up. It is a picture that lies behind the claim of several authors that, while specially relativistic theories can be made generally covariant in the sense just described, they are nevertheless *not* diffeomorphism invariant.

Take the *kinematically possible models* (KPMs) of the theory to be suitably smooth functions from some given manifold equipped with a Minkowski metric, $\langle M, \eta_{ab} \rangle$ into $\mathbb{R}$. That is, they are objects of the form $\langle M, \eta_{ab}, \Phi \rangle$, where η_{ab} is held *fixed*—it is *identically* the same in every model.[15] The *dynamically possible models* (DPMs) are then the proper subset of these objects picked out by the requirement that Φ satisfies the Klein–Gordon equation relative to the η_{ab} common to all the KPMs. So understood, equation (3) is not an equation for η_{ab} and Φ together. Rather, it is an equation for Φ alone, *given* η_{ab} (*cf.* [34], 107). For ease of future reference, call this version of the specially relativistic theory of the scalar field **SR1**.

---

[15]This means that the concept of a fixed field is not equivalent to the concept of an absolute object in the Anderson–Friedman sense. In using "fixed" in this quasi-technical sense, I follow Belot (see, e.g., [7], 197, fn 137). The distinction is explored more fully in Section 7.

Our initial definition of diffeomorphism invariance runs as follows:

Diffeomorphism Invariance (version 1). A theory T is *diffeomorphism invariant* iff, if $\langle M, O_1, O_2, \ldots \rangle$ is a solution of T, then so is $\langle M, d^*O_1, d^*O_2, \ldots \rangle$ for all $d \in \text{Diff}(M)$.[16]

So defined, diffeomorphism invariance corresponds to what has sometimes simply been identified as general covariance in the post-Hole Argument philosophical literature.[17] Friedman is explicit in taking general covariance as defined above (*cf.* [32], 51) to be equivalent to diffeomorphism invariance as just defined (*cf.* [32], 58). In arguing for this equivalence [32, 52–4], he appears to overlook the crucial possibility, exploited here, that a coordinate-free equation relating two geometric objects A and B, can nonetheless be interpreted as an equation for B alone, given a fixed A. (We shall see in Section 9 that Earman [21] seems to be guilty of a similar oversight.)

Returning to **SR1**, it is clear that, with the KPMs and DPMs defined as suggested, the theory does *not* satisfy the definition of diffeomorphism invariance just given. If $\langle M, \eta_{ab}, \Phi \rangle$ is a model of the theory, $\langle M, d^*\eta_{ab}, d^*\Phi \rangle$ will be a model only if $d^*\eta_{ab} = \eta_{ab}$, for only in that case will $\langle M, d^*\eta_{ab}, d^*\Phi \rangle$ correspond to a KPM, let alone a DPM!

Contrast **SR1** to the generally relativistic theory of the scalar field. To make the analogy as close as possible, consider the sector of the theory defined on the same manifold M mentioned in **SR1**. Call this theory **GR1**. Superficially, the KPMs and the DPMs of **GR1** are the same type of objects as those of **SR1**: triples of the form $\langle M, g_{ab}, \Phi \rangle$, where g_{ab}, like η_{ab}, is a Lorentzian metric field. But now one does not have the option of taking g_{ab} to be fixed.[18] Rather the KPMs of the theory are *all possible* triples of the form $\langle M, g_{ab}, \Phi \rangle$, subject only to g_{ab} and Φ satisfying suitable differentiability (and perhaps boundary) conditions. The DPMs are picked out as a proper subset of the KPMs by two equations:

$$g^{ab}\nabla_a\nabla_b\Phi = 0, \tag{4}$$
$$G_{ab} = 8\pi T_{ab}. \tag{5}$$

---

[16]In this statement of the condition, O_i and d^*O_i are distinct mathematical objects; one is not contrasting different coordinate representations of the very same objects.

[17] See, e.g., Earman [17, 47]. As mentioned, Norton distinguishes active and passive general covariance. His statement of the former [42, 1226] is almost identical to the statement of diffeomorphism invariance just given, save that he considers diffeomorphisms between distinct manifolds. (His statement of passive general covariance [42, 1230] differs, however, from the characterisation of general covariance given above, in focusing on the closure properties of the set of coordinate representations of a theory's models, rather than on the nature of the equations that pick out such models.)

[18]Strictly speaking, one could interpret equations (4) and (5), given below, as describing a theory of a single field Φ propagating on a fixed g_{ab}. The resulting space of DPMs would consist of a single point in this cut-down space of KPMs! What, exactly, would be wrong with such a set-up? We take ourselves to have evidence for the (approximate) truth of our theory (GR) even though we have not pinned down a specific model. But on this variant of the theory, pinning down the theory requires pinning down a unique model.

Equation (5) is the Einstein field equation, relating the Einstein tensor G_{ab}, encoding certain curvature properties of g_{ab}, to the energy momentum tensor T_{ab}.[19] Equation (4) might look superficially like equation (3), but now it is no longer an equation for Φ given g_{ab}. Rather (4) and (5) together form a coupled system of equations—the "Einstein–Klein–Gordon equations"—for g_{ab} and Φ together. This generally relativistic theory is, of course, diffeomorphism invariant: if $\langle M, g_{ab}, \Phi \rangle$ satisfies equations (4) and (5), so does $\langle M, d^*g_{ab}, d^*\Phi \rangle$ for any diffeomorphism d.

The rather dramatic way in which **SR1** fails to meet our definition of diffeomorphism invariance—that for a generic diffeomorphism d, $\langle M, d^*\eta_{ab}, d^*\Phi \rangle$ is not even a KPM when $\langle M, \eta_{ab}, \Phi \rangle$ is a DPM—suggests a modification of our definition. Rather than considering the effect of a diffeomorphism on all of the fields of a theory's models, we can exploit the distinction, built into the very construction of the theory, between fixed fields and dynamical fields. Letting F stand for the solution-independent fixed fields common to all KPMs, and letting D stand for the dynamical fields, we can consider the effect of acting only on the latter. This leads to the following amended definition:

Diffeomorphism Invariance (final version). A theory T is *diffeomorphism invariant* iff, if $\langle M, F, D \rangle$ is a solution of T, then so is $\langle M, F, d^*D \rangle$ for all $d \in \text{Diff}(M)$.

More generally, one can say that a theory T is G-invariant, for some subgroup $G \subseteq \text{Diff}(M)$ iff, if $\langle M, F, D \rangle$ is a solution of T, then so is $\langle M, F, g^*D \ldots \rangle$ for all $g \in G$.

Since **GR1** involves no fixed fields, acting only on the dynamical fields *just is* to act on all the fields. Our amendment to the definition of diffeomorphism invariance therefore makes no material difference in this case. For this reason, focus on theories like **GR1** tends to obscure the difference between our two definitions. Turning to the case of **SR1**, this theory still fails to be diffeomorphism invariant under the new definition: for an arbitrary diffeomorphism d, if $\langle M, \eta_{ab}, \Phi \rangle$ is a solution of **SR1**, then $\langle M, \eta_{ab}, d^*\Phi \rangle$, in general, will not be. However, assuming no boundary conditions are being imposed, $\langle M, \eta_{ab}, d^*\Phi \rangle$ will nonetheless be a KPM of the theory. This becomes significant when considering the definition of the invariance of the theory under proper subgroups of $\text{Diff}(M)$.

Suppose T has models of the form $\langle M, F, D \rangle$ and that d is a symmetry of the fixed, background structure, i.e. $d^*F = F$. In this case, $\langle M, d^*F, d^*D \rangle = \langle M, F, d^*D \rangle$ and so, for this subgroup of $\text{Diff}(M)$, an invariance principle that asks us to consider transformations of all fields, background and dynamical, will give the same verdict as those that consider transformations only of the dynamical fields. Further, it follows from the general covariance of the theory, i.e. from the fact that its defining equation can be give a coordinate-free expression, that when d is a symmetry of F, $\langle M, d^*F, d^*D \rangle = \langle M, F, d^*D \rangle$ will be a DPM whenever $\langle M, F, D \rangle$ is.[20] We can

---

[19]For our massless real scalar field, $T_{ab} = (\nabla_a \Phi)(\nabla_b \Phi) - \frac{1}{2} g_{ab} g^{mn} (\nabla_m \Phi)(\nabla_n \Phi)$.

[20]Note that this claim is not identical to Earman's claim that it follows from general covariance that a diffeomorphism that is symmetry of a theory's spacetime structure will also be what he calls a "dynamical symmetry" [17, 46–7]. The reason is that Earman's "general covariance" corresponds to the (unmodified) definition of diffeomorphism invariance given above.

therefore define G-invariance either by analogy with the first definition of diffeomorphism invariance or (as advocated) by analogy with the final version, and we will get the verdict that if G is a subgroup of the automorphism group of F, then the theory is G-invariant.

The definitions give different verdicts, however, when we consider the opposite implication: if T is a G-invariant theory, does it follow that G is a subgroup of the automorphism group of its fixed fields F? If G-invariance requires that if $\langle M, F, D \rangle$ is a DPM then so is $\langle M, g^*F, g^*D \ldots \rangle$, for all $g \in G$, then no diffeomorphism that is not also an automorphism of F could be a member of G. Such a diffeomorphism does not map KPMs to KPMs. However, if G-invariance only requires that if $\langle M, F, D \rangle$ is a DPM then so is $\langle M, F, g^*D \ldots \rangle$, then the automorphisms of F can be a *proper* subgroup of G. In fact, this is exactly the situation in the case of **SR1**. Let d correspond to a conformal transformation of η_{ab}. Since we are considering the *massless* Klein–Gordon field, if $\langle M, \eta_{ab}, \Phi \rangle$ is a DPM, then so is $\langle M, \eta_{ab}, d^*\Phi \rangle$, even though $d^*\eta_{ab} \neq \eta_{ab}$. We can only capture this fact in terms of the statement that the theory is invariant under the relevant group if we define such invariance in the modified manner.[21]

Let us take a step back and recall the wider project. We are interested in assessing the claim that diffeomorphism invariance is intimately linked to background independence. I contend that the distinction drawn in this section between general covariance and diffeomorphism invariance, and exemplified by **SR1**'s satisfaction of the first but not the second, is the right one for this purpose, for it makes good sense of several remarks by the claim's defenders.

For example, Smolin [57, §6] offers an extended discussion of diffeomorphism invariance and its connection to background independence. His focus is on the interpretational consequences of diffeomorphism invariance, rather than on providing a positive characterisation of the property as such, so no direct comparison with the definition proposed here can be made. (He is also particularly concerned to stress the *gauge* status of diffeomorphisms in the context of a diffeomorphism-invariant formulation of a theory, a topic I return to in Section 9.) However, his contrasting diffeomorphism invariance with general coordinate invariance is fully consonant with the distinction of this section

> it can be asserted—indeed it is true—that with the introduction of explicit background fields any field theory can be written in a way that is generally coordinate invariant. This is not true of diffeomorphisms [sic] invariance, which relies on the fact that in general relativity there are no non-dynamical background fields. [57, 233]

It is natural to read the second half of this passage as committing Smolin to the claim that **SR1** cannot be made diffeomorphism invariant because the theory involves a non-dynamical background, η_{ab}.

---

[21] Similar, historically inspired examples are Galilean-invariant classical mechanics set in full Newtonian spacetime and, more interestingly, Newtonian gravitational theory set in Galilean spacetime [see, e.g., 38]. What these examples should remind one is that such theories are epistemologically problematic. The background structure that they postulate introduces allegedly meaningful properties (e.g. absolute velocities) that are undetectable in principle. This motivates the search for formulations with weaker background structure (see, e.g. [48], §3 and §6).

Consider, now, a revealing passage from Rovelli. Having summarised what he takes to be the philosophical implications of GR's lack of non-dynamical background structures, he states that these implications are "coded in the active diffeomorphism invariance (diff invariance) of GR" [52, 108]. He goes on to elaborate in a footnote

> Active diff invariance should not be confused with passive diff invariance, or invariance under change of co-ordinates...A field theory is formulated in [a] manner invariant under passive diffs (or change of co-ordinates), if we can change the co-ordinates of the manifold, re-express all the geometric quantities (dynamical *and non-dynamical*) in the new coordinates, and the form of the equations of motion does not change. A theory is invariant under active diffs, when a smooth displacement of the dynamical fields (*the dynamical fields alone*) over the manifold, sends solutions of the equations of motion into solutions of the equations of motion. [52, 122]

I take it that **SR1** is precisely a theory formulated in a manner invariant under passive diffs, but not active diffs, whereas **GR1** is a theory invariant under active diffs. In other words, Rovelli's "passive diffeomorphism invariance" is what I called above general covariance. Identifying Rovelli's "non-dynamical" fields with fixed fields, his "active diffeomorphism invariance" corresponds to our (amended) definition of diffeomorphism invariance.

Finally, Giulini [34] offers equivalent definitions, although he adopts a rather different approach to characterising general covariance. He schematically represents a theory's equations of motion as

$$\mathcal{F}[\gamma, \Phi, \Sigma] = 0 \tag{6}$$

Here γ goes proxy for structures given by maps *into* the manifold M (representing particle worldlines, strings, etc.) and Φ goes proxy for the dynamical fields: maps from spacetime into some value space (or, more generally, structures given by sections in some bundle over M). Finally, Σ stands for the fixed ("background") structures.[22]

He then distinguishes what he calls the notion of *covariance* from *invariance* as follows (see [34, 108]). Equation (6) is said to be *covariant* under diffeomorphisms iff

$$\mathcal{F}[\gamma, \Phi, \Sigma] = 0 \quad \text{iff} \quad \mathcal{F}[d \cdot \gamma, d \cdot \Phi, d \cdot \Sigma] = 0 \quad \forall d \in \text{Diff}(M). \tag{7}$$

It is *invariant* under diffeomorphisms iff:

$$\mathcal{F}[\gamma, \Phi, \Sigma] = 0 \quad \text{iff} \quad \mathcal{F}[d \cdot \gamma, d \cdot \Phi, \Sigma] = 0 \quad \forall d \in \text{Diff}(M). \tag{8}$$

The only difference between these conditions is that in the former but not in the latter case one allows the diffeomorphism to act on the fixed fields. In absence of fixed fields, therefore, the distinction between the conditions collapses: covariance implies invariance.

---

[22]In both our examples theories, γ is empty and the scalar field Φ belongs to Giulini's category Φ. In the case of **SR1**, η_{ab} belongs to Σ; in **GR1**, g_{ab} belongs to (Giulini's) Φ, and Σ is empty.

The distinction between the γ and Φ, on the one hand, and the Σ on the other is crucial in understanding these conditions. Consider, first, condition (8). The statement that $\mathcal{F}[\gamma, \Phi, \Sigma] = 0$ iff $\mathcal{F}[d \cdot \gamma, d \cdot \Phi, \Sigma] = 0$ simply means that $\langle \gamma, \Phi \rangle$ and $\langle d \cdot \gamma, d \cdot \Phi \rangle$ stand or fall together as solutions of (6). The condition is therefore this section's (modified) statement of diffeomorphism invariance.

Now consider condition (7). The fact that $\mathcal{F}[\gamma, \Phi, \Sigma] = 0$ is only an equation for γ and Φ (but not Σ) means that $\mathcal{F}[\gamma, \Phi, \Sigma] = 0$ and $\mathcal{F}[d \cdot \gamma, d \cdot \Phi, d \cdot \Sigma] = 0$ are *distinct* equations. The condition states that if $\langle \gamma, \Phi \rangle$ is a solution to (6), then $\langle d \cdot \gamma, d \cdot \Phi \rangle$ must be a solution of a structurally similar equation involving the *different* field(s) $d \cdot \Sigma$. The condition (7), therefore, says nothing about whether d maps a solution of (6) to another solution *of the same equation*. Given that Σ represents fixed fields, (7) does not collapse into our original, unmodified statement of diffeomorphism invariance. All that it requires is that (6) be well defined in the differential-geometric sense. It is therefore equivalent to the requirement that the equation have a generally covariant expression in the sense given earlier.

5 Diffeomorphism-Invariant Special Relativity

The previous section described a generally covariant but non-diffeomorphism-invariant formulation of an intuitively background-dependent theory, **SR1**. This was contrasted with a generally covariant and diffeomorphism-invariant formulation of an intuitively background-independent theory, **GR1**.[23] What should one make of **SR1**'s failure to be diffeomorphism invariant? Does it support Smolin's contention that diffeomorphism invariance "relies on" the absence of background fields? In this section and the next, I suggest that it does not. At the very least, whether it does depends on what counts as a "background field."

We need to consider yet another formulation of a theory, which I will call **SR2**. This theory's space of KPMs is the very same set of objects that formed the space of KPMs of the generally relativistic **GR1**. But, rather than being picked out via equations (4) and (5), the subspace of DPMs is defined via

$$g^{ab} \nabla_a \nabla_b \Phi = 0, \tag{4}$$

$$R^a{}_{bcd} = 0, \tag{1}$$

[23]From here on, when I refer simply to "diffeomorphism invariance" I am referring to the property captured by the second (final) definition given in the previous section. The merits, or otherwise, of the first definition will not be discussed further.

where R^a_{bcd} is the Riemann curvature tensor of g_{ab}.[24] Several comments are in order before we assess the interpretational dilemmas that **SR2** presents.

First, the contrast between **SR1** and **SR2** highlights something of a contrast between the philosophy literature, including the post-Hole Argument literature, and discussions of background independence arising from attempts to quantise GR. Crudely put, philosophers have tended to have a formulation of a theory like **SR2** in mind when they have considered 'generally covariant' formulations of special relativity (see, e.g., [22, 518]), whereas physicists have tended to have something like **SR1** in mind. This is not unrelated to the fact, noted in the previous section, that Friedman, Earman, and even Norton (used to) identify (active) general covariance with diffeomorphism invariance (as initially characterised in the previous section).

This is not to say that the physics literature has not discussed theories like **SR2**—we shall shortly see that it has—but it is possible to mistake a discussion of an **SR1**-type theory for that of a **SR2**-type theory. One does not arrive at **SR2** simply by stipulating that equation (1) is to be satisfied. One must also indicate how g_{ab}, as it occurs in (4) and (1), is to be interpreted. After all, the field η_{ab} of **SR1** satisfies a formally identical equation to (1). It is just that, in this context, the equation does not function to pick out a class of DPMs from a wider class of KPMs. Instead it characterises a fixed field common to all the KPMs. In **SR2**, it is important that (4) and (1), just like (4) and (5) in **GR1**, are understood as coupled equations for both Φ and g_{ab}.

Finally, of course, we should note the crucial fact that **SR2**, like **GR1** and unlike **SR1**, is diffeomorphism invariant.

6 Connecting Diffeomorphism Invariance and Background Independence

What does the diffeomorphism invariance of **SR2** tell us about the alleged link between diffeomorphism invariance and background independence? A proper answer to this question will require disentangling various meanings of "background", but here is the obvious moral: **SR2** is a diffeomorphism-invariant but intuitively background-dependent theory. Diffeomorphism invariance therefore cannot be equated with—or be seen as a formal expression of, or sufficient condition for—background independence. Diffeomorphism invariance is not, *per se*, what differentiates GR from pre-relativistic theories.

Here is one way that this conclusion might be resisted. Consider the following questions. (Q1) Is **SR2** a background-independent theory? (Q2) Are **SR1** and **SR2**

---

[24] As with those of **GR1**, the theory's KPMs are restricted to fields defined on a given manifold M. In the previous section, this restriction served to allow as direct as possible a comparison between **GR1** and **SR1**. When comparison with **SR1** is not at issue, the restriction is arbitrary. One can (and should) generalise the formulations of **SR2** and **GR1** further, not least to allow for different global topologies.

merely different ways of formulating the same theory? Suppose that one answers (Q1) in the affirmative, on the grounds that g_{ab} in a model of the theory is a *solution to an equation*. It therefore counts as a 'dynamical field'; it is not 'fixed a priori'. This, in effect, is to treat 'background field' as synonymous with 'solution-independent fixed field' in the sense highlighted in Section 4. One then goes on to answer question (Q2) in the negative. Precursors of GR were not background independent, period, and so only **SR1** is faithful to the pre-GR understanding of the spacetime structure of special relativity.

I take it that this package is a highly implausible cocktail of views. First, one should ask: on what basis can one assert that **SR1** and **SR2** constitute genuinely distinct theories, rather than merely different formulations of the same theory? On the face of it, since their models involve the same types of geometric object, and since all objects in any solution of one theory are diffeomorphic to the corresponding objects in some solution of the other, the two formulations appear to be, not merely empirically equivalent, but equivalent in a thoroughgoing sense. The DPMs of one theory are isomorphic to the DPMs of the other; it is just that, for each solution of one of the theories, the other theory has an infinite set of diffeomorphic copies.

Second, the classification of **SR2** as relevantly similar to **GR1**, and so background independent, focuses on a minor similarity between the theories at the expense of a more significant contrast. True, the g_{ab}s of both theories are treated as 'solutions of equations' and *in this sense* they are not fixed, but this fact seems much less interesting than their obvious differences. Recall the intuitive characterisation of the differences between the spacetime structures of GR and pre-relativistic theories given in Section 1: in GR, the curvature of spacetime varies, not just in time and space, but across models, and the material content of spacetime influences how it does so. The fact that the g_{ab} of **SR2** is the solution of an equation is not a sufficient condition for either of these features. The g_{ab} of **SR2** is not affected by matter, because it is wholly determined (up to isomorphism) by equation (1). Relatedly, in the sense that matters, the metric structure of spacetime does not differ from DPM to DPM: the g_{ab}s in any two DPMs are isomorphic to one another.[25]

These features of **SR2** mean that, if one wishes to remain faithful to the natural pre-theoretic sense of "background", it should be classified as a background-*dependent* theory. They further suggest that one should regard **SR1** and the diffeomorphism-invariant **SR2** as different formulations of the same, background-dependent theory. In contrast, **GR1** is (a diffeomorphism-invariant formulation of) a background-independent theory. This situation might bring to mind Bergmann's claim, noted in Section 3, that the distinctive feature of GR is its lack of a non-generally covariant formulation. This feature of GR could not be equated with its background independence: a background-dependent theory might lack a non-generally covariant formulation because its background structures lack symmetries. However, now we have the distinction between general covariance and diffeomorphism invariance on the table, the general approach might appear more promising.

---

[25] Strictly, the global topology of the manifold M might allow for infinitely many non-isomorphic flat metric fields. Even so, these will all be locally isomorphic.

The idea is that it is the *lack* of a *non*-diffeomorphism-invariant formulation, rather than the existence of a diffeomorphism-invariant formulation, that is the mark of a background-independent theory. A non-diffeomorphism-invariant formulation of a theory requires that some elements of its models are regarded as fixed, identically the same from model to model. If a theory is background dependent, in the sense that it involves non-dynamical fields that (intuitively) do not vary from model to model, then those fields can be represented by fixed structures in a non-diffeomorphism-invariant formulation of the theory. But if the theory is background independent, in the sense that all of its fields can vary from model to model, it lacks elements that can be represented by fixed structures. Of necessity, it will be diffeomorphism invariant.[26] The background fields of a theory are to be identified with those fields that appear as fixed elements in *some* non-diffeomorphism-invariant formulation that theory. So, for example, the metric field, g_{ab}, of **SR2** represents background structure because it represents the same structure that is represented in the alternative formulation of the theory, **SR1**, by η_{ab}.

There is clearly a close connection between identifying a background field in this way and Anderson's notion of an *absolute object* [1, 2]. I will return to this connection at the end of the next section, after reviewing one more complication.

7 Absolute Objects and the Action–Reaction Principle

Assume that background-independent theories can only be formulated in a diffeomorphism-invariant manner. That leaves open the issue of whether every theory that must be formulated in a diffeomorphism-invariant manner lacks background fields. Whether one endorses this further claim in part depends on a subtlety concerning what it takes to be a background field.

When the metric field of GR is presented as an example of field that, unlike its precursors in pre-relativistic theories, is not a background field, two of its features are often run together: (i) like other fields in the theory, the metric is *dynamical*; (ii) it also obeys *the action–reaction principle*: it is affected by every field whose evolution it constrains. The second feature entails the first (assuming the entity in question is not entirely dynamically redundant); a field obviously cannot be dynamically affected and yet not be dynamical. However, the converse implication does not hold. A field might affect without being affected and yet have non-trivial dynamics of its own.

Consider, for example, the theory (call it **GR2**) given by the following equations:

$$g^{ab}\nabla_a\nabla_b\Phi = 0, \tag{4}$$

---

[26]This proposal fits with some of the more careful claims from the quantum gravity community concerning the link between background independence and diffeomorphism invariance. For example, in an informal website article on the meaning of background independence, Baez claims: "making the metric dynamical instead of a background structure leads to the fact that all diffeomorphisms are gauge symmetries in general relativity" [5].

$$R_{ab} = 0. \tag{1}$$

Here R_{ab} is the Ricci tensor associated with g_{ab}. In other words, equation (1) is the vacuum Einstein equation, even though the theory's models contain a material scalar field. In this theory the metric is clearly dynamical; it varies from DPM to DPM. Since it is constrained to obey equation (4), the matter field 'feels' the metric. However, in contrast to the situation in GR, matter does not act back on the metric. The action–reaction principle is violated. To adapt Einstein's terminology, as quoted in Section 2, the metric of **GR2** is a *causal absolute* even though it is a thoroughly dynamical field.

Should g_{ab} count as a background field in this theory? One might naturally characterise the metric as a background *relative to* the dynamics of Φ. It is a kind of "dynamical background field". But it does not seem correct to classify *the theory as a whole* as background dependent on this account. After all, in those models where Φ vanishes, the theory just is vacuum GR. This verdict matches that reached if one sticks with the criterion proposed in the previous section (necessary diffeomorphism invariance), for **GR2** lacks a non-diffeomorphism-invariant formulation in just the way **GR1** does.

GR2 serves another illustrative purpose. At the end of the previous section, I suggested that there is a link between whether a field can appear as a fixed field in a non-diffeomorphism-invariant formulation of a theory and whether that field is an absolute object in Anderson's sense. Although Anderson informally introduces absolute objects in terms of their violation of the action–reaction principle, the definition he goes on to give characterises them in terms of a notion of sameness in all DPMs of the theory.[27] What the metric field g_{ab} of **GR2** illustrates is that a field can be an action–reaction violating causal absolute without being an absolute object in the Andersonian sense.

Let us return to the connection between absolute objects and fixed fields. How exactly, are they related? The answer is not entirely straightforward, partly because different authors define absolute objects slightly differently.

Anderson's formal definition of absolute objects does not characterise them directly. Instead he defines them in terms of conditions intended to determine when a subset of the dynamical variables of a theory constitute the components of the theory's absolute objects [2, 83]. Friedman [32, 56–60] later advocated a coordinate-free characterisation, according to which a geometric object field counts as absolute if there exist the right kind of maps between any two models of the theory that preserve the object in question (more details shortly). According to Friedman's set-up, the metric fields of both **SR1** and **SR2** count as absolute objects, even though the metric

---

[27] The values of the absolute objects are said to determine the values non-absolute objects but not vice versa ([2, 83]; see also [4] 1658, fn 6). In Anderson [1, 192], he says that "an absolute element in a theory indicates a lack of reciprocity". This is consistent with absolute objects being sufficient, but not necessary, for a violation of the action–reaction principle.

is a fixed field only in **SR1**.[28] This is not true according to Anderson's definitions. On his way of setting things up, in a non-covariant coordinate presentation of **SR1**, there are *no* absolute elements, because the metric field is not explicitly represented (*cf.* [2], 87). *In this formulation of the theory*, all of the variables required to characterise a solution (in this case, the values of Φ relative to some inertial coordinate system) are the components of a genuinely dynamical object. Nevertheless, it is clear that the metric of **SR2** counts as an absolute object according to Anderson's definition. I suggested above that one should regard **SR1** and **SR2** as different formulations of the same theory, and thus regard their metric fields as representing the same element of physical reality. Generalising this move, one can say that an object that features as a fixed field in one formulation of a theory will appear as an absolute object in reformulations of the theory in which that object is no longer treated as fixed.

So far we have noted that fields that are (or can be represented as) fixed are (or can be represented as) absolute objects. What about the converse? If a diffeomorphism-invariant theory contains an absolute object, can it be given a non-diffeomorphism-invariant formulation in which that object features as a fixed field? Here, again, the way Friedman and Anderson define "absolute object" makes a difference. While both, in different ways, formalise a notion of "sameness in every model", Anderson's notion of sameness is global whereas Friedman's is local. More specifically, Fried-man holds that, if the models of a theory take the form $\langle M, O_1, \ldots, O_n \rangle$, then object O_i is an absolute object just if, for any two models $\mathcal{M}_1 = \langle M, O_1, \ldots, O_n \rangle$ and $\mathcal{M}_2 = \langle M, O_1', \ldots, O_n' \rangle$, and for every $p \in M$, there are neighbourhoods A and B of p, and a diffeomorphism $h : A \rightarrow B$ such that $O_i' = h^* O_i$ on $A \cap B$. Fried-man's absolute objects can therefore possess "global degrees of freedom": differences between such objects might distinguish between classes of DPMs even though the objects are (in the sense just characterised) everywhere locally indistinguishable.[29] The upshot is that a theory that involves absolute objects in Friedman's sense may not have a (natural) non-diffeomorphism-invariant formulation in terms of fixed fields.

A popular move is to equate background fields and absolute objects, and so to treat background independence as the lack of absolute objects. Giulini [34] offers a careful recent development of this strategy. As Giulini notes, and as is discussed in depth by Pitts [45], several "counterexamples" suggest that neither Anderson's proposal nor Friedman's get things just right. The counterexamples come in three categories. (1) There are cases where structure that, intuitively, should count as background is not classified as an absolute. (2) There are cases where structure that, intuitively,

---

[28]Effectively, we are distinguishing two senses of "dynamical". The metric of **SR2** counts as dynamical in a liberal sense, because it varies non-trivially in the space of KPMs and is constrained to be what it is in any DPM via the "equation of motion" (1). But in a stricter sense it is not dynamical, because (up to a diffeomorphism) it is the same in every model of the theory. The stricter sense takes "dynamical" to mean "not absolute"; the liberal sense takes "dynamical" to mean "not fixed".

[29]Consider, for example, flat Lorentzian metrics on a manifold with non-trivial global topology. Such metrics need not be globally isometric even though they are everywhere flat. Some models might be temporally finite whereas others are temporally infinite but spatially finite in a preferred spatial direction.

should not count as background is classified as an absolute. Finally, (3), it is noted that, on Anderson's definition (suitably localised), GR itself turns out to have an absolute object (and so should count as background dependent).

Torretti's [64] example of a theory set in classical spacetimes of arbitrary but constant spatial curvature is of type (1). Pitts observes that if one decomposes the spatial metric into a conformal spatial metric density and a scalar density, then the former is an absolute object while the latter, while constant in space and time, counts as a genuine, global degree of freedom.

The best-known case of type (2) is the Jones–Geroch example of the "dust" four-velocity in GR coupled to matter that is characterised by only a four-velocity field and a mass density. Pitts sees both Friedman's own suggestion—that one take the 4-momentum field of the dust as primitive [32, 59]—and the option of defining the "4-velocity" so that it vanishes in matter-free regions, as motivated by an Andersonian ban on formulations of a theory that contain physically redundant variables [45, 361–2].[30] My own view is that both of these "solutions" miss the central problem posed by the example. In the context of this theory, the non-vanishing velocity field is, intuitively, as dynamical as the 4-momentum. The trouble arises not because we mistook as indispensable an object that Anderson's definition correctly classifies as absolute. The trouble is that Anderson's definition, intuitively, misclassifies that object.

The example suggests that the notion of absolute objects might not, in fact, be a better candidate than the notion of fixed fields for articulating the sense of "dynamical" relevant to characterising background structure. Consider, for example, a diffeomorphism-invariant formulation of a theory set in Minkowski spacetime and involving matter characterised, in part, by a (non-vanishing) four-velocity. One can define two distinct proper subsets of the KPMs (and, correspondingly, the DPMs) of this theory. The first is obtained by specialising to a particular metric field on the manifold, and retaining all and only those KPMs (and DPMs) that include this metric field. The second is obtained by specialising to a particular representation of the four-velocity. If we view each set of models as determining some theory, then both theories involve (in some sense) a fixed field. However, in the case of the theory obtained by specialising to a particular metric, the solution set is identifiable, as a subspace of the KPMs, via some differential equations for the truly dynamical objects given the fixed field (the metric). In the case of the "theory" with the fixed velocity field, in contrast, it seems highly doubtful that we will be able to view the particular (flat) metrics occurring in the DPMs as all and only the solutions of an equation for the metric *given* the velocity field. (Imagine specialising to coordinates in which the velocity field takes the value $(1, 0, 0, 0)$ and consider how likely it is that the set of admissible components of the metric field in such coordinates are picked out via an equation.)

A similar strategy might be pursued in the case of ((3)). The candidate absolute object in question is the determinant of the metric, $\sqrt{-g}$. One might accept this

---

[30]Pitts pursues the topic further in Pitts [46].

verdict without accepting that this automatically means that GR should count as background dependent. The latter might be held to further require that $\sqrt{-g}$ be interpretable as a fixed field.[31]

Suppose, however, that one sticks with the proposal that the lack of absolute objects is equivalent to background independence. What light does that shed on the relationship between background independence and diffeomorphism invariance? Does a theory lack a non-diffeomorphism-invariant formulation just if it lacks absolute objects? We have seen that, not only are fixed fields not absolute objects (on either Anderson's definition or Friedman's), but *being representable in terms of a fixed field* is also not equivalent to being an absolute object. Since the presence of fixed fields would seem to be necessary for the failure of diffeomorphism invariance, this means that necessary diffeomorphism invariance cannot be equivalent to background independence understood as lack of absolute objects.

There is a rather desperate way to reconnect the question of whether Diff(M) is a symmetry group with background independence: redefine symmetry! For example, one might try stipulating that Diff(M) is a symmetry* group of a theory T iff, if $\langle M, A, D \rangle$ is a model of T, then so is $\langle M, A, d^*D \rangle$ for all $d \in$ Diff(M). (Formally this looks just the definition of diffeomorphism invariance from Section 4, with "F", for "fixed field" replaced by "A", for "absolute object".) The proposal is problematic, on at least three grounds.

First, the notion of symmetry* is transparently ad hoc. When our theory contained fixed fields, restricting the action of Diff(M) to the dynamical (i.e. *non-fixed*) fields was natural. Only by doing so could one define a natural group action on the space of KPMs. The symmetry group is then naturally defined to be the subgroup of this group that fixes the space of DPMs. When one has a diffeomorphism-invariant theory that includes absolute objects, one (obviously!) does not need to stipulate that Diff(M) acts only on the dynamical (i.e. *non-absolute*) fields in order for its action on the space of KPMs to be well defined.

Second, defining the action of Diff(M) on the space of KPMs in such a way that it does not act on the As breaks the natural definition of symmetry. The definition yields, as intended, that a theory with, say, a flat Lorentzian metric as its absolute object will fail to have Diff(M) as a symmetry* group. But it will *also* fail to have the Poincaré group as a symmetry* group. For any *given* solution $\langle M, A, D \rangle$, the maximal group G such that, for all $g \in G$, $\langle M, A, g^*D \rangle$ is a solution, will be isomorphic to the Poincaré group (or, possibly, a supergroup of the Poincaré group). But for

---

[31] Can the equations of the theory be interpreted as equations *for* the other variables *given* fixed $\sqrt{-g}$? This seems to be the correct verdict for unimodular GR, but not (or not clearly so) for GR itself. For further discussion of this case, although not in terms of the notion of fixed fields, see Earman [19]; Pitts [45]; Sus [61, 62].

two arbitrary solutions $\langle M, A, D \rangle$ and $\langle M, A', D' \rangle$, the groups so defined need not coincide. In fact, in general, they will coincide only when $A = A'$.[32]

Suppose one circumvents these problems by adding some epicycles to the definition of symmetry*. There remains a third reason to be dissatisfied with the proposal that background independence is equivalent to Diff(M)'s being a symmetry* group. At bottom, what is doing all the work is the notion of absolute object, in terms of which the gerrymandered notion of symmetry is defined. If our interest is in characterising background independence, why not simply characterise it as the lack of absolute objects and be done with it? In particular, the detour via symmetry* does not give us a better handle on GR's background independence versus SR's background dependence.

8 Diff(M) as a Variational Symmetry Group

When physicists talk of a generally covariant formulation of a specially relativistic theory, they typically have in mind a formulation like **SR1**. Undue focus on such examples, at the expense of examples like **SR2**, might explain why the connection between background independence and diffeomorphism invariance is sometimes taken to be tighter than it really is. However, theories along the lines of **SR2** do get considered by those who defend a diffeomorphism invariance/background independence link. As we have seen, the possibility of such formulations of specially relativistic theories is central to Anderson's thinking (and explains the idiosyncrasies of his definition of symmetry). The option is also considered by Rovelli, who concedes

> even full diffeomorphism invariance, should probably not be interpreted as a rigid selection principle, capable of selecting physical theories *just by itself*. With sufficient acrobatics, any theory can perhaps be re-expressed in a diffeomorphism invariant language. ...

> But there are prices to pay. First, [**SR2**]...has a "fake" dynamical field, since g is constrained to a single solution up to gauges, by the second equation of the system. Having no physical degrees of freedom, g is physically a fixed background field, in spite of the trick of declaring it a variable and then constraining the variable to a single solution. Second, we can insist on a lagrangian formulation of the theory...[59], but to do this we must introduce an additional field, and it can then be argued that the resulting theory, having an additional field is different from [the original] [17]. [54]

---

[32]Invariance, as I defined it in Section 4, is called *covariance* by Anderson [2, 75]. He defines a theory's symmetry, or "invariance" group as the "largest subgroup of the covariance group...which is simultaneously the symmetry group of its absolute objects" [2, 87]. It would seem, therefore, that Anderson's symmetry group is related to the notion of symmetry* in exactly the way the group of automorphisms of the fixed fields of a theory is related the symmetry group (as defined in Section 4) of that theory. In both cases one should expect the former to be a (possibly proper) subset of the latter. But we have just seen that, without some finessing, the symmetry* group of a theory will be trivial. The same trouble afflicts a flatfooted reading of Anderson's definition. Consider **SR2**. The symmetry group of any *particular* absolute $g_{\mu\nu}$, occurring in a particular DPM, will be (isomorphic to) the Poincaré group (*cf.* [2, 87]), but the only diffeomorphism that belongs to every such group is the identity map.

Several comments are in order. First, reference to "sufficient acrobatics" seems like hyperbole, given the relatively straightforward nature of the transition from a theory like **SR1** to a reformulation along the lines of **SR2**.

Second, it is true that, in **SR2**, g_{ab} is a "fake" dynamical field. It *should* be classified as background structure. Despite our treating it as dynamical in the liberal sense, it remains non-dynamical in a stricter sense. The previous sections have reviewed apparatus that allows us to draw precisely these distinctions, and to differentiate **GR1** and **SR2**, despite both theories being equally diffeomorphism invariant. So, it is not clear why there is a "price to pay" in adopting such a formulation, particularly since we are regarding **SR2** as merely a reformulation of **SR1**. Rovelli, perhaps, would question this last stance. The diffeomorphism invariance of any theory might be taken to have significant implications for the nature of the true physical magnitudes of the theory, and thus require that one distinguish **SR2** from (the non-diffeomorphism-invariant) **SR1**. If so, I disagree, for reasons I explain in the final section of this paper.

Third, and most interestingly, Rovelli's description of the second cost suggests a quite different way to connect the question of whether diffeomorphisms are symmetries to background independence. *Prima facie*, there is a formal difference between **SR2** and **GR1** that I have not so far mentioned. The two theories are defined on the same space of KPMs. In the case of **GR1**, the space of solutions picked out by its equations can also be fixed via a variational problem defined in terms of the action $S_{GR1} = \int d^4x(\mathscr{L}_G + \mathscr{L}_\Phi)$.[33] On the face of it, the same is not true of **SR2**. One can pick out the solution space of **SR1** in terms of a variational problem, defined via the action $S_{SR1} = \int d^4x\mathscr{L}_\Phi$, where $\mathscr{L}_\Phi$ depends on the *fixed* metric field η_{ab}. In the context of the space of KPMs common to **GR1** and **SR2**, however, elements in the solution space of **SR2** are not stationary points of $\int d^4x\mathscr{L}_\Phi$. The latter can identified by considering the Euler–Lagrange equations one obtains by applying Hamilton's principle to both Φ *and* g_{ab}. From the first, one gets the Klein–Gordon equation, but from the second one gets the trivialising condition that the stress-energy tensor for Φ vanishes.

These reflections might suggest that background independence could be linked to the symmetry status of Diff(M) in the following way:

Background Independence (version 1). A theory T is background independent if and only if it can be formulated in terms of a variational problem for which Diff(M) is a variational symmetry group.

Although one can write an action for **SR1** in a generally covariant or coordinate-independent manner, Diff(M) is not a symmetry group of the variational problem that defines the theory's models.[34] Recall that the action of Diff(M) on the **SR1**'s space of KPMs acts on Φ but not on η_{ab}, and does not leave the space of DPMs invariant. A useful alternative way of stating the proposed condition is as follows:

---

[33]The "gravitational" part of the Lagrangian is the Einstein–Hilbert Lagrangian $\mathscr{L}_G = \sqrt{-g}\kappa R$, where R is the curvature scalar and κ is a suitable constant. The "matter" term is the standard Lagrangian for the massless Klein–Gordon field: $\mathscr{L}_\Phi = \sqrt{-g}g^{ab}\nabla_a\Phi\nabla_b\Phi$.

[34]See Belot [7, 161–2] for further discussion of the notion of a variational symmetry.

Background Independence (version 2). A theory T is background independent if and only if its solution space is determined by a generally covariant action *all of whose dependent variables are subject to Hamilton's principle.*

This rules out the generally covariant version of the **SR1** action principle, since in this case only Φ and not η_{ab} is subject to Hamilton's principle. It will also rule out **SR2** if the solution space of this theory really is not obtainable from an appropriately formulated action principle.

Despite these promising results, the proposal does not work. In the quotation above, Rovelli refers to Sorkin [59]. In that paper, Sorkin, rediscovering a procedure originally employed by Rosen [50], shows how one can derive equations (4) and (1) from a diffeomorphism-invariant action. One obtains a Sorkin-type action by replacing $\mathscr{L}_G$ in S_{GR1} with a different "gravitational" term, $\mathscr{L}_S = \sqrt{-g}\Theta^{abcd}R_{abcd}$. The theory therefore involves a Lagrange multiplier field, Θ^{abcd}, in addition to the fields common to **SR2** and **GR1**. In this new action, all the dependent variables are to be subject to Hamilton's principle. For ease of reference, let us call the resulting theory (so formulated) **SR3**. Varying Θ^{abcd} leads to equation (1). Since Φ does not occur in $\mathscr{L}_S$, varying this field has the same effect as in **GR1**, and leads to the Klein–Gordon equation (4). (One also needs to consider variations of g_{ab}. Rather than the EFE, this leads to an equation that relates Θ^{abcd}, g_{ab} and Φ.)[35]

Let us assume, for the moment, that in **SR3** we have yet another way to formulate the specially relativistic theory that has been our example throughout this paper. Since its models are determined by a diffeomorphism-invariant action, all of whose dependent variables are subject to Hamilton's principle, the theory counts as background independent according to our latest proposal. The proposal therefore needs to be revised. A natural thought is to amend it as follows:

Background Independence (version 3). A theory T is background independent if and only if its solution space is determined by a generally covariant action: (i) all of whose dependent variables are subject to Hamilton's principle, and (ii) all of whose dependent variables represent physical fields.

The idea is that **SR3** fails to satisfy the second of these conditions because the dynamics of the additional field Θ^{abcd} strongly suggest that it is not a physical field. It makes no impact on the evolution of g_{ab} and Φ and hence, were it a genuine element of reality, it would be completely unobservable (on the natural assumption that our empirical access to it would be through its effect on "standard" matter fields such as Φ). Indeed, it is only on the basis of interpreting Θ^{abcd} as a mere mathematical device that one can view **SR3** as a reformulation of **SR2**.

In the quotation at the start of this section, Rovelli suggests that one might instead regard **SR3** as a different theory from **SR2**, on the grounds that **SR3** involves an

---

[35]Note that the evolution of Θ^{abcd} is constrained by, but does not affect the evolutions of g_{ab} and Φ. The action–reaction principle is therefore violated by Φ, with respect to Θ^{abcd}, and not just by g_{ab}. The theory illustrates that requiring that all of the dependent variables in an action be subject to Hamilton's principle does not entail that the resulting theory satisfies the action–reaction principle, *pace* Baez [5].

additional field (presumably because one views this field as representing a genuine element of reality, the points just made notwithstanding). This might seem to provide an alternative way to argue that our revised proposal does not classify **SR2** as background independent on the basis of **SR3**'s satisfying its conditions: if **SR3** is a different theory, it clearly does not show that the solutions of **SR2** can be derived from a diffeomorphism-invariant action.

While this might get the classification of **SR2** correct, it does so at the cost of misclassifying **SR3**. According to the current suggestion, **SR3** now *is* a theory that meets the conditions for being background independent. But this is not the right result. The fact the equation of motion for its metric field is derived from a diffeomorphism-invariant action expressed only in terms of physical fields, hardly makes that metric more dynamical than the metric of **SR2**. After all, they both obey exactly the same equation of motion. And once this problem is recognised, reclassifying Θ^{abcd} as unphysical does not seem like enough to salvage the proposal. Even if **SR3** is no longer a counterexample, might there not be a relevantly similar theory that the proposal incorrectly classifies as background independent? The Rosen–Sorkin method is not the only way to construct a diffeomorphism-invariant variational problem for a theory that involves non-dynamical fields. These alternative procedures arguably provide examples of exactly the type envisaged.

One such procedure, developed by Karel Kuchař, is *parameterization*. In the simplest case one starts with the *Lorentz-covariant* expression for the action, defined with respect to inertial frame coordinates. Note that the field η_{ab} does not explicitly occur in this expression. One then treats the four coordinate fields X^μ of this formulation as themselves dependent variables ("clock fields"), writes them as functions of arbitrary coordinates, $X^\mu = X^\mu(x^\nu)$, and re-expresses the Lagrangian in terms of these new variables. Hamilton's principle is applied to the original dynamical variables, now conceived of as functions of x^ν, *and* to the coordinate fields, X^μ. In our simple example of **SR1**, stationarity under variations of Φ leads to an equation for Φ and X^μ that is satisfied just if Φ satisfies the standard Lorentz-covariant Klein–Gordon equation (1) with respect to the X^μ. Stationarity under variations of the X^μ yields equations that are automatically satisfied if the first equation is satisfied (see, e.g. §II.A [66]). Let us call the resulting theory **SR4**.

Another technique is described by Lee and Wald [40, 734].[36] Let the KPMs of **SR5** be defined in terms of two maps from the spacetime manifold, M. One is our familiar scalar field Φ. The other is a diffeomorphism y into a copy of spacetime, $\tilde{M}$, that is equipped with a particular flat Lorentzian metric field. One can use the diffeomorphism y to pull back the metric on $\tilde{M}$ onto M, and use the result, $g_{ab}(y)$, to define the standard Lagrangian, $\mathcal{L}_\Phi(y, \Phi) = \sqrt{-g(y)}g(y)^{ab}(\nabla_a\Phi)(\nabla_b\Phi)$, and action functional $S = \int d^4x\mathcal{L}_\Phi$. To determine the theory's solutions we require that S is stationary under variations in both of the theory's fundamental variables, y and Φ. Φ variations give us that Φ satisfies the Klein–Gordon equation with respect to $g_{ab}(y)$. Variations in y give equations that involve the vanishing of terms that are

---

[36]See Belot [7, 206–9] for an extended discussion of this example.

proportional to $\nabla_n T^n{}_b$, where T^{ab} is the stress-energy tensor for Φ. Since $\nabla_n T^n{}_b = \mathbf{0}$ follows from the Klein–Gordon equation, these equations are automatically satisfied.

Both **SR4** and **SR5** are examples of theories defined by diffeomorphism-invariant actions all of whose dependent variables are subject to Hamilton's principle. They will therefore be counterexamples to our latest proposal just if (i) they are background dependent and (ii) all of their fields are physical fields. One way to explore whether (i) and (ii) are satisfied is to consider how the theories relate to **SR2**. In particular, if they count as reformulations of **SR2**, then they are formulations of a background-dependent theory.

First, recall that a model of **SR2** is a triple of the form $\langle M, g_{ab}, \Phi \rangle$, where g_{ab} is flat. A model of **SR4**, is of the form $\langle M, \Phi, X^0, X^1, X^2, X^3 \rangle$. That is, it lacks a (primitive) field g_{ab}, and includes instead four scalar fields. Finally, models of **SR5** are of the form $\langle M, y, \Phi \rangle$, where y is a diffeomorphism into $\tilde{M}$, a copy of M equipped with a fixed metric.

For both **SR4** and **SR5**, there is a natural map from that theory's solution space to the solution space of **SR2**. For **SR4**, one first defines the unique flat metric field g_{ab}^X associated with the fields X^μ (the metric for which the X^μ are every-where Riemmann–normal coordinates). One then requires that the map associates $\langle M, \Phi, X^0, X^1, X^2, X^3 \rangle$ with $\langle M, g_{ab}, \Phi \rangle$ just if $g_{ab}^X = g_{ab}$. For **SR5**, $\langle M, y, \Phi \rangle$ maps to $\langle M, g_{ab}, \Phi \rangle$ just if $g(y)_{ab} = g_{ab}$. In the first case, the map is many-one. The solution space of **SR4** is intuitively 'bigger' than that of **SR2**. In the case of **SR5**, however, the map is a bijection.

This machinery helps articulate how both **SR4** and **SR5** can naturally be viewed as reformulations of **SR2**.[37] First, consider **SR4**. For any model of **SR2** one can choose special coordinates that encode its metric via the requirement that, in these coordinate systems, $g_{ab} = \mathrm{diag}(-1, 1, 1, 1)$. In order to understand **SR4** as a reformulation of **SR2**, one interprets the fundamental fields of **SR4** to be such coordinate fields. So interpreted, **SR4** is a formulation of a background-dependent theory, since **SR2** is. Do the X^μ count as "physical fields"? Unlike the Θ^{abcd} of **SR3**, they certainly encode something physical, since they encode the metrical facts. But there is also a sense in which they do not themselves directly represent something physical: coordinate systems are not physical objects. Note also that encoding a flat metric via special coordinates in the manner proposed does not uniquely determine the coordinates. If $\{X^\mu\}$ corresponds to one such set of fields, then so will any set $\{X'^\mu\}$ where the X'^μ are related to the X^μ by a Poincaré transformation. This is the source of the fact that the map from models of **SR4** to those of **SR2** is many-one. This means that (on the suggested interpretation our formalism) the $\{X^\mu\}$ contain some redundancy; "internal" Poincaré transformations $X^\mu \mapsto X'^\mu$ should be regarded as mere gauge re-descriptions.

---

[37]A similar observation can be made concerning **SR3**. Its models are of the form $\langle M, g_{ab}, \Phi, \Theta^{abcd} \rangle$ and the map from its solution space to that of **SR2** simply involves throwing away Θ^{abcd}: $\langle M, g_{ab}, \Phi, \Theta^{abcd} \rangle \mapsto \langle M, g_{ab}, \Phi \rangle$. This map is many-one, but the differences between **SR3** models mapped to the same **SR2** model concern differences in the non-physical field Θ^{abcd}.

The nature of the bijection between the solution space of **SR5** and that of **SR2** makes their interpretation as reformulations of the same background-dependent theory even more straightforward. Are **SR5**'s basic variables physical fields? The dynamical role of y is exhausted by its use to define the pull-back metric on M. It is only through this metric that y enters into the Lagrangian of the theory. Nonetheless, there is again a clear sense in which the machinery involves arbitrary elements that do not represent the physical facts directly. In particular, we might have set up the theory in terms of a different (but still flat) metric on the target manifold. As a mathematical object, this would constitute a different formulation of the theory, and yet the difference does not show up at the level of the pulled-back metrics on M: the same range of metrics for M is surveyed, just via different maps to a different object.

The upshot is that it is not clear whether **SR4** and **SR5**, interpreted as reformulations of **SR2**, constitute counterexamples to the proposed criterion for background independence. All hinges on whether the relevant fields count as physical fields. They clearly encode physical facts but, equally clearly, they do not do so in the most perspicuous manner. One might seek to solve this dilemma via further proscriptive modifications to the proposal. This, of course, risks creating further problems.[38] More importantly, one should recognise that we are now far past the point where one might hope to articulate a simple and illuminating connection between diffeomorphism invariance and background independence.

Rovelli writes

> Diffeomorphism invariance is the key property of the mathematical language used to express the key conceptual shift introduced with GR: the world is not formed by a fixed non-dynamical spacetime structure, which defines localization and on which the dynamical fields live. Rather, it is formed solely by dynamical fields in interactions with one another. Localization is only defined, relationally, with respect to the fields themselves. [54, 1312]

The moral of our investigation so far is that diffeomorphism invariance cannot be taken to express the shift from non-dynamical to only dynamical spacetime structures. Theories with non-dynamical structure can be formulated in a fully diffeomorphism-invariant manner. But note that Rovelli's description of the key conceptual shift introduced with GR involves two elements. In addition to the move from non-dynamical to dynamical spacetime, there is the claim that, in GR, "localization is only defined, relationally, with respect to the fields themselves". I agree that this is how one should understand diffeomorphism-invariant theories. What the existence of diffeomorphism-invariant formulations of theories with non-dynamical structure indicates, however, is that this feature of a theory is not peculiar to theories that lack non-dynamical fields. A diffeomorphism-invariant, relational approach to "localization" is as appropriate in the context of Newtonian physics and special relativity as it is in GR. A defence of this claim is the task of the last two sections.

---

[38]For example, does the metric field of **GR1** represent the physical facts in the most perspicuous manner? If **GR1** is not to count as fully background independent, it should not be on account of this type of failure.

9   An Aside on the Gauge Status of Diff(M)

My central claim is this: the observable content of, and the nature of the genuine physical magnitudes of, a specially relativistic theory, whether formulated along the lines of **SR1** or **SR2**, are identical in nature to those of an analogue generally relativistic theory, such as **GR1**. In the next section, I will spell out how this can be so. In this section, I say a little about when one should interpret diffeomorphisms as gauge transformations.

In the previous section, we saw that Rovelli claimed that **SR3** might be distinguished from **SR2** on the grounds that the former involves an additional field. In the passage quoted above, he cites Earman, who does indeed argue that one should distinguish **SR3** from more standard formulations of specially relativistic Klein–Gordon theory. Earman's reasoning, however, is rather different from Rovelli's.

Earman [21] defines (massive variants of) **SR1**, **SR2** and **SR3**, via the analogues of the equations considered earlier in this paper.[39] (To ease exposition, I use this paper's labels to refer to Earman's theories.) He is primarily concerned with the comparison between **SR1** (as obtained from an action principle) and **SR3**. Earman's reasons for differentiating the theories, unlike Rovelli's, have nothing directly to do with the presence of an additional field. He views the theories as distinct because he believes that, in the context of **SR1**, Φ can be treated as an observable but, in **SR3**, it cannot because: (i) only gauge-invariant quantities are observable and (ii) one should regard the Diff(M) symmetry of **SR3** as a gauge symmetry. Earman takes (ii) to be justified by the fact that Diff(M) is both a local *and a variational* symmetry group in the context of **SR3**. In reaching this judgement in this way, he takes himself to be applying a "uniform method for getting a fix on gauge that applies to any theory in mathematical physics whose equations of motion/field equations are derivable from an action principle" and that is "generally accepted in the physics community" [18, 19].

As I have argued elsewhere [47], the fact that this apparatus tells us that Diff(M) is not a gauge group of **SR1** is not surprising. Diff(M) *is not a symmetry group of* **SR1** and so *a fortiori* it is not a gauge symmetry group. What one really wishes to know is whether one should view Diff(M) as a gauge group of **SR2**. Earman does not address this question head-on, but one suspects that his answer would be in the negative, for he argues that the solution sets of **SR1** and **SR2** are the same [21, 455]. This, of course, simply cannot be correct. It cannot be the case that (i) Diff(M) is not a symmetry group of **SR1**; (ii) Diff(M) is a symmetry group of **SR2**; and (iii) the solution sets of **SR1** and **SR2** are the same. It is (iii) that should be given up, and it will be instructive to see where Earman's argument goes wrong.

---

[39]His equation (3) [21, 451] is (once corrected) the massive analogue of my (3), and defines his **SR1**-type theory. His equations (5) and (6) [21, 455] are the analogues of (4) and (1), and define his **SR2**-type theory.

Here is what he says

> The solution sets for [SR1] and for [SR2] are the same, at least on the assumption that the spacetime manifold is $\mathbb{R}^4$. For then there is a global coordinate system $\{x^\mu\}$ such that $g_{\mu\nu} = \eta_{\mu\nu}$ (where $\eta_{\mu\nu}$ is the Minkowski matrix) solves [(1)]. Moreover, in this coordinate system [(4)] reduces to [(3)[40]]. And every solution of [(1)] can be transformed, by a suitable coordinate transformation, into a solution of the form $g_{\mu\nu} = \eta_{\mu\nu}$. Thus, every solution of [SR2] is a solution of [SR1]. Similar reasoning shows that the converse is also true. [21, 455, 466, n 26]

This argument, effectively, ignores the distinction between fields that are solutions to equations and fields that feature in equations as fixed fields. Here is one way to see the error. Fix a coordinate system K on M (of the kind Earman considers). Relative to K, η_{ab} always has the same components in the coordinate representation of every solution of **SR1**. Every one of these coordinate descriptions is also a description with respect to K of a solution of **SR2**. But, *in addition to these*, every possible set of coordinate functions that one can obtain from the original sets by acting by a diffeomorphism on $\mathbb{R}^4$ also describes—*still relative to K*—a solution of **SR2**. Note, too, that each of these additional sets of coordinate functions corresponds (relative to K) to a representation of a (mathematically, though not necessarily physically) *distinct* solution of **SR2**. But these new coordinate functions are not descriptions of solutions of **SR1** relative to K (the components of the metric tensor have been changed, so they no longer describe η_{ab}).[41]

I conclude that Earman's claims do not speak against the natural interpretation of Diff(M) as a gauge group of **SR2**. His own favoured apparatus is simply silent on the question. When physicists themselves justify the use of the apparatus to identify gauge freedom, they take the deterministic nature of the theories in question as a premise (see, e.g. [16, 20]). In the context of **SR2**, this premise also leads to the conclusion that Diff(M) is a gauge group. In fact, Belot [8] shows how one can regiment the intuitions that are arguably behind such arguments in order to define a notion of gauge equivalence that matches Earman's favoured notion in its verdicts concerning Lagrangian theories but which applies more widely. Unsuprisingly, Belot's definition tells us that Diff(M) is a gauge group of **SR2**. There remains just one task. We need to see how this interpretative stance with respect to **SR2** can be reconciled with a relatively orthodox account of the nature of the observables of both background-dependent SR and background-independent GR.

---

[40] Since Earman refers to $\eta_{\mu\nu}$ as the Minkowski *matrix*, and since he has switched from Roman indices—which I interpret as signalling coordinate-free, abstract index notation—to Greek indices, it would seem more appropriate to refer to his equation (2), i.e., to equation (1), rather than to his (3).

[41] They can be understood as descriptions of solutions of **SR1**, but only if we allow ourselves to describe things with respect to coordinate systems other than K (in fact, we need to consider one coordinate system for each class related by Poincaré transformations). And when we do this, each solution of **SR1** is, of course, multiply represented.

10   On the Meaning of Coordinates

Recall, again, the similarities between **GR1** and **SR2**. The two theories share a space of KPMs. They differ only in terms of which subsets of this space are picked out as dynamically possible. The DPMs of each theory, although distinct sets of mathematical objects, are sets of the same *kind* of objects. That much is mathematical fact. These similarities, I submit, make plausible the following interpretative stance: one should treat the two theories uniformly. On this view, the physical magnitudes of the two theories describe the same types of physical objects. The theories postulate the same kind of stuff; they just differ over which configurations of this stuff are physically possible.

Why might one reject such a view? The reason, I think, has to do with a popular, but potentially misleading, way of thinking about the coordinates of non-generally covariant formulations of pre-relativistic theories. As I will describe in a moment, this way of thinking about the coordinates of, for example, Lorentz-invariant theories has implications for how one conceives of the content of those theories. It leads to a way of thinking about the theory's physical content that does not transfer to theories without special coordinates. The lack of non-dynamical background fields entails (though, as we saw, cannot be equated with) the lack of such coordinates. It is therefore natural to see the shift from SR to GR, in which background structures are excised, as heralding a radical change in the nature of the content of our physical theories. Against this, I want to highlight an alternative way of conceiving of the special coordinates of a non-covariant physics. This alternative way is perfectly compatible with the fundamental nature of the content of our physics remaining unchanged in the passage from background dependence to background independence. It also provides an independently plausible account of the content of background-dependent theories, such as SR.

The influence of the problematic view might well flow from the following passage in Einstein's groundbreaking paper on special relativity:

> The theory to be developed—like every other electrodynamics—is based upon the kinematics of rigid bodies, since *the assertions of any such theory concern relations between rigid bodies (systems of coordinates), clocks, and electromagnetic processes.* [23, 38, my emphasis]

Einstein seems here to be claiming that the meaning of the theoretical claims of Lorentz-invariant electromagnetism—that is, what those claims are fundamentally about—concerns the relationships between electromagnetic phenomena and rods and clocks. In other words, the content of the theory's claims is held to be about relationships between electromagnetic phenomena and *material bodies outside of the electromagnetic system under study.*

Versions of this type of view, as an interpretation of the special coordinates of specially relativistic and Newtonian physics, are explicitly endorsed by, for example, Stachel [60, 141–2], Westman and Sonego [67, 1592–3] and, in several places, Rovelli. To give a flavour of the importance of the view for Rovelli, I quote at length

For Newton, the coordinates **x** that enter his main equation

$$\mathbf{F} = m \frac{d^2\mathbf{x}(t)}{dt^2} \tag{2.152}$$

are the coordinates of absolute space. However, since we cannot directly observe space, the only way we can coordinatize space points is by using physical objects. The coordinates **x**...are therefore defined as distances from a chosen system O of objects, which we call a "reference frame"...

In other words, the physical content of (2.152) is actually quite subtle:

There exist reference objects O with respect to which the motion of any other object A is correctly described by (2.152)...

Notice also that for this construction to work it is important that the objects O forming the reference frame are not affected by the motion of the object A. There shouldn't be any dynamical interaction between A and O. [53, 87–8][42]

The similarity with Einstein's claim is clear. The "physical content" of an equation of restricted covariance turns out to involve claims about relations between the dynamical quantities that are explicitly represented in the equations and other material bodies that are only implicitly represented via the special coordinates. There is one difference worth noting. For Einstein, the important role of external bodies is to make meaningful spatial and temporal intervals; the bodies in question are rods and clocks. Rovelli, in contrast, emphasises two other roles played by the bodies of his reference system: they fix a particular coordinate system (define its origin) and, more importantly, they define same place over time. In fact, in spelling out his notion of a material reference system, Rovelli seems to take the notion of spatial distance as primitive and empirically unproblematic.

Now contrast this Einstein–Stachel–Rovelli (ESR) way of understanding special coordinates to what I will call the Anderson–Trautman–Friedman (ATF) perspective (recall footnote 12), which has already been adopted throughout in this paper. According to this latter view, a generally covariant formulation of a theory has the advantage over formulations of limited covariance of making the physical content of the theory fully explicit. This content includes certain spatiotemporal structures, such as those encoded by the Minkowski metric field η_{ab}. In cases where these structures are highly symmetric, one can encode certain physical quantities (e.g. spatiotemporal intervals) via special choices of coordinates adapted to these structures. Newton's special coordinates are not fundamentally defined in terms of, and Newton's equations do not make implicit reference to, external material bodies. Rather they are

---

[42] A similar claim is found in Rovelli [51, 187–9]. There Rovelli combines the claim that in pre-relativistic physics "reference system objects are not part of the dynamical system studied, their motion...is independent from the dynamics of the system studied" with the further assertion that the "mathematical expression" of the failure of this condition in GR is "the invariance of Einstein's equations under active diffeomorphisms."

equations that encode physically meaningful chronometric and inertial structure, via certain "gauge fixing" coordinate conditions.[43]

In order to avoid confusion, let me stress that according to both the ESR view and the ATF view the special coordinates of a non-covariant form of pre-relativistic physics have a different meaning to arbitrary coordinates in GR (or a generally covariant form of the pre-relativistic theory). On both views the special coordinates have physical meaning. The accounts just differ over what that physical meaning is.

To help further clarify the differences between two views, let me highlight three distinct features that concrete applications of coordinate systems must or may have.

1. The coordinate system must be anchored to the world in some way. If it is to be concretely applied, and predictively effective, we must be able to practically determine which coordinate values' particular observable events are to be assigned.
2. The coordinate system might be anchored to the world by observable material objects outside the system under study. (The system under study might be a proper subsystem of the universe.)
3. The coordinate system might partially encode, or be partially defined in terms of, physically meaningful spatiotemporal quantities (spacetime intervals; inertial trajectories, etc.). In order for this to be applied in concrete cases, we require physical systems that disclose these facts. Further, these systems may or may not be external to the system being modelled by our theory.

The ATF perspective wholly concerns the third point: the special coordinates of non-generally covariant formulations of theories encode physical magnitudes. It is simply silent on the issues raised in the first two points. The ESR perspective assumes such encoding too, but it makes various further commitments concerning how such coordinate systems are anchored to the world, and what kind of systems disclose the magnitudes that the coordinate systems encoded. It is important to see that these additional claims are not necessary concomitants of the idea that there is such encoding.

To see this, consider how one might in practice get one's hands on an ATF special coordinate system. The coordinates encode spatial intervals and temporal intervals. So one needs to be able to measure spatial and temporal intervals. But without further argument, one's ability to measure these should not be taken to require that the rods and clocks one uses are outside the system that one is describing, much less outside the scope of the theory one is using. Note that such spatiotemporal measurement is equally essential to the concrete application of GR, not now to give meaning to special coordinates, but to give empirical content to one of the dynamical fields that is explicitly described.

The ESR idea that, necessarily, special coordinates in pre-relativistic physics gain their meaning from material systems outside the system being studied, blurs the distinction between (i) coordinates encoding physical magnitudes that are disclosed by

---

[43]Specifically, one imposes $\Gamma^{\mu}_{\nu\rho} = 0$, $t_{\mu} = (1, 0, 0, 0)$ and $h^{\mu\nu} = \mathrm{diag}(0, 1, 1, 1)$, where $\Gamma^{\mu}_{\nu\rho}$ are the components of the connection, t_{μ} are the components of the one-form that defines the temporal metric and $h^{\mu\nu}$ are the components of the spatial metric.

systems not covered by the theory in question and (ii) the coordinates being anchored to the world via material systems outside the system under study. Rovelli's idea that "localisation" is inherently non-relational in pre-relativistic physics really only relies on (ii). However, it is easy to see that (ii) is not an intrinsic feature of the special coordinates of pre-relativistic physics. Even if in practice we often use physical systems to measure spatiotemporal intervals (and thereby fix the "magnitude-encoding" aspect of the coordinate system) that we do not (or cannot) actually model in our theory, the anchoring of particular coordinates to the world might simply involve the stipulation that some qualitatively characterisable components of the system under study are to be given such-and-such coordinate values.

Consider the case of a Lorentz-covariant formulation of our theory of the specially relativistic scalar field, for which $\Phi(x)$ is supposed to be an "observable", in contrast to the analogous quantity in GR. If the special coordinate system in terms of which Φ is being described is anchored to the world by some reference system not described by the theory, and if the coordinates are understood as encoding objective spatiotemporal quantities, then it is clear what physical meaning $\Phi(x_0)$ is supposed to have (for any given, particular x_0) and what the difference in meaning is between the quantities $\Phi(x_0)$ and $\Phi(x_0 + \Delta x)$. However—and this is the absolutely crucial observation—such coordinate representations of Φ can also be understood to be physically meaningful (in essentially the same way) *without* understanding them in terms of "non-relational localisation" thought of as provided by an external anchor for the coordinate system.

Imagine, for example, that one measures Φ to take a certain value (at one's location). One stipulates that this value is to be given coordinate values x_0.[44] One then asks what value the theory predicts that the field will take at a certain spatiotemporal distance away from the observed value. Since such spatiotemporal distances are encoded in the coordinates of the Lorentz-covariant formulation of the theory, this is to ask what the theory predicts the value of $\Phi(x_0 + \Delta x)$ will be, *given the value of* $\Phi(x_0)$, where the coordinate difference Δx encodes the spatiotemporal interval we are interested in. Note that, conceived of in this way, $\Phi(x)$ and $\Phi(x + \Delta x)$ specify, not two independently predictable quantities ultimately defined in terms of the relationship of Φ to an unstated reference object, but a single diffeomorphism-invariant coincidence quantity, involving how the variation of Φ is related to the underlying metric field η_{ab}.

If one considers Newtonian physics or special relativity as potentially providing complete cosmological theories, then any anchoring of special coordinate systems has to be done, ultimately, in this second way. Moreover, any systems that disclose the metric facts are, by hypothesis, describable by the theory. Of course, this is not

---

[44]In reality, in order both to provide a uniquely identifying description of the field that allows us to anchor the coordinate system, and to provide sufficient initial data that a prediction can be extracted from the theory, one should really consider the observation of a certain qualitatively characterisable and spatially extended continuum of field values. This complication does not alter the basic structure of the story given in the text.

how we now understand the empirical applicability of Newtonian physics or special relativity in the actual world. But the point is that there is no logical incoherence in so conceiving of them. Indeed, it was the interpretation each was assumed to have prior to 1905 and 1915 respectively. A theory's including non-dynamical background fields does not, per se, preclude such a cosmological interpretation.

To summarise, the additional commitments of the ESR interpretation of coordinates, over those of the ATF view, are not necessary consequences of a theory's being background dependent in the sense of involving non-dynamical structure. The conditions that ESR write into the very meaning of all special coordinate systems might correctly characterise some concrete applications of such systems, but they need not do so. In fact, sometimes, they do not do so. Consider, for example, a case whose philosophical importance is stressed by Julian Barbour: the use of Newtonian mechanics by astronomers to determine ephemeris time and the inertial frames.[45] Here certain facts about simultaneity and spatial distances are determined "externally", but the way the coordinate system is anchored to the world, and the way *some* of the spatiotemporal quantities encoded by the coordinate system are determined (time intervals and an inertial standard of equilocality) are not.

There is, perhaps, one qualification to be made. I have argued that, in the context of classical background-dependent physics, the ESR story about special coordinate systems does not provide an analysis of their fundamental meaning. This, however, does not rule out something like the story being correct for background-dependent quantum theory. In this context, the suggestion would be that certain (non-quantum) background structure in the theory, namely, Minkowski spacetime geometry, really does acquire physical meaning via an implicit appeal to physical systems outside the scope of the theory. Even if something along these lines were correct (and I register my scepticism), the point to be stressed is that its correctness is not to be understood as flowing from the necessary meaning of such coordinate systems in classical background-dependent physics.

Acknowledgements This material began to take something close to its current shape during a period of sabbatical leave spent at UC San Diego. I thank the members of the UCSD philosophy department for their generous welcome. I have benefitted from correspondence and/or discussions with John Dougherty, Carl Hoefer, Dennis Lehmkuhl, Thomas Moller-Nielsen, Matt Pead, Brian Pitts, Carlo Rovelli, Adán Sus, David Wallace and Chris Wüthrich. I have also received helpful comments from numerous audience members at talks on related material, in Leeds, Konstanz, Oxford, Les Treilles, Wuppertal, at the Second International Conference on the Ontology of Spacetime in Montreal, at the Southern California Philosophy of Physics group, and at the Laboratoire SPHERE Philosophy of Physics Seminar at Paris Diderot. Research related to this paper was supported during 2008–10 by a Philip Leverhulme Prize. Finally, special thanks are due to the editor for his encouragement, and to Sam Fletcher and Neil Dewar for numerous comments on an earlier draft.

---

[45]For a popular account that stresses the philosophical morals, see Barbour [6, Ch. 6].

References

1. Anderson, J. L. (1964). Relativity principles and the role of coordinates in physics. In H.-y. Chiu and W. F. Hoffmann (Eds.), *Gravitation and Relativity*, pp. 175–194. New York: W. A. Benjamin.
2. Anderson, J. L. (1967). *Principles of Relativity Physics*. New York: Academic Press.
3. Anderson, J. L. (1996). Answer to question 22. ["is there a gravitational force or not?," Barbara S. Andereck, Am. J. Phys. 63(7), 583 (1995)]. *American Journal of Physics 64*(5), 528–529.
4. Anderson, J. L. and R. Gautreau (1969). Operational formulation of the principle of equivalence. *Physical Review 185*(5), 1656–1661.
5. Baez, J. C. (2000). What is a background-free theory? http://math.ucr.edu/home/baez/background.html. Accessed 9 2009.
6. Barbour, J. B. (1999). *The End of Time: The Next Revolution in Our Understanding of the Universe*. London: Weidenfeld & Nicholson.
7. Belot, G. (2007). The Representation of Time and Change in Mechanics. See [14], pp. 133–227.
8. Belot, G. (2008). An elementary notion of gauge equivalence. *General Relativity and Gravitation 40*, 199–215.
9. Belot, G. (2011). Background-independence. *General Relativity and Gravitation 43*(1), 2865–2884.
10. Bergmann, P. G. (1942). *An Introduction to the Theory of Relativity*. New York: Prentice-Hall.
11. Brown, H. R. (2005). *Physical Relativity: Space-time Structure from a Dynamical Perspective*. Oxford: Oxford University Press.
12. Brown, H. R. and D. Lehmkuhl (2013). Einstein, the reality of space, and the action-reaction principle. http://philsci-archive.pitt.edu/id/eprint/9792.
13. Brown, H. R. and O. Pooley (2006). Minkowski space-time: A glorious non-entity. See [15], pp. 67–88.
14. Butterfield, J. N. and J. Earman (Eds.) (2007). *Philosophy of Physics* (1 ed.), Volume 2 of *Handbook of the Philosophy of Science*. Amsterdam: Elsevier.
15. Dieks, D. (Ed.) (2006). *The Ontology of Spacetime*, Volume 1 of *Philosophy and Foundations of Physics*. Amsterdam: Elsevier.
16. Dirac, P. A. M. (2001 [1964]). *Lectures on Quantum Mechanics*. New York: Dover Publications, Inc.
17. Earman, J. (1989). *World Enough and Space-Time: Absolute versus Relational Theories of Space and Time*. Cambridge, MA: MIT Press.
18. Earman, J. (2002). Response by John Earman. *Philosophers' Imprint 2*, 19–23.
19. Earman, J. (2003). The cosmological constant, the fate of the universe, unimodular gravity, and all that. *Studies In History and Philosophy of Modern Physics 34*(4), 559–577.
20. Earman, J. (2006a). The implications of general covariance for the ontology and ideology of spacetime. See [15], pp. 3–24.
21. Earman, J. (2006b). Two challenges to the requirement of substantive general covariance. *Synthese 148*(2), 443–468.
22. Earman, J. and J. D. Norton (1987). What price spacetime substantivalism? the hole story. *The British Journal for the Philosophy of Science 38*, 515–525.
23. Einstein, A. (1905). Zur Elektrodynamik bewegter Körper. *Annalen der Physik 322*(1), 891–921. Reprinted in Einstein et al. [30, 37–65].
24. Einstein, A. (1916). The foundation of the general theory of relativity. *Annalen der Physik 49*, 769–822. Reprinted in Einstein et al.[30, 109–64].
25. Einstein, A. (1918). Prinzipielles zur allgemeinen Relativitätstheorie. *Annalen der Physik 360*, 241–244.
26. Einstein, A. (1920). Antwort auf Ernst Reichenbächer, "Inwiefern läßt sich die moderne Gravitationstheorie ohne die Relativität begründen?". *Die Naturwissenschaften 8*, 1010–1011. Reprinted in [29 Doc. 49]; pages references are to the accompanying translation volume.
27. Einstein, A. (1922). *The Meaning of Relativity: Four Lectures Delivered at Princeton University, May, 1921*. Princeton: Princeton University Press.

28. Einstein, A. (1924). Über den Äther. *Schweizerische naturforschende Gesellschaft, Verhanflungen 105*, 85–93. Translated by S. W. Saunders in [56, 13–20]; page references are to this translation.
29. Einstein, A. (2002). *The Berlin Years: Writings 1918-1921*, Volume 7 of *The Collected Papers of Albert Einstein*. Princeton, NJ: Princeton University Press.
30. Einstein, A., H. A. Lorentz, H. Weyl, and H. Minkowski (1952). *The Principle of Relativity: A Collection of Original Papers on the Special and General Theory of Relativity*. New York: Dover. Translated by W. Perrett and G. B. Jeffrey.
31. Fletcher, S. C. (2013). Light clocks and the clock hypothesis. *Foundations of Physics 43*(11), 1369–1383.
32. Friedman, M. (1983). *Foundations of Space-Time Theories: Relativistic Physics and Philosophy of Science*. Princeton University Press.
33. Gaul, M. and C. Rovelli (2000). Loop quantum gravity and the meaning of diffeomorphism invariance. In J. Kowalski-Glikman (Ed.), *Towards Quantum Gravity: Proceeding of the XXXV International Winter School on Theoretical Physics Held in Polanica, Poland, 2–11 February 1999*, Volume 541 of *Lecture Notes in Physics*, Berlin, Heidelberg, pp. 277–324. Springer.
34. Giulini, D. (2007). Remarks on the notions of general covariance and background independence. In I.-O. Stamatescu and E. Seiler (Eds.), *Lecture Notes in Physics*, Volume 721, pp. 105–120. Berlin, Heidelberg: Springer.
35. Gryb, S. B. (2010). A definition of background independence. *Classical and Quantum Gravity 27*(2), 5018.
36. Hoffmann, B. (Ed.) (1966). *Perspectives in Geometry and Relativity: Essays in Honor of Václav Hlavatý*. Bloomington: Indiana University Press.
37. Janssen, M. (2012). The twins and the bucket: How Einstein made gravity rather than motion relative in general relativity. *Studies In History and Philosophy of Modern Physics 43*(3), 159–175.
38. Knox, E. (2014). Newtonian spacetime structure in light of the equivalence principle. *The British Journal for the Philosophy of Science 65*, 863–880.
39. Kretschmann, E. (1917). Über den physikalischen Sinn der Relativitätspostulate. *Annalen der Physik 53*, 575–614.
40. Lee, J. and R. M. Wald (1990). Local symmetries and constraints. *Journal of Mathematical Physics 31*(3), 725–743.
41. Misner, C., K. S. Thorne, and J. A. Wheeler (1973). *Gravitation*. San Francisco: W. H. Freeman and Company.
42. Norton, J. D. (1989). Coordinates and covariance: Einstein's view of space-time and the modern view. *Foundations of Physics 19*(1), 1215–1263.
43. Norton, J. D. (1993). General covariance and the foundations of general relativity: Eight decades of dispute. *Reports on Progress in Physics 56*(7), 791–858.
44. Ohanian, H. C. and R. Ruffini (2013). *Gravitation and Spacetime* (3rd ed.). Cambridge: Cambridge University Press.
45. Pitts, J. B. (2006). Absolute objects and counterexamples: Jones–Geroch dust, Torretti constant curvature, tetrad-spinor, and scalar density. *Studies in History and Philosophy of Modern Physics 37*(2), 347–371.
46. Pitts, J. B. (2009). Empirical equivalence, artificial gauge freedom and a generalized Kretschmann objection. http://philsci-archive.pitt.edu/4995/.
47. Pooley, O. (2010). Substantive general covariance: Another decade of dispute. In M. Suárez, M. Dorato, and M. Rédei (Eds.), *EPSA Philosophical Issues in the Sciences: Launch of the European Philosophy of Science Association*, Volume 2, pp. 197–209. Dordrecht: Springer.
48. Pooley, O. (2013). Substantivalist and relationalist approaches to spacetime. In R. W. Batterman (Ed.), *The Oxford Handbook of Philosophy of Physics*, pp. 522–586. Oxford: Oxford University Press.
49. Rickles, D. (2008). Who's afraid of background independence? In D. Dieks (Ed.), *The Ontology of Spacetime II*, Volume 4 of *Philosophy and Foundations of Physics*, pp. 133–152. Amsterdam: Elsevier.

50. Rosen, N. (1966). Flat space and variational principle. See [36], Chapter 33, pp. 325–327.
51. Rovelli, C. (1997). Halfway through the woods: Contemporary research on space and time. In J. Earman and J. D. Norton (Eds.), *The Cosmos of Science: Essays of Exploration*, Volume 6 of *Pittsburgh–Konstanz Series in the Philosophy and History of Science*, pp. 180–223. Pittsburgh: University of Pittsburgh Press.
52. Rovelli, C. (2001). Quantum spacetime: What do we know? In C. Callender and N. Huggett (Eds.), *Physics Meets Philosophy at the Planck Scale: Contempory Theories in Quantum Gravity*, pp. 101–122. Cambridge: Cambridge University Press.
53. Rovelli, C. (2004). *Quantum Gravity*. Cambridge: Cambridge University Press.
54. Rovelli, C. (2007). Quantum gravity. See [14], pp. 1287–1329.
55. Rozali, M. (2009). Comments on background independence and gauge redundancies. *Advanced Science Letters 2*(2), 244–250.
56. Saunders, S. W. and H. R. Brown (Eds.) (1991). *The Philosophy of the Vacuum*. Oxford: Oxford University Press.
57. Smolin, L. (2003). Time, structure and evolution in cosmology. In A. Ashtekar, R. S. Cohen, D. Howard, J. Renn, S. Sarkar, and A. Shimony (Eds.), *Revisiting the Foundations of Relativistic Physics: Festschrift in Honor of John Stachel*, Volume 234 of *Boston Studies in the Philosophy of Science*, pp. 221–274. Dordrecht: Kluwer.
58. Smolin, L. (2006). The case for background independence. In D. Rickles, S. French, and J. Saatsi (Eds.), *The Structural Foundations of Quantum Gravity*, Chapter 7, pp. 196–239. Oxford: Oxford University Press.
59. Sorkin, R. D. (2002). An example relevant to the Kretschmann–Einstein debate. *Modern Physics Letters A 17*(11), 695–700.
60. Stachel, J. (1993). The meaning of general covariance. In J. Earman, A. I. Janis, G. J. Massey, and N. Rescher (Eds.), *Philosophical problems of the internal and external worlds: essays on the philosophy of Adolf Grünbaum*, Volume 1 of *Pittsburgh–Konstanz series in the philosophy and history of science*, pp. 129–160. Pittsburgh: University of Pittsburgh Press.
61. Sus, A. (2008). *General Relativity and the Physical Content of General Covariance*. Ph. D. thesis, Universitat Autònoma de Barcelona.
62. Sus, A. (2010). Absolute objects and general relativity: Dynamical considerations. In M. Suarez, M. Dorato, and M. Rédei (Eds.), *EPSA Philosophical Issues in the Sciences: Launch of the European Philosophy of Science Association*, Volume 2, Chapter 23, pp. 239–249. Springer.
63. Synge, J. L. (1960). *Relativity: The General Theory*. Amsterdam: North-Holland.
64. Torretti, R. (1984). Space-time physics and the philosophy of science. *The British Journal for the Philosophy of Science 35*(3), 280–292.
65. Trautman, A. (1966). Comparison of Newtonian and relativistic theories of space-time. See [36], Chapter 42, pp. 413–425.
66. Varadarajan, M. (2007). Dirac quantization of parametrized field theory. *Physical Review D 75*(4), 44018.
67. Westman, H. F. and S. Sonego (2009). Coordinates, observables and symmetry in relativity. *Annals of Physics 324*(8), 1585–1611.

Gauge Theory of Gravity and Spacetime

Friedrich W Hehl

Abstract The advent of general relativity in 1915/1916 induced a paradigm shift: since then, the theory of gravity had to be seen in the context of the geometry of spacetime. An outgrowth of this new way of looking at gravity is the gauge principle of Weyl (1929) and Yang–Mills–Utiyama (1954/1956). It became manifest around the 1960s (Sciama–Kibble) that gravity is closely related to the Poincaré group acting in Minkowski space. The gauging of this external group induces a Riemann–Cartan geometry on spacetime. If one generalizes the gauge group of gravity, one discovers still more involved spacetime geometries. If one specializes it to the translation group, one finds a specific Riemann–Cartan geometry with teleparallelism (Weitzenböck geometry).

1 Apropos a Theory of Spacetime Theories

In this workshop, we are supposed to move "Towards a theory of spacetime theories." The idea seems to be that there are many spacetime theories around: the Riemannian spacetime theory in the framework of general relativity (GR), the Weitzenböck spacetime theory in teleparallelism approaches to gravity, the Riemann–Cartan spacetime theory withing the Poincaré gauge theory of gravity (PG), the superspace(time) theory within supergravity, the Weyl(–Cartan) spacetime theory within a gauge theory of the Weyl group, etc. The list could be continued with spacetime theories emerging in quantization approaches to gravity where spacetime becomes mostly a discrete structure. There is a plethora of different spacetime theories around and it is hardly possible to view all of them from some kind of a unifying principle, let alone from one theory encompassing these spacetime theories as specific subcases.

F.W. Hehl (✉)
Institute for Theoretical Physics, University of Cologne, Cologne, Germany
e-mail: hehl@thp.uni-koeln.de

F.W. Hehl
Department of Physics & Astronomy, University of Missouri, Columbia,
MO, USA

© Springer Science+Business Media, LLC 2017   145
D. Lehmkuhl et al. (eds.), *Towards a Theory of Spacetime Theories*,
Einstein Studies 13, DOI 10.1007/978-1-4939-3210-8_5

Orientation in this seemingly chaotic landscape of spacetime theories can be provided by looking at the successful theories of our days that are able to predict and describe correctly fundamental phenomena occurring in nature. There is the standard model of particle physics, based on the Poincaré group (also known as inhomogeneous Lorentz group) and the internal groups $SU(3)$, $SU(2)$, $U(1)$. The Poincaré group is the group of motion in the Minkowski spacetime of special relativity (SR), and it classifies the particles according to their masses and their spins. The internal groups describe the strong and the electroweak interactions by means of the respective gauge (or Yang–Mills) theory.

A book on the centennial of the discovery of SR was called [1]: "Special Relativity. Will it survive the next 100 years?" When I read this title in 2005, I thought for a moment that I must have been in a time machine and in reality I am living in 1905. Hadn't SR already been superseded in 1915/1916 by GR, I wondered? I pointed this out to the editors that this title looks anachronistic to me and is hardly appropriate for editors who both are known to subscribe to GR. It turned out that both wanted to ask whether SR survives *locally* as a valid theory. But they did not want to change the title since this fact was, as they told me, known to everybody anyway. I gave up since I realized that in a time when in the tabloid press a title is more for catching one's attention than for spreading the truth, the scientific literature cannot stand aside.

But what is my point? Well, we all seem to agree that at least presently SR is universally valid locally *in a freely falling frame*. So far no deviations therefrom have been found. Only at very high accelerations, the principle of locality, inherent in SR, may need to be amended [2]. In any case, our march toward a theory of all spacetime theories has at least a definitive starting point.

But was SR superseded by GR? Yes, of course—in spite of the title of reference [1]. The abstraction of a Minkowski space can only be uphold when gravitational effects can safely be neglected. If you measure Planck's constant or the elementary charge by a conventional laboratory experiment, then this assumption is justified. But if you go down the stairs, you had better not neglect gravity, otherwise you may fall downwards; or if you measure the deflection angle of a light ray gracing a star, you also had better not neglect gravity. From the laboratory to at least the scale of the planetary system, GR is in excellent agreement with experiment. On the galactic scale this is taken for granted by most physicists, but this is disputed by supporters of MOND, of TeVeS, of f(R)-theory, or of nonlocal gravity,[1] for example, compare the presentations in [6]. Anyway, GR is mostly accepted for the global description of the cosmos and if the cosmological principle is assumed, namely homogeneity and isotropy of space, Einstein's field equation predicts a Friedmann cosmos. The cosmos started with the Big Bang and it is usually assumed to be equipped with a scalar inflationary field providing a sufficiently fast expansion. Needless to say that this framework is based on a number of extreme extrapolations.

---

[1]Mashhoon and the author [3, 4] formulated a *nonlocal* translational *gauge theory* of gravity that can account for the observed rotation curves of spiral galaxies without invoking any dark matter; in fact, this nonlocal theory of gravitation appears to be consistent with gravitational physics from the scale of the solar system to the scale of clusters of galaxies—see the most recent results by Rahvar and Mashhoon [5].

The message is then that the Minkowski spacetime picture is replaced by the Riemannian one. But this does not rest on the same strong experimentally well-confirmed basis as the local presence of the Minkowski spacetime of SR.

2   Is the Gauge Idea the Underlying Principle for All Interactions?

Since the advent of GR, it was clear that a spacetime theory is inextricably linked to gravity. One cannot be understood without the other. Coming back to the topic of our workshop, it is then clear that gravity has to be considered in this general context willy nilly. Accordingly, a spacetime theory is at the same time, at least in some of its parts, a theory of gravity.

Let us then turn to gravity: Is GR all we have? Well, by some people GR is declared to be sacrosanct and you may touch it only by superimposing some abstract mathematical framework supposedly quantizing GR, see [7]. But practitioners of this method increasingly become aware that they have to amend the Hilbert–Einstein Lagrangian of the free gravitational field by non-Riemannian supplementary terms, thereby dissolving to a certain extend the Riemannian structure they started with [8–10]. Hence alternatives to GR gain credibility even if GR is left fixed at first.

Is GR the only reasonable theory of gravity? No, it is not. Already in 1956, Utiyama began to formulate gravity as a gauge theory, for a selection of classical papers, see [11]. The strong and electro-weak gauge theories are based on internal symmetry groups—mathematically semi-simple Lie groups—linked to conserved currents. The gauge idea basically requires that the *rigid* (or global) symmetry group related to the conserved current under consideration has to be made *local*; without giving up the invariance of the Lagrangian, this is only possible by the introduction of a gauge potential $A = A_i dx^i$ (a covector or an 1-form) that transform under this group suitably; for each parameter of the group one needs one covector field. Thus, the group dictates the interaction emerging from that scheme: a new interaction is created from a conserved current via the (reciprocal) Noether theorem and the symmetry group attached to it.

In the standard model of particle physics all gauge groups are internal, that is, they act in some internal space. In the original Yang–Mills theory, for example, it was the isospin space. But the gauge idea of *localizing a symmetry* does not seem to be restricted to internal groups. An external group affects by definition spacetime. If we have a conserved current and a corresponding group, nothing prohibits us to apply the gauge principle.

How does gravity come into this framework? The source of Newtonian gravity is the mass of a body. In classical physics, mass is a conserved quantity, as has been experimentally demonstrated by Lavoisier (around 1790). In SR mass conservation is no longer valid—as has been shown in the 1930s by more accurate experimental techniques—and is superseded by energy-momentum conservation, as has been most vividly demonstrated in Alamogordo in 1945. Clearly then, the Poisson equation controlling Newton's gravitational potential ϕ, namely $\Delta\phi(\mathbf{r}, t) = 4\pi G\rho(\mathbf{r}, t)$, with Δ as the Laplace operator, G as the gravitational constant, and ρ as the mass density, has to be replaced by an equation that carries on its right-hand side the energy density of matter (and/or radiation). However, according to SR, the energy density is the time–time component of the symmetric energy-momentum current $t_{ij} = t_{ji}$ of matter (and/or radiation).

For an isolated physical system, the energy-momentum current t_{ij} is conserved: $\partial_j t_i{}^j = 0$. This is an expression of the fact that the action of the system is invariant under translations in time and space. Consequently, the conserved energy-momentum current together with the translation group $T(4)$ acting in Minkowski space should underlie gravity. Since the translation group has four parameters, one describing a time translation and three describing space translations, we expect four potential one-forms ϑ^α, for $\alpha = 0, 1, 2, 3$. As we will see further down, this framework leads to a teleparallelism theory of gravity and back to a theory that is equivalent to GR for conventional (bosonic) matter. Accordingly, GR can be understood as a gauge theory of the translation group $T(4)$, which is an *external* group.

Ergo, all interactions, including gravity, are governed by gauge field theories. But let us now turn back to the history of the gauge idea:

3   The Gauging of the Poincaré Group

As we mentioned before, Utiyama [12] first attacked the problem of understanding gravity as a gauge theory by means of gauging the Lorentz group $SO(1, 3)$. In this way, Utiyama supposedly derived general relativity. However, the problematic character of his derivation is apparent. First of all, he had to introduce in an ad hoc way tetrads $e_i{}^\alpha$ (or coframes $\vartheta^\alpha = e_i{}^\alpha dx^i$), first holonomic (natural), and later anholonomic (arbitrary) ones. Secondly, he has to assume the connection $\Gamma_i{}^{\alpha\beta}$ of spacetime to be Riemannian, without any convincing argument.

But thirdly, perhaps the strongest reason, the current linked to the (homogeneous) Lorentz group is the *angular momentum current* $\mathfrak{J}_{ij}{}^k = -\mathfrak{J}_{ji}{}^k$, which is conserved, $\partial_k \mathfrak{J}_{ij}{}^k = 0$. However, as we have seen in the last section, gravity is coupled to the conserved and symmetric energy-momentum current t_{ik}. Accordingly, Einstein in 1915 took in general relativity the symmetric energy-momentum current t_{ik} as the source of gravity in his field equation and *not* the angular momentum current. Hence Utiyama was not on the right track. Interestingly enough, in numerous publications even today, the Lorentz group is incorrectly thought of as gauge group of GR; usually the conserved current coupled to it is not even mentioned.

This can be also viewed from the translational gauge group of gravity, at which we arrived above. In a Minkowski space, as in any Euclidean space, the group of motions consists of translations *and* rotations. In fact, the semidirect product of the translation group and the Lorentz group, $T(4) \rtimes SO(1, 3)$, is the Poincaré group $P(1, 3)$ with its $4 + 6$ parameters (and its $4 + 6$ gauge potentials ϑ^α and $\Gamma^{\alpha\beta} = -\Gamma^{\beta\alpha}$, respectively).

In a Euclidean or Minkowskian space the translations do not live alone, they are accompanied, in a nontrivial way, by the (Lorentz) rotations. Accordingly, since we find reasons to gauge the translations in a Minkowski spacetime, it is hardly avoidable to gauge also the rotations. If one has spinless matter, this argument may be skipped. However, if we have fermionic matter, its rotational behavior is closely linked to the translational behavior. Kibble, who was the first to gauge the Poincaré group [13], poses the following question [14]:

> ... Is it possible that starting from a theory with rigid symmetries and applying the gauge principle, we can recover the gravitational field? The answer turned out to be yes, though in a subtly different way and with an intriguing twist. Starting from special relativity and applying the gauge principle to its Poincaré-group symmetries leads most directly not precisely to Einstein's general relativity, but to a variant, originally proposed by Élie Cartan, which instead of a pure Riemannian space-time uses a space-time with torsion. In general relativity, curvature is sourced by energy and momentum. In the Poincaré gauge theory, in its basic version, additionally torsion is sourced by spin.

This is also the basic message of our seminar: Gauging an external group, here the Poincaré group, leads directly to a new geometry of spacetime, here the Riemann–Cartan geometry of spacetime. To an external gauge group a certain geometry of spacetime is attached, the Minkowski space is deformed in accordance with the gauged symmetries. Moreover, without a conserved current, there can be no real gauge procedure in the sense of Weyl and Yang–Mills. If somebody tries to sell you a gauge theory without mentioning the associated conserved current, do not believe her or him a word. Gauging the Weyl group *without* considering the scale current and gauging the conformal group *without* considering the conformal currents are procedures that may lead to something, but certainly not to gauge theories à la Weyl–Yang–Mills, see the discussion in [11].

Often I have heard the argument that gravity can have no relation to the translation group since GR takes place in a Riemannian space and therein the translations are an ill-defined concept since they are not integrable, for example. However, this argument rests on a misunderstanding. In a gauge approach, at the start of the procedure, that is, before the rigid symmetry is made local, we consider the gravity-free case. Accordingly, we are in Minkowski space where a translation is part of the group of motion. Only after we localized the symmetry, we lose the underlying Minkowski space, it gets deformed, and one has to reconstruct the emerging geometry. This is the radicality of the gauge principle: an interaction is created by a symmetry. The translation group $T(4)$, a subgroup of the Poincaré group $P(1, 3)$, which acts in a Minkowski space, creates the gravitational potential ϑ^α. The Lorentz subgroup $SO(1, 3)$ creates another gravitational potential $\Gamma^{\alpha\beta} = -\Gamma^{\beta\alpha}$, the consequences of which we will have to discuss.

4 Einstein's Discussion of the Transition from Special to General Relativity

Before we turn to the subject of the gauging of the Poincaré group, we remind ourselves how Einstein "derived" gravity [15]. When Einstein developed GR, he could take a classical mass point with mass m as a starting point for his investigations. He studied its behavior in an accelerated reference system. Technically, in order to switch on acceleration, he transformed the original Cartesian coordinate system X^i to a curvilinear coordinate system x^i. Let us look at this in more detail. The points in the Minkowski space of SR can be described with the help of Cartesian coordinates X^i, with $i = 0, 1, 2, 3$. In these coordinates, the line element reads

$$ds^2 = (dX^0)^2 - (dX^1)^2 - (dX^2)^2 - (dX^3)^2 = o_{ij}dX^i \otimes dX^j, \tag{1}$$

with $o_{ij} = \text{diag}(1, -1, -1, -1)$ and summation over repeated indices. The equation of motion of a *force-free* mass in an inertial frame K,

$$\frac{d^2X^k}{ds^2} = 0, \tag{2}$$

leads for the particle trajectory to a straight line with constant velocity.

The same motion, as viewed from the accelerated frame K', can be derived by a transformation of (2) to curvilinear coordinates,

$$\frac{D^2x^k}{Ds^2} := \frac{d^2x^k}{ds^2} + \widetilde{\Gamma}_{ij}{}^k \frac{dx^i}{ds}\frac{dx^j}{ds} = 0, \tag{3}$$

with the Riemannian connection (Christoffel symbols of the 2nd kind):

$$\widetilde{\Gamma}_{ij}{}^k := \tfrac{1}{2}g^{k\ell}\left(\partial_i g_{j\ell} - \partial_j g_{i\ell} + \partial_\ell g_{ij}\right) = \widetilde{\Gamma}_{ji}{}^k; \tag{4}$$

here we abbreviated the partial differentiation $\partial/\partial x^i$ as ∂_i. The massive particle accelerates with respect to the non-inertial frame K' in such a way that this acceleration is independent of its mass. But an observer in K' cannot tell whether this motion is accelerated or induced by a homogeneous gravitational field of strength $\widetilde{\Gamma}_{ij}{}^k$. In other words, the reference system K' can be alternatively considered as being at rest with respect to K, but a homogeneous gravitational field is present that is described by the *Christoffel symbols* $\widetilde{\Gamma}_{ij}{}^k$.

Nothing has happened so far. We are still in a Minkowski space in which—as is shown in geometry—the *Riemann curvature tensor* belonging to the Christoffel symbols

$$\widetilde{R}_{ijk}{}^\ell := 2\partial_{[i}\widetilde{\Gamma}_{j]k}{}^\ell + 2\widetilde{\Gamma}_{[i|m|}{}^\ell\,\widetilde{\Gamma}_{j]k}{}^m \tag{5}$$

vanishes, that is $\widetilde{R}_{ijk}{}^{\ell} = 0$; brackets around indices denote antisymmetrization: $[ij] := \{ij - ji\}/2$. This is the ingenuity of Einstein's approach: He considers force-free motion from two different reference frames and identifies thereby the Christoffels as describing—according to the equivalence principle—a homogeneous gravitational field. Of course, this gravitational field in Minkowski space is fictitious, it is simulated, it does not really exist since the Riemann curvature vanishes.

Besides massive point particles, we have light rays ("photons") that can be considered in a similar way. For light propagation we have $ds^2 = 0$, but the geodesic line (3) can be reparametrized with the help of a suitable affine parameter. Then, from the point of view of reference frame K', a light ray that propagates in a straight line in the inertial frame K appears to be deflected in K'. According to Einstein [16], "... the principle of the constancy of the *velocity of light in vacuo* must be modified, since we easily recognize that the path of the light ray with respect to K' must in general be curvilinear." Thus, the gravitational field deflects light. This is one of Einstein famous and successful predictions.

In order to create a real gravitational field—this is Einstein's assumption—we must relax the rigidity of Minkowski space and allow for Riemannian curvature, inducing in this way a "deformed" spacetime carrying nonvanishing curvature $\widetilde{R}_{ijk}{}^{\ell} \neq 0$. A prerequisite for this procedure to work is the fact that the Christoffels depend at most on first derivatives $\partial_k g_{ij}(x)$ of the metric $g_{ij}(x)$. These first derivatives appear even in a flat space in an accelerated frame. Only nonvanishing second derivatives tell us about real gravitational fields.

There is one more thing to be seen from (3). If we multiply it with a slowly varying scalar mass density ρ of dust matter, then we recognize that the Christoffels are coupled to the (symmetric) energy-momentum tensor density of dust,[2]

$$\rho \frac{d^2 x^k}{ds^2} + \mathfrak{t}^{ij} \, \widetilde{\Gamma}_{ij}{}^k = 0 \qquad \text{with} \qquad \mathfrak{t}^{ij} := \rho u^i u^j \qquad (6)$$

and $u^i := dx^i/ds$ as velocity of the dust. The fictitious nontensorial force density $\mathfrak{f}^k := \mathfrak{t}^{ij} \, \widetilde{\Gamma}_{ij}{}^k$, as observed by Weyl [18], is somewhat analogous to the Lorentz force acting on a charged particle in electrodynamics $\mathfrak{f}^k_{\text{Lor}} := \mathfrak{J}^i F_i{}^k$, with $\mathfrak{J}^i = \rho_{\text{el}} u^i$ as electric current density and F_{ik} as electromagnetic field strength, the difference being that here the force density $\mathfrak{f}^k$ is quadratic in u^i, whereas the Lorentz force density $\mathfrak{f}^k_{\text{Lor}}$ is linear in u^i; note also that the electromagnetic field is antisymmetric $F_{ik} = -F_{ki}$ and the gravitational field symmetric $\widetilde{\Gamma}_{ij}{}^k = +\widetilde{\Gamma}_{ji}{}^k$. Thus, as a by-product, we have identified the energy-momentum tensor density of matter as the source of gravity.

---

[2]A more detailed discussion can be found in Adler et al. [17], p. 351.

5 Neutron Interferometer Experiments

However, in the meantime, I mean since 1916, we have learned that there are fermions in nature. Besides mass m, they carry half-integer spin s. Instead of a mass point, we will then study the simplest massive fermion, the Dirac field in an inertial and a non-inertial reference frame thus taking care of Synge's verdict *"Newton successfully wrote apple = moon, but you cannot write apple = neutron."* This is what, in fact, Kibble [13] has done in 1961.

But even better, experimentally it has been clear since 1975 that the Colella–Overhauser–Werner (COW) experiment [19] is the "modern" archetypal experiment for a fermion in a gravitational field: A monochromatic neutron beam, extracted from a nuclear reactor, falls freely in the gravitational field of the earth. The phase shift of its wave function $\Psi(x)$, caused by the gravitational field, is measured by means of an interferometer built from a silicon crystal, see also [20]. Accordingly, the single-crystal interferometer is at rest with respect to the laboratory, whereas the neutrons are subject to the gravitational potential. Bonse and Wroblewski (BW) [21] compared this with the effect of *acceleration* relative to the laboratory frame by letting the interferometer oscillate horizontally. With these experiments of BW and COW the effect of local acceleration and local gravity on matter waves has been shown to be equivalent. Later, with atomic beam interferometry, the accuracy of these type of results were appreciably improved.

It is strange, but in most textbooks on gravitation—and in most philosophical discussions on gravity—these successful experiments on the behavior of Dirac fields under acceleration (BW) and in a gravitational field (COW) are simply not mentioned. Most textbook authors and philosophers rather restrict themselves to Einstein's 1916 discussion and to experiments related therewith. In writing a textbook on gravitation, is it indecent to refer to experiments that have a certain quantum flavor? Is it appropriate to be silent about experiments that provide new insight into the structure of the gravitational field?

The neutrons in the COW and BW experiments have spin $\frac{1}{2}$; they are fermions. At the energies prevalent in the COW and the BW experiments, the neutron (including its spin) can be supposed to be elementary, its composition out of three quarks can be neglected. Accordingly, if the neutron is force-free, it can be described by a Dirac spinor $\Psi(x)$ obeying the free Dirac equation[3] $(i\gamma^k \partial_k - m)\Psi(x) = 0$. Thus, the neutron obeys approximately a classical one-particle equation, namely the Dirac or, in the nonrelativistic limit, the Pauli–Schrödinger equation and, if the spin can be neglected, the Schrödinger equation. That this evaluation is correct has been borne out by experiments of the COW and BW type [20]: the neutrons of the COW and the BW experiments obey a Schrödinger equation including a Newtonian gravitational potential energy or a corresponding acceleration term, respectively.

---

[3]Here $\hbar = 1$, $c = 1$, the imaginary unit is denoted by i, the Dirac gamma matrices by γ^k, and the mass of the neutron by m. If an electromagnetic field is present, the Dirac equation has to be coupled minimally to it and a Pauli-term added that takes into account the non-standard magnetic moment of the neutron.

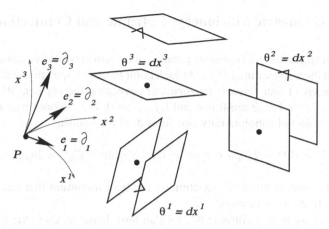

Fig. 1 Natural frame $e_b = \delta_b^j \partial_j$ and natural coframe $\vartheta^a = \delta_i^a dx^i$ at a point P of a three-dimensional manifold ($a, b = 1, 2, 3$). The coordinates of P are denoted by x^i, $i = 1, 2, 3$, whereas δ_a^b is the Kronecker symbol. The coframe ϑ^a is supposed to be also at the same point P, but the three one-forms ϑ^a are shifted for better visibility in three different directions. Note that $\vartheta^1(e_1) = 1$, $\vartheta^1(e_2) = 0$, etc., that is, ϑ^a is dual to e_b according to $e_b \lrcorner \vartheta^a \equiv \vartheta^a(e_b) = \delta_b^a$; for the figure, see [22]

The basic difference between the mass point and the Dirac field is that the latter requires an *orthonormal* reference frame for its description. A Dirac spinor is a half-integer representation of the [covering group $SL(2, C)$ of the] Lorentz group $SO(1, 3)$, that is, it is intrinsically tied to the Lorentz group. In Minkowski space, it is simple to introduce an orthonormal frame. On starts with Cartesian coordinates and takes the tangent vectors of the coordinate lines as "natural" frame $e_\beta = \delta_\beta^j \partial_j$, compare Figure 1. If one translates and Lorentz rotates such a frame, one can find an arbitrary frame $e_\beta = e^j{}_\beta \partial_j$ that, in general, cannot any longer be derived from coordinate lines. Before we discuss this from a more general point of view, let us first make a general remark:

Pitts [23] argues, using work of Ogievetsky and Polubarinov of the 1960s, that one does not require orthonormal frames for introducing spinors in curved spacetime and that coordinate systems are sufficient. Frames are very useful for Fermi-Walker transport and for gravitomagnetism already in GR. For the gauge theory of gravity, frames were used by Sciama and Kibble, see [11], and we can hardly see a benefit for kicking them out. The price one has to pay for the removal of frames is to go to nonlinear group representations and to other complications. We do not know whether this prevention of frames is really conclusive and leave the answer to this question to the future.

6 Some Geometric Machinery: Coframe and Connection

Suppose that spacetime is a four-dimensional continuum in which we can distinguish one time and three space dimensions. At each point P, we can span the local cotangent space by means of four *linearly independent* covectors, the *coframe* $\vartheta^\alpha = e_i{}^\alpha dx^i$. Here $\alpha, \beta, \ldots = 0, 1, 2, 3$ are frame and $i, j, \ldots = 0, 1, 2, 3$ coordinate indices. In general, the object of anholonomity two-form does not vanish,

$$C^\alpha := d\vartheta^\alpha = \tfrac{1}{2}C_{ij}{}^\alpha dx^i \wedge dx^j \neq 0, \qquad \text{with} \qquad C_{ij}{}^\alpha = 2\partial_{[i}e_{j]}{}^\alpha, \tag{7}$$

see [24]. This specification of spacetime is the bare minimum that one needs for applications to classical physics.

As soon as we have a coframe ϑ^α, we can also define its dual, the frame composed of four likewise linearly independent vectors $e_\alpha = e^i{}_\alpha \partial_i$ by the duality relation $e_\beta \rfloor \vartheta^a = \vartheta^\alpha(e_\beta) = \delta_\beta^\alpha$. Geometrically speaking, frame and coframe are equivalent as reference frames for physical quantities. For physical reasons, the coframe turns out to be the translational gauge type potential and thus does fit more smoothly into a gauge formalism.

Having now a reference coframe ϑ^α, we want to do physics in such a spacetime. We need a tool to express, for instance, that a certain field is constant. If the field is a scalar ϕ, there is no problem, the gradient $d\phi = (\partial_i\phi)dx^i$, if equated to zero, will do the job. However, if the field is a vector or, more generally, a spinor or an arbitrary tensor field ψ, we need a law that specifies the parallel transfer of ψ from one point P to a neighboring point P'. Let us see how Einstein in 1955 looked in retrospect at the development of GR [25]:

> ... the essential achievement of general relativity, namely to overcome "rigid" space (ie the inertial frame), is *only indirectly* connected with the introduction of a Riemannian metric. The directly relevant conceptual element is the "displacement field" (Γ^l_{ik}), which expresses the infinitesimal displacement of vectors. It is this which replaces the parallelism of spatially arbitrarily separated vectors fixed by the inertial frame (ie the equality of corresponding components) by an infinitesimal operation. This makes it possible to construct tensors by differentiation and hence to dispense with the introduction of 'rigid' space (the inertial frame). In the face of this, it seems to be of secondary importance in some sense that some particular Γ field can be deduced from a Riemannian metric...[4]

Einstein's "displacement field" can be implemented by means of a *linear connection* $\Gamma_\alpha{}^\beta = \Gamma_{i\alpha}{}^\beta dx^i$ ("affinity"). The one-form field $\Gamma_\alpha{}^\beta(x)$, with its 64 independent com-

---

[4]When I showed this quotation during my seminar, E. Scholz (Wuppertal) immediately remarked that the fact of the importance of the connection as guiding field was already clear to Weyl in 1918, or at least in the 1920s. And D. Rowe (Mainz) added that also Einstein was aware of the importance of the concept of a connection since at least the late 1920s. Both remarks are certainly true. However, there is a subtle difference: Weyl referred to a symmetric connection since he was concerned with coordinates and not with frames. When, in 1929, he introduced frames [26], Weyl's connection still remained symmetric, and only in 1950 he considered also *a*symmetric connections in the context of gravity [27]. In contrast, Einstein was concerned with *a*symmetric connections at least since 1925, when he formulated a unified theory of gravity and electricity and introduced what is nowadays called incorrectly the Palatini variational principle [28].

ponents, has to be prescribed before the parallel transport of a spinor or a tensor field ψ can be performed and, associated with it, a covariant derivative be defined (whose vanishing would imply that the field is constant). The linear connection $\Gamma_\alpha{}^\beta(x)$, shortly after the advent of general relativity, was recognized as a fundamental ingredient of spacetime physics, for more details see [11], for instance. The law of parallel transport embodies the *inertial properties* of matter.

The connection $\Gamma_\alpha{}^\beta$ represents 4×4 potentials of the four-dimensional group of general linear transformations $GL(4, R)$. Very similar to the Yang–Mills potential of the $SU(3)$, for example.

Coframe and connection ϑ^a, $\Gamma_\alpha{}^\beta$—still the metric is not involved—provide a good arsenal for further geometrical battles. Having a connection, we can covariantly differentiate. We define straightforwardly the "field strengths" torsion T^α and curvature $R_\alpha{}^\beta$ as

$$T^\alpha := d\,\vartheta^\alpha + \Gamma_\beta{}^\alpha \wedge \vartheta^\beta = \tfrac{1}{2} T_{ij}{}^\alpha dx^i \wedge dx^j, \tag{8}$$

$$R_\alpha{}^\beta := d\,\Gamma_\alpha{}^\beta - \Gamma_\alpha{}^\gamma \wedge \Gamma_\gamma{}^\beta = \tfrac{1}{2} R_{ij\alpha}{}^\beta dx^i \wedge dx^j. \tag{9}$$

One recognizes that T^α and $R_\alpha{}^\beta$ are the gauge field strengths of the affine group $A(4, R) = T(4) \rtimes GL(4, R)$.

Let us look at the torsion in components. From (8) we find

$$T_{ij}{}^\alpha = 2\partial_{[i}e_{j]}{}^\alpha + 2\Gamma_{[i|\beta}{}^\alpha e_{|j]}{}^\beta = C_{ij}{}^\alpha + 2\Gamma_{[ij]}{}^\alpha. \tag{10}$$

In a holonomic (coordinate) frame, $C_{ij}{}^\alpha = 0$. Thus, $T_{ij}{}^\alpha \overset{*}{=} 2\Gamma_{[ij]}{}^\alpha$; incidentally, a "star equal" $\overset{*}{=}$ is used, see [24], if a formula is only valid for a restricted class of frames or coordinates. In such a frame—and only in a holonomic one—the vanishing of the torsion translates into the *symmetry of the connection*. It is now obvious why this symmetry is called a "bastard symmetry." In $\Gamma_{[ij]}{}^\alpha = \Gamma_{[i|\beta}{}^\alpha e_{|j]}{}^\beta$, the index 'i' originates from the one-form character of the connection, whereas the index 'j' is related to the Lie-algebra index 'β'. Only in a holonomic frame the symmetry of a connection looks natural. In an anholonomic frame, here $C_{ij}{}^\alpha \neq 0$, it is nothing trivial. It is a fundamental assumption that has to be justified similar as the vanishing of the curvature.

A space with $T^\alpha \neq 0$, $R_\alpha{}^\beta \neq 0$, we call an *affine* space. If $T^\alpha = 0$, we have a *symmetric* affine space, if $R_\alpha{}^\beta = 0$, we have a *teleparallel* affine space (or of a space with teleparallelism). Should we require $T^\alpha = 0$ and $R_\alpha{}^\beta = 0$, we have a symmetric flat affine space.

We followed here the lead of Schrödinger [29] and introduced first the connection before we will turn to the metric.

7 More Geometry: Metric and Orthonormal Coframe

However, our experience in Minkowski space tells us that there must be more structure on the spacetime manifold than the symmetric flat affine space possesses. Locally at least, we are able to measure time and space intervals and angles. A pseudo-Riemannian (or Lorentzian) metric[5] $g_{ij} = g_{ji}$ is sufficient for accommodating these measurement procedures. If $g_{\alpha\beta}$ denotes the components of the metric with respect to the coframe, we have $g_{ij} = e_i{}^\alpha e_j{}^\beta g_{\alpha\beta}$ and $\mathbf{g} = g_{\alpha\beta}\,\vartheta^\alpha \otimes \vartheta^\beta$. In an orthonormal coframe we recover

$$g_{\alpha\beta} \stackrel{*}{=} o_{\alpha\beta} := \begin{pmatrix} 1 & 0 & 0 & 0 \\ 0 & -1 & 0 & 0 \\ 0 & 0 & -1 & 0 \\ 0 & 0 & 0 & -1 \end{pmatrix}. \tag{11}$$

Now, in analogy to the procedures in equations (8) and (9), we can derive the field strength, the *nonmetricity* one-form, corresponding to the potential $g_{\alpha\beta}$, by differentiation:

$$Q_{\alpha\beta} := -Dg_{\alpha\beta} = -dg_{\alpha\beta} + \Gamma_\alpha{}^\gamma g_{\gamma\beta} + \Gamma_\beta{}^\gamma g_{\alpha\gamma} = Q_{i\alpha\beta}dx^i. \tag{12}$$

Accordingly, the coframe $\vartheta^\alpha(x)$, the linear connection $\Gamma_\alpha{}^\beta(x)$, and the metric $g_{\alpha\beta}(x)$ control the geometry of spacetime. The metric determines the distances and angles, the coframe serves as translational gauge potential, whereas the connection provides the guidance field for matter reflecting its inertial properties and it is the $GL(4, R)$ gauge potential. The space equipped with these $10 + 16 + 64$ potentials $(g_{\alpha\beta}, \vartheta^\alpha, \Gamma_\alpha{}^\beta)$ we call a *metric-affine* space, the corresponding field strength are the $40 + 24 + 96$ fields $(Q_{\alpha\beta}, T^\alpha, R_\alpha{}^\beta)$, for reviews and the corresponding formalism, see [11, 31, 32].

In a metric-affine space, we can lower the second index of the connection according to $\Gamma_{\alpha\beta} := \Gamma_\alpha{}^\gamma g_{\gamma\beta}$. Then we can compare it with the Riemann (Levi-Civita) connection $\widetilde{\Gamma}_{\alpha\beta}$. After some algebra, see [24], we find in terms of components:

$$\Gamma_{\alpha\beta\gamma} = \widetilde{\Gamma}_{\alpha\beta\gamma} + \tfrac{1}{2}(T_{\alpha\beta\gamma} - T_{\beta\gamma\alpha} + T_{\gamma\alpha\beta}) + \tfrac{1}{2}(Q_{\alpha\beta\gamma} + Q_{\beta\gamma\alpha} - Q_{\gamma\alpha\beta}). \tag{13}$$

It should be stressed that this decompositions are useful if a direct comparison is made with the Riemannian piece $\widetilde{\Gamma}$. However, in the variational formalism of a gauge theory of gravity, besides $g_{\alpha\beta}$ and ϑ^α, the connection $\Gamma_\alpha{}^\beta$ is considered as *independent* variable. Then such a decomposition is unwarranted under those circumstances.

Can we give a satisfactory justification for the emergence of three different gravitational gauge potentials? We take the Minkowski space of SR as basis for our

---

[5]Nowadays there exists a definite hint that the conformally invariant part of the metric, the light cone, is electromagnetic in origin (see [22, 30]), that is, it can be derived from premetric electrodynamics together with a linear constitutive law for the empty spacetime (vacuum). Hence the metric, or at least its conformally invariant part, doesn't appear as a fundamental structure, it rather emerges in an electromagnetic context.

considerations. It is a fact of life that the geometry of a Minkowski (or a Euclidean) space consists of an interplay between properties that relate to parallel displacement and those that relate to distance and angle measurements. In Minkowski space this duality between affine (inertial) and metric properties is solved in that the affine properties are exclusively expressed in terms of metric properties: the metric properties dominate the affine ones.

If we "liberate" the affine properties, we are immediately led, in four dimensions, to the affine group $A(4, R) = T(4) \rtimes GL(4, R)$ and, gauging it, to the coframe ϑ^α and the linear connection $\Gamma_\alpha{}^\beta$ as gauge potentials. The metric properties, expressed by the metric g_{ij}, are then left behind.

Since macroscopic gravity in GR is so successfully described by means of the metric g_{ij} as (Einstein's) gravitational potential, it suggests itself to add the metric—in its anholonomic form $g_{\alpha\beta}$—as third member to the gravitational potentials. There are two procedures possible: We pick, instead of an arbitrary coframe, an *orthonormal* one, which is constructed with the help of the metric; in this way the metric is absorbed and, besides this orthonormal coframe, only the connection remains as variable. However, since this restricts the freedom of choosing also non-orthonormal coframes, we take all three potentials as independent variables. The Lagrangian formalism of the corresponding field theory will then provide the relation between the coframe and the metric, and it will turn out that there is, indeed, a close link between both variables, see [32]. At the same time—and this is a real progress in understanding—we find that the metric energy-momentum current of matter $t^{\alpha\beta}$ couples to the metric and the canonical one $\mathfrak{T}_\alpha$ couples to the coframe. Their interdependence is beautifully displayed in the three-potentials' approach.

In a metric-affine space, as shown by Hartley [33], *normal frames* can be found: *locally* it is possible to find suitable coordinates and suitable frames such that

$$(\vartheta^\alpha, \Gamma_\alpha{}^\beta) \stackrel{*}{=} (\delta_i^\alpha dx^i, 0) . \tag{14}$$

This is the new type of Einstein elevator. In GR, the Einstein elevator was described by a holonomic reference frame ϑ^α with $C^\alpha = 0$. Then, in the Riemann spacetime of GR, one could introduce Riemannian normal coordinates. Here, in the gauge theoretical approach, the constraint of holonomicity is dropped and this new degree of freedom, which expresses itself in a rotational acceleration, admits to introduce normal frames. The equivalence principle can then be applied in this new context. For new developments of this notion, see Nester [34] and Giglio and Rodrigues [35].

As soon as we require in a metric-affine space integrability of length and angle measurements, we have to postulate[6] $Q_{\alpha\beta} = 0$. Then we arrive at a *Riemann–Cartan* space (RC-space), which was mentioned in the context of the gauging of the Poincaré group in Sec. 3. In such a space, if we choose *orthonormal* frames, the connection

---

[6]If one wants to keep the angles integrable, but not the length, one can postulate only the vanishing of the tracefree part of the nonmetricity, $Q_{\alpha\beta} - \frac{1}{4}g_{\alpha\beta}Q_\gamma{}^\gamma = 0$. This results in a Weyl–Cartan space with nonvanishing Weyl covector $\frac{1}{4}Q_\gamma{}^\gamma$, see the contribution of Scholz [36]; however, in this approach also the torsion is put to zero.

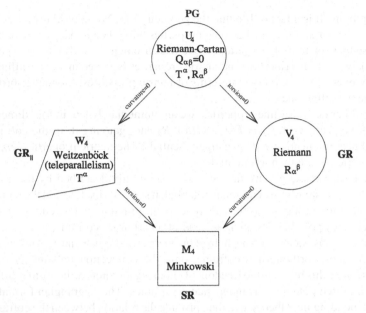

Fig. 2 The Riemann–Cartan space (or U_4), a metric-affine space with vanishing nonmetricity, is the arena for the Poincaré gauge theory (PG). It can either become a *Weitzenböck* space W_4, if its curvature vanishes, or a *Riemann* space V_4, if the torsion happens to vanish. GR acts in a V_4, teleparallelism theories of gravity in a W_4

becomes antisymmetric; then we have the $6 + 24$ potentials $(\vartheta^a, \Gamma^{\alpha\beta} = -\Gamma^{\beta\alpha})$ as gravitational variables. Again normal frames like in (14) can be found

$$(\vartheta^\alpha, \Gamma^{\alpha\beta}) \overset{*}{=} (\delta_i^\alpha dx^i, 0) \,, \tag{15}$$

as has been first shown by von der Heyde [37]. This geometrical fact shows clearly that the Lorentz connection $\Gamma^{\alpha\beta}$ is, besides the orthonormal frame ϑ^α, the appropriate gauge field variable. After this geometrical detour we are back to where we started from. In Figure 2 different subcases of a RC-space are displayed.

8 Dirac Field $\Psi(x)$ in Minkowski Space in a Non-inertial Reference Frame

After this rather long geometrical interlude, we come back to physics and consider again the Dirac field $\Psi(x)$. A mass point in an inertial frame moves according to equation (2), one in an accelerated frame, according to equation (3). The inertial forces are represented by the Christoffels in equation (4). Let us execute the analogous process for the Dirac electron. Since the Dirac electron is referred to an orthonormal (co)frame, we have to study its behavior under translational and rotational accelerations, see [38].

In Minkowski space in Cartesian coordinates, we have the force-free Dirac equation as analog of equation (17),

$$(i\gamma^i\partial_i - m)\Psi \overset{*}{=} 0,$$ (16)

and in a non-inertial frame in flat Minkowski space we find

$$\left[i\gamma^\alpha e^i{}_\alpha(\partial_i + \frac{i}{4}\sigma_{\beta\gamma}\widetilde{\Gamma}_i{}^{\beta\gamma}) - m\right]\Psi = 0, \qquad \sigma_{\beta\gamma} := i\gamma_{[\beta}\gamma_{\gamma]}.$$ (17)

These two equations correspond to the Einsteinian equations (2) and (3). Namely, in the nonrelativistic WKB-approximation, when the spin can be neglected, equation (16) becomes (2). You may wonder whether this is true since (16), in contrast to (2), is mass dependent. For this reason, some people argued that this violates the equivalence principle in the sense that the motion of a force-free particle (field) must be independent of m. However, what they overlooked is that also in classical mechanics the Hamilton–Jacobi equation for a force-free particle *is* mass dependent—and the classical nonrelativistic analog of the Dirac equation is the Hamilton–Jacobi equation. Accordingly, all is fine and in the desired approximation the mass will drop out.

The new potentials, emerging in a non-inertial frame, are $(e^i{}_\alpha, \widetilde{\Gamma}_i{}^{\beta\gamma})$. The latter one, in Minkowski space, can be expressed in terms of derivatives of the former: $\widetilde{\Gamma}_i{}^{\beta\gamma} = \widetilde{\Gamma}_i{}^{\beta\gamma}(\partial_j e^k{}_\delta)$. However, we will not substitute $\widetilde{\Gamma}_i{}^{\beta\gamma}$ in terms of the frame since we will relax the constraint $T_{ij}{}^\alpha = 0$ subsequently.

This is what we will do now. Einstein relaxed the constraint $\widetilde{R}_{ijk}{}^\ell = 0$, since that is all he found for a point particle, we relax the constraints $T_{ij}{}^\alpha = 0$ and $\widetilde{R}_{ij}{}^{\alpha\beta} = 0$, since a Dirac field has a more involved structure as displayed in particular in a non-inertial frame. This relaxation of both constraints leads directly to a *Riemann–Cartan spacetime* as the arena appropriate for a Poincaré gauge theory (PG).

Why could not we do by only relaxing the curvature constraint, $\widetilde{R}_{ij}{}^{\alpha\beta} \neq 0$, but keeping the torsion constraint, $T_{ij}{}^\alpha = 0$? Well, this is possible. However, it is not in the sense of local field theory. Why should we keep the nonlocal constraint[7] $T_{ij}{}^\alpha = 0$, which corresponds to 24 partial differential equations of first order, when we know that its relaxation does away with these PDEs and still allows locally to get rid of gravity according to (15)?

Whereas Einstein discussed the equivalence principle on the level of the equations of motion, in gauge theories, because of the application of the Noether theorem for rigid and local symmetries, the discussion takes place on the level of Lagrangians.

---

[7]Explicitly, this constraint reads $T_{ij}{}^\alpha = 2(\partial_{[i}e_{j]}{}^\alpha + \Gamma_{[i|\beta|}{}^\alpha e_{j]}{}^\beta) = 0$. These are $6 \times 4 = 24$ PDEs for the coframe components $e_i{}^\alpha$. For their solution, we not only have to know the local values of $e_i{}^\alpha$, but also their values in the infinitesimal neighborhood. In this sense, the constraint is nonlocal and contrived, see [37] for a more detailed discussion.

If we multiply $D_\alpha \Psi = (\partial_\alpha + \frac{i}{4}\sigma_{\beta\gamma}\Gamma_\alpha{}^{\beta\gamma})\Psi$ from the left by $i\overline{\Psi}\gamma^\alpha$, average with its Hermitian conjugate, and add a mass term, we find the (real) Dirac Lagrangian density:[8]

$$\frac{\mathfrak{L}}{e} = \frac{i}{2}e^i{}_\alpha\left[\overline{\Psi}\gamma^\alpha\left(\partial_i + \frac{i}{4}\sigma_{\beta\gamma}\Gamma_i{}^{\beta\gamma}\right)\Psi\right] + \text{herm. conj.} + m\overline{\Psi}\Psi$$

$$= \frac{i}{2}(\overline{\Psi}\gamma^\alpha D_\alpha\Psi - \Psi\gamma^\alpha\overline{D_\alpha\Psi}) + m\overline{\Psi}\Psi. \tag{18}$$

The action is $W = \int d^4x\mathfrak{L}$. The Lagrangian in an inertial frame in Cartesian coordinates can be read off by making the substitutions $e^i{}_\alpha \to \delta^i_\alpha$, $\Gamma_i{}^{\beta\gamma} \to 0$.

9 Some Results of the Lagrange–Noether Formalism

To identify the currents that couple to the gravitational potentials $(e_i{}^\alpha, \Gamma_i{}^{\alpha\beta})$, some formalism is necessary that may disturb the philosophically minded reader. We try to simplify these considerations and will, instead of working in a RC-spacetime (for a rigorous treatment see [32]), restrict ourselves to the Minkowski space in Cartesian coordinates.

The action W is invariant under 4 rigid spacetime *translations* of 6 rigid Lorentz *rotations* (3 boosts plus 3 spatial rotations). As a consequence, we have (see Corson [39]) energy-momentum and angular momentum conservation,

$$\partial_k\mathfrak{T}_i{}^k \overset{*}{=} 0, \qquad \partial_k\mathfrak{J}_{ij}{}^k \overset{*}{=} 0, \tag{19}$$

with the *canonical energy-momentum* tensor density

$$\mathfrak{T}_i{}^k := \overset{*}{} \delta_i^k\mathfrak{L} - \frac{\partial\mathfrak{L}}{\partial\partial_k\Psi}\partial_i\Psi, \tag{20}$$

the total canonical angular momentum tensor density, consisting of an intrinsic and an orbital part,

$$\mathfrak{J}_{ij}{}^k := \overset{*}{} \mathfrak{S}_{ij}{}^k + x_i\mathfrak{T}_j{}^k - x_j\mathfrak{T}_i{}^k = -\mathfrak{J}_{ji}{}^k, \tag{21}$$

and the *canonical spin* angular momentum tensor density (l_{ij} = Lorentz generators)

$$\mathfrak{S}_{ij}{}^k := \overset{*}{} \frac{\partial\mathfrak{L}}{\partial\partial_k\Psi}l_{ij}\Psi = -\mathfrak{S}_{ji}{}^k. \tag{22}$$

From this straightforward consideration in Minkowski space alone, we recognize that the canonical energy-momentum $\mathfrak{T}_i{}^k$ and the canonical angular momentum $\mathfrak{J}_{ij}{}^k$

---

[8]$e := \det e_i{}^\alpha$, $\partial_\alpha = e^i{}_\alpha\partial_i$, $D_\alpha = e^i{}_\alpha D_i$.

are the translational and the Lorentz currents of matter. Only the intrinsic spin part $\mathfrak{S}_{ij}{}^k$ of the angular momentum is a tensor; the orbital part is only a tensor under Cartesian coordinate transformations. For the Dirac field we find ($\mathfrak{T}_\alpha{}^k = e^i{}_\alpha \mathfrak{T}_i{}^k$, etc.):

$$\mathfrak{T}_\alpha{}^i \overset{*}{=} \frac{i}{2}\left(\overline{\Psi}\gamma^i\partial_\alpha\Psi - \Psi\gamma^i\partial_\alpha\overline{\Psi}\right), \tag{23}$$

$$\mathfrak{S}_{\alpha\beta}{}^i \overset{*}{=} \frac{1}{8}\overline{\Psi}\left(\sigma_{\alpha\beta}\gamma^i + \gamma^i\sigma_{\alpha\beta}\right)\Psi. \tag{24}$$

The spin is totally antisymmetric: after some algebra, we can put (24) into the form

$$\mathfrak{S}_{\alpha\beta\gamma} = \frac{1}{4}\varepsilon_{\alpha\beta\gamma\delta}\overline{\Psi}\gamma^\delta\gamma_5\Psi, \tag{25}$$

with $\gamma_5 := -\frac{i}{4!}\varepsilon_{\alpha\beta\gamma\delta}\gamma^\alpha\gamma^\beta\gamma^\gamma\gamma^\delta$.

We compare (23) and (24) with the Lagrangian (18) and consider small deviations from the inertial case, that is, $e^i{}_\alpha = \delta^i{}_\alpha + \varepsilon^i{}_\alpha$, with $\varepsilon^i{}_\alpha \ll 1$, then we find after some algebra and some rearrangements to linear order in $\varepsilon^i{}_\alpha$,

$$\mathfrak{L} \sim e_i{}^\alpha\mathfrak{T}_\alpha{}^i + \Gamma_i{}^{\alpha\beta}\mathfrak{S}_{\alpha\beta}{}^i - m\overline{\Psi}\Psi. \tag{26}$$

There is some resemblance to the structure in (6) even though we work here on a Lagrangian level. This coupling of geometry to matter displayed in (26) suggests the following representation of the canonical currents:

$$\mathfrak{T}_\alpha{}^i = \frac{\delta\mathfrak{L}}{\delta e_i{}^\alpha}, \qquad \mathfrak{S}_{\alpha\beta}{}^i = \frac{\delta\mathfrak{L}}{\delta\Gamma_i{}^{\alpha\beta}}. \tag{27}$$

Of course, this was a heuristic consideration, but with the full Lagrange-Noether machinery acting in RC-spacetime, it can be made rigorous [32]: The canonical currents $\mathfrak{T}_\alpha{}^i$, $\mathfrak{S}_{\alpha\beta}{}^i$, defined via the Noether theorem according to (20) and (22), can be shown to be equal to the "dynamical" currents that couple to the gravitational potentials according to (27). These currents should also play a decisive role in quark and gluon physics, see [40].

A short summary of the formalism in this section

For those of you who were lost in this formalism, a short bird eye's view on the results: In order to compactify our notation, we change to exterior calculus. We introduce the matrix-valued one-form $\gamma := \gamma_\alpha\vartheta^\alpha$ and the Hodge star operator $*$. Then the Dirac equation in an arbitrary orthonormal frame in a RC-space can be rewritten as

$$i^*\gamma \wedge D\Psi + {}^*m\Psi = 0, \tag{28}$$

with the covariant exterior derivative $D\Psi := (d + \frac{i}{4}\sigma_{\alpha\beta}\Gamma^{\alpha\beta})\Psi$. Let us then formulate the twisted Lagrange four-form of the Dirac field,

$$L = L(\vartheta^\alpha, \Psi, D\Psi) = \frac{i}{2}\left(\overline{\Psi}\,^*\gamma \wedge D\Psi + \overline{D\Psi} \wedge {}^*\gamma\,\Psi\right) + {}^*m\,\overline{\Psi}\Psi\,, \qquad (29)$$

which is minimally coupled to the RC-spacetime via the gauge potentials ϑ^α [contained in $\gamma = \gamma_\alpha\vartheta^\alpha$] and $\Gamma^{\alpha\beta} = -\Gamma^{\beta\alpha}$ [contained in D]. Note that only the potentials themselves enter the Lagrangian, but not their derivatives. Thus, the Lagrangian (29), formulated in a RC-spacetime, because of (15), looks *locally* special-relativistic. This attests to the validity of the relaxation process discussed above. The currents, as we saw above in (27), are then defined as follows:

$$\mathfrak{T}_\alpha = \frac{\delta L}{\delta\vartheta^\alpha}\,, \qquad \mathfrak{S}_{\alpha\beta} = \frac{\delta L}{\delta\Gamma^{\alpha\beta}}\,. \qquad (30)$$

These innocently looking equations (29) and (30), all living in a RC-spacetime, are the net outcome of our considerations so far.

It was then Sciama [41] and Kibble [13] in the early 1960s who added the Hilbert–Einstein type Lagrangian of the RC-spacetime to (18) and formulated the corresponding simplest field equations of the gauge theory of gravity; for a historical view see O'Raifeartaigh [42] and the reprint volume [11], for a modern representation Blagojević [43] and Ryder [44].

10   Field Equations of Sciama and Kibble

The Ricci tensor in a RC-spacetime is defined according to $\mathrm{Ric}_i{}^\alpha := e^j{}_\beta R_{ji}{}^{\alpha\beta}$. A corresponding scalar density $ee^i{}_\alpha\mathrm{Ric}_i{}^\alpha$ is the simplest nontrivial gravitational Lagrangian. The total action is (Λ = cosmological constant)

$$W_{\text{tot}} = \int d^4x \left[\frac{1}{2\kappa}e(e^i{}_\alpha\mathrm{Ric}_i{}^\alpha - 2\Lambda) + \mathfrak{L}(e_k{}^\gamma, \Psi, D\Psi)\right], \qquad (31)$$

with Einstein's gravitational constant κ. Variation with respect to $e_i{}^\alpha$ and $\Gamma_i{}^{\alpha\beta}$ yields the gravitational field equations of Sciama [41] and Kibble [13]:

$$\mathrm{Ric}_\alpha{}^i - \tfrac{1}{2}e^i{}_\alpha\,\mathrm{Ric}_\gamma{}^\gamma + \Lambda e^i{}_\alpha = \frac{\kappa}{e}\,\mathfrak{T}_\alpha{}^i\,, \qquad (32)$$

$$\mathrm{Tor}_{\alpha\beta}{}^i + e^i{}_\alpha\mathrm{Tor}_{\beta\gamma}{}^\gamma - e^i{}_\beta\mathrm{Tor}_{\alpha\gamma}{}^\gamma = \frac{\kappa}{e}\,\mathfrak{S}_{\alpha\beta}{}^i\,. \qquad (33)$$

We made here the torsion a bit more visible. Please note that Ric and $\mathfrak{T}$ have both 16 independent components, whereas Tor and $\mathfrak{S}$ have both 24 independent components. These field equations are just linear algebraic equations between Ric and Tor on the geometrical side and $\mathfrak{T}$ and $\mathfrak{S}$ on the matter side, respectively. The Dirac case is particularly simple, there (33) collapses to just four equations.

The first equation can be easily recognized as an Einstein type field equation. However, the Ricci tensor is here asymmetric as well as the canonical energy-momentum tensor of matter. The second equation relates the torsion linearly to the spin of matter. If we consider matter *without* spin, the torsion vanishes and the first field equation reduces just to the Einstein field equation of GR, for a review see [45].

In exterior calculus, these field equations, given first in this form by Trautman, see [46], look even a bit more transparent:[9]

$$\tfrac{1}{2}\eta_{\alpha\beta\gamma} \wedge R^{\beta\gamma} - \Lambda\eta_\alpha = \kappa\,\mathfrak{T}_\alpha \,, \tag{34}$$

$$\tfrac{1}{2}\eta_{\alpha\beta\gamma} \wedge T^\gamma = \kappa\,\mathfrak{S}_{\alpha\beta} \,. \tag{35}$$

The two equations (32), (33) or (34), (35) are the field equations of the Einstein–Cartan(–Sciama–Kibble) theory of gravity or, in short, of the *Einstein–Cartan theory* (EC). This is a special case of a Poincaré gauge theory, namely that which has the curvature scalar of the RC-spacetime as gravitational Lagrangian. EC is a viable gravitational theory.

The Maxwell field carries helicity, that is, spin projected along its wave vector, but is does not carry spin proper as a gauge covariant quantity. Therefore, there is no electromagnetic contribution to the material spin on the right-hand side of (33) or (35). Light is insensitive to torsion; torsion cannot be "seen."[10]

Torsion effects in EC-theory are minute. Besides the Einsteinian gravitational field, we have additionally a very weak *spin–spin contact interaction* that is proportional to the gravitational constant, which is measurable in principle. For a particle of mass m and reduced Compton wave length $\lambda_{Co} := \hbar/mc$ (with $\hbar$ = reduced Planck constant, c = speed of light), there exists in EC a critical density and, equivalently, a critical radius of ($\ell_{P\ell}$ = Planck length)

$$\rho_{EC} \sim m/(\lambda_{Co}\ell_{P\ell}^2) \quad \text{and} \quad r_{EC} \sim (\lambda_{Co}\ell_{P\ell}^2)^{1/3} \,, \tag{36}$$

respectively, see [45]. For a nucleon we have $\rho_{EC} \approx 10^{54}$ g/cm^3 and $r_{EC} \approx 10^{-26}$ cm. Whereas those densities are extremely high from a usual lab perspective or even from the point of view of a neutron star ($\approx 10^{16}$ g/cm^3), in cosmology they are standard. It may be sufficient to recall that inflation is believed to set in around the Planck density of 10^{93} g/cm^3.

At densities higher than ρ_{EC}, EC-theory is expected to overtake GR. There is no reason why GR should survive under those conditions, since *for fermions* the gauge theoretical framework seems more trustworthy. Some cosmological models of EC can be found in [11].

---

[9]Here we have: Hodge star *, $\eta_\alpha = {}^\star\vartheta_a$, $\eta_{\alpha\beta} = {}^\star(\vartheta_a \wedge \vartheta_\beta)$, $\eta_{\alpha\beta\gamma} = {}^\star(\vartheta_a \wedge \vartheta_\beta \wedge \vartheta_\gamma)$. Moreover, $\mathfrak{T}_\alpha = \tfrac{1}{e}\mathfrak{T}_\alpha{}^\gamma\eta_\gamma$ and $\mathfrak{S}_{\alpha\beta} = \tfrac{1}{e}\mathfrak{S}_{\alpha\beta}{}^\gamma\eta_\gamma$.

[10]Only a nonminimal coupling of the electromagnetic field to torsion-square pieces is conceivable, see [47].

It is probably fair to say that EC has been established as a consistent and viable theory of gravity and the Riemann–Cartan geometry of spacetime has won solid support so that its study should not be skipped in philosophical circles as undesirable complication of the Riemann geometry of GR.

11   Quadratic Poincaré Gauge Theory of Gravity (qPG)

Let me first express a word of caution: In a fairly recent paper, Mao et al. [48] believe to have shown "… that Gravity Probe B is an ideal experiment for further constraining nonstandard torsion theories, …" Nothing could be further away from the truth. Following the guiding principle that nothing is more practical than a good theory, Puetzfeld et al. [49, 50] have shown that the measurement of torsion requires elementary particle spins as test objects whereas in Gravity Probe B the rotating quartz balls carry orbital angular momentum only, but do not carry uncompensated elementary particle spin. Thus, the results in [48] are simply incorrect in spite of the wide publicity that this paper has won.

But back to Einstein–Cartan theory (EC). It is in many ways a very degenerate theory. A contact interaction in physics cries for a generalization to a propagating interaction, as has been the way things developed in the Fermi theory of weak interaction—which was a contact interaction par excellence—to the theory of the propagating W and Z. The recipe is very simple: The EC-Lagrangian is linear in the Lorentz field strength, add terms that are quadratic in the translational field strength (torsion), and the Lorentz field strength (curvature).

Instead of boring you with all the details of this development to quadratic Lagrangians in a RC-spacetime and who did what and when and why, I shock you again with a messy formula. This is the most general quadratic Lagrangian including parity violating pieces (see [9] and the explanations in the subsequent paragraphs):

$$V = \frac{1}{2\kappa}[(a_0 R + b_0 X - 2\Lambda)\,\eta \tag{37}$$

$$+ \tfrac{1}{3}a_2 \mathscr{V} \wedge {}^*\mathscr{V} - \tfrac{1}{3}a_3 \mathscr{A} \wedge {}^*\mathscr{A} - \tfrac{2}{3}\sigma_2 \mathscr{V} \wedge {}^*\mathscr{A} + a_1{}^{(1)}T^\alpha \wedge {}^{*(1)}T_\alpha]$$

$$- \frac{1}{2\rho}\left[(\tfrac{1}{12}w_6 R^2 - \tfrac{1}{12}w_3 X^2 + \tfrac{1}{12}\mu_3 RX)\,\eta + w_4{}^{(4)}R^{\alpha\beta} \wedge {}^{*(4)}R_{\alpha\beta}\right.$$

$$\left. + {}^{(2)}R^{\alpha\beta} \wedge (w_2{}^{*(2)}R_{\alpha\beta} + \mu_2{}^{(4)}R_{\alpha\beta}) + {}^{(5)}R^{\alpha\beta} \wedge (w_5{}^{*(5)}R_{\alpha\beta} + \mu_4{}^{(5)}R_{\alpha\beta})\right].$$

The first two lines represent *weak* gravity, with the conventional gravitational constant κ, the last two lines speculative *strong* gravity with the dimensionless strong gravity constant ρ. The unknown constants $(a_0; a_1, a_2, a_3; b_0, \sigma_2)$, weigh the different terms of weak gravity, the unknown constants $(w_2, w_3, w_4, w_5, w_6; \mu_2, \mu_3, \mu_4)$ those of strong gravity. What a mess!

But let us discuss the formula line by line: In the *first line* R is the EC-term, $X := \frac{1}{4}\eta_{\alpha\beta\gamma\delta}R^{[\alpha\beta\gamma\delta]}$ is the (parity violating) curvature pseudoscalar, which vanishes in Riemannian space, but is nonvanishing in RC-space. This term is presently very popular in the quantum gravity scene, Λ is the cosmological constant, and η the "volume element."

The *second line* houses all torsion-square pieces. We have a tensor torsion $^{(1)}T^\alpha$, a vector torsion $\mathscr{V}$, and an axial vector torsion $\mathscr{A}$. They can enter in the combinations shown. The remarkable fact is that for dimensional reasons the first line and the second line give rise to similar effects. Instead of the EC-theory with R, you can select a suitable linear combination of torsion-square pieces acting in a RC-space with vanishing curvature (Weitzenböck space), see, for example, Itin [51] or [11] and the historical article of Sauer [52]. On the first two lines there are literally hundreds of published papers studying different properties. Numerous printed pages could be saved, if our colleagues would start with the first two lines right away and just motivate their choice of the unknown constants.

Now we turn to the remaining more speculative pieces, which are, however, fairly plausible due to their Yang–Mills type structure. After all, C.N. Yang himself proposed such a theory [53]. We are not in bad company! In the *third line* we turn our attention immediately to the first three pieces: They are just squares built from the curvature scalar and/or the curvature pseudoscalar. The curvature in a RC-space $R^{\alpha\beta}$ decomposes into six irreducible pieces $^{(I)}R^{\alpha\beta}$: they are numbered by I, running from one to six. The pseudoscalar X is number three, the scalar R number six. The last term in the third line is then a square piece of number four. In the *fourth line* we have the remaining curvature square pieces. The term with number one drops out due to certain identities.

This is only algebra. Where is the physics? you may ask. Well, we have to find out. It will be a task of the future to single out of this set of quadratic Lagrangians (37) the physically acceptable one. How such possible developments may look like, I will illustrate with one example. Shie–Nester–Yo [54] developed a fairly realistic cosmological model of Friedmann type with *propagating connection* by picking the Lagrangian

$$V_{\text{SNY}} = \underbrace{\frac{1}{2\kappa}\left(a_0 R\,\eta + a_1{}^{(1)}T^\alpha \wedge {}^{\star(1)}T_\alpha\right)}_{\text{weak Newton-Einstein gravity}} - \underbrace{\frac{w_6}{24\rho}R^2\,\eta}_{\text{strong YM-type gravity}} \quad . \tag{38}$$

They found two conventional graviton helicities, as in GR, and this, for $a_0 \neq a_1$, combined with a torsion mode of mass of $\mu := a_1 - a_0$ and spin 0^+ (spin zero with positive parity, that is, an ordinary scalar), which has many attractive features. Of course, equation (38) is a subcase of equation (37). In the meantime this paper has been generalized by including parity violating pieces, inter alia, and it has been numerically evaluated. This paper has about 45 follow-up papers. In this way one collects more and more insight into the possible physics behind the most general quadratic PG-Lagrangian.

12 Outlook

What is the benefit of all of that for the theory of spacetime? Well, it is a small but decisive step beyond the established Riemannian spacetime structure of GR. Cartan's torsion has been incorporated into the body of knowledge of classical spacetime geometry. At the same time it has been demonstrated that the Poincaré group $P(1, 3) = T(4) \rtimes SO(1, 3)$, acting in the Minkowski space, and the behavior of the Dirac field in non-inertial frames leads, via the gauge principle, to the Riemann–Cartan geometry of spacetime. That is, the $P(1, 3)$ symmetry induced the Riemann–Cartan geometry.

The generalization of this procedure seems to be straightforward. If we add the group of dilations to $P(1, 3)$, assuming scale covariance in addition to the $P(1, 3)$ covariance, we arrive at the 11 parametric Weyl group. Gauging it, requires one more potential, namely the Weyl covector Q, defined in terms of the nonmetricity according to $Q := \frac{1}{4} Q_\gamma{}^\gamma = \frac{1}{4} g^{\alpha\beta} Q_{\alpha\beta}$, see equation (12). Associated with it comes a conservation law and the Noether current $\Delta = \delta L / \delta Q$, the dilation or scale current, which Weyl had mistaken for the electric current. If we turn the crank, a Weyl–Cartan spacetime emerges together with a gauge field equation that has the dilation current as source. This is standard Weyl lore from a contemporary point of view, see [11], Chapter 8.

I hope it does not take you by surprise that I cannot see much common ground with the theory of E. Scholz presented during this workshop [36]. In his approach, spacetime is governed by a Weyl geometry with vanishing torsion, but the dilation current is not an inhabitant of the Weyl space of Scholz—or, at least, this current has not been identified as such and lives anonymously and drifts around uncontrolled by any field equation.

Instead, one can add to the $P(1, 3)$ simple supersymmetry (symmetry between fermions and bosons) by extending the Poincaré algebra with anticommuting fermionic generators thus being led to a Poincaré superalgebra. The corresponding gauge procedure creates a so-called superspace(time) geometry. The field equations of simple supergravity can be immediately written down by using the EC-field equations (34) and (35); as sources one takes the energy-momentum and the spin currents of the massless Rarita–Schwinger field, which carries spin $\frac{3}{2}$. The Rarita–Schwinger field conspires with the effective spin two of the EC-field to build up a super multiplet $(2, \frac{3}{2})$, compare [11], Chapter 12.

In this way, we see that also in supersymmetry the gauge concept of Weyl and Yang–Mills–Utiyama is successful. And the geometry of spacetime turned out to have a potential "super" structure beyond Riemann–Cartan geometry.

Mielke [55] generalized the Poincaré group $T(4) \rtimes SO(1, 3)$ to the $SL(5, R)$ and recovered by symmetry breaking reasonable 4-dimensional gravitational gauge structures. This could be a future-pointing approach.

Acknowledgements I would like to thank Dennis Lehmkuhl, Erhard Scholz, and Gregor Schiemann most sincerely for the invitation to this workshop, for the hospitality, and for the lively discussions. I am also grateful to Peter Baekler (Düsseldorf), Milutin Blagojević (Belgrade), Yakov Itin

(Jerusalem), Claus Kiefer (Cologne), Bahram Mashhoon (Columbia, Missouri), Eckehard Mielke (Mexico City), James Nester (Chung-li), Yuri Obukhov (Cologne/Moscow), and Dirk Puetzfeld (Bremen) for many oral or email discussion on the subject of gravity and gauging. The criticism of two referees also helped to improve this article. I am particularly grateful to J. Brian Pitts (Cambridge, UK) for detailed remarks and suggestions. This work was supported by the German-Israeli Foundation for Scientific Research and Development (GIF), Research Grant No. 1078-107.14/2009.

References

1. J. Ehlers and C. Lämmerzahl (eds.), *Special Realtivity. Will it survive the next 100 years?* (Springer, Berlin, 2006).
2. B. Mashhoon, *Necessity of acceleration-induced nonlocality,* Annalen der Physik (Berlin) **523**, 226–234 (2011) [http://arXiv.org/pdf/1006.4150].
3. F. W. Hehl and B. Mashhoon, *Nonlocal gravity simulates dark matter,* Phys. Lett. B **673**, 279–282 (2009) [http://arXiv.org/pdf/0812.1059].
4. F. W. Hehl and B. Mashhoon, *A Formal framework for a nonlocal generalization of Einstein's theory of gravitation,* Phys. Rev. D **79**, 064028 (2009) [13 pages] [http://arXiv.org/pdf/0902.0560].
5. S. Rahvar and B. Mashhoon, *Observational tests of nonlocal gravity: galaxy rotation curves and clusters of galaxies,* Phys. Rev. D **89**, 104011 (2014) [27 pages] [http://arXiv.org/pdf/1401.4819].
6. B. Famaey et al. (organizers), *Modified Gravity Approaches to the Dark Sector,* Conference at the Observatoire Astronomique de Strasbourg, 28 June to 1 July 2010; for the talks see [http://astro.u-strasbg.fr/MGAtotheDARK/Home.html].
7. C. Kiefer, *Quantum Gravity,* 3rd ed. (Oxford University Press, Oxford, UK, 2012).
8. D. Diakonov, A. G. Tumanov and A. A. Vladimirov, *Low-energy General Relativity with torsion: A Systematic derivative expansion,* Phys. Rev. D **84**, 124042 (2011) [16 pages] [http://arXiv.org/pdf/1104.2432].
9. P. Baekler and F. W. Hehl, *Beyond Einstein–Cartan gravity: Quadratic torsion and curvature invariants with even and odd parity including all boundary terms,* Class. Quant. Grav. **28**, 215017 (2011) [11 pages] [http://arXiv.org/pdf/1105.3504].
10. Y. N. Obukhov and F. W. Hehl, *Extended Einstein–Cartan theory à la Diakonov: the field equations,* Phys. Lett. B **713**, 321–325 (2012) [http://arXiv.org/pdf/1202.6045].
11. M. Blagojević and F. W. Hehl (eds.), *Gauge Theories of Gravitation,* a reader with commentaries (Imperial College Press, London, 2013).
12. R. Utiyama, *Invariant theoretical interpretation of interactions,* Phys. Rev. **101**, 1597–1607 (1956).
13. T. W. B. Kibble, *Lorentz invariance and the gravitational field,* J. Math. Phys. **2**, 212–221 (1961).
14. T. W. B. Kibble, Excerpt from the Foreword in Ref. [11].
15. A. Einstein: *The Meaning of Relativity,* 5th ed. (Princeton University Press, Princeton, NJ, 1955) [first published in 1922].
16. A. Einstein, Reprint 1.1 in Ref. [11].
17. R. Adler, M. Bazin, and M. Schiffer, *Introduction to General Relativity,* 2nd ed. (McGraw-Hill, New York, 1975).
18. H. Weyl, *Space-Time-Matter,* translated from the fourth German edition of 1921 by H. Brose (Dover, New York, 1952).
19. R. Colella, A. W. Overhauser, and S. A. Werner: *Observation of gravitationally induced quantum interference,* Phys. Rev. Lett. **34**, 1472–1474 (1975).
20. H. Rauch and S. A. Werner, *Neutron Interferometry,* Lessons in Experimental Quantum Mechanics (Clarendon Press, Oxford, UK, 2000).

21. U. Bonse and T. Wroblewski, *Measurement of neutron quantum interference,* Phys. Rev. Lett. **51**, 1401–1404 (1983).

22. F. W. Hehl and Y. N. Obukhov, *Foundations of Classical Electrodynamics:* Charge, Flux, and Metric (Birkhäuser, Boston, MA, 2003).

23. J. B. Pitts, *The nontriviality of trivial general covariance: How electrons restrict 'time' coordinates, spinors (almost) fit into tensor calculus, and $\frac{7}{16}$ of a tetrad is surplus structure,* Stud. Hist. Philos. Mod. Phys. **43**, 1–24 (2012) [http://arXiv.org/pdf/1111.4586].

24. J. A. Schouten, *Tensor Analysis for Physicists,* 2nd ed. reprinted (Dover, Mineola, NY, 1989).

25. A. Einstein, Excerpt from the Preface (dated 04 April 1955) in: *Cinquant'anni di Relatività 1905–1955.* M. Pantaleo (ed.) (Edizioni Giuntine and Sansoni Editore, Florence, 1955) (translation from the German original by F. Gronwald, D. Hartley, and F. W. Hehl).

26. H. Weyl, *Elektron and Gravitation. I,* Zeitschrift für Physik **56**, 330–352 (1929), translation into English in Ref. [42], pp. 121–144.

27. H. Weyl, *A remark on the coupling of gravitation and electron,* Phys. Rev. **77**, 699–701 (1950).

28. A. Einstein, *Einheitliche Feldtheorie von Gravitation und Elektrizität,* Sitzungsber. Preuss. Akad. Wiss. Berlin, Phys.-math. Klasse (1925) pp. 414–419.

29. E. Schrödinger, *Space-Time Structure,* reprinted with corrections (Cambridge University Press, Cambridge, UK, 1960).

30. F. W. Hehl and Y. N. Obukhov, *Spacetime metric from local and linear electrodynamics: A new axiomatic scheme,* Lect. Notes Phys. (Springer) **702**, 163–187 (2006) [http://arXiv.org/pdf/gr-qc/0508024].

31. F. Gronwald and F. W. Hehl, *On the gauge aspects of gravity,* in: *Proc. Int. School of Cosm. & Gravit.* 14th Course: Quantum Gravity. Held in Erice, Italy. Proceedings, P.G. Bergmann et al. (eds.) (World Scientific, Singapore, 1996) pp. 148–198 [http://arXiv.org/pdf/gr-qc/9602013].

32. F. W. Hehl, J. D. McCrea, E. W. Mielke, and Y. Ne'eman, *Metric-affine gauge theory of gravity: Field equations, Noether identities, world spinors, and breaking of dilation invariance,* Phys. Rept. **258**, 1–171 (1995).

33. D. Hartley: *Normal frames for non-Riemannian connections.* Class. Quantum Grav. **12**, L103–L105 (1995) [http://arXiv.org/pdf/gr-qc/9510013].

34. J. M. Nester, *Normal frames for general connections,* Annalen der Physik (Berlin) **19**, 45–52 (2010).

35. J. F. T. Giglio and W. A. Rodrigues, Jr, *Locally inertial reference frames in Lorentzian and Riemann–Cartan spacetimes,* Annalen der Physik (Berlin) **524**, 302–310 (2012) [http://arXiv.org/pdf/1111.2206].

36. E. Scholz, *Paving the way for transitions—a case for Weyl geometry,* this workshop [49 pages] [http://arXiv.org/pdf/1206.1559].

37. P. von der Heyde, *The equivalence principle in the U_4 theory of gravitation,* Nuovo Cim. Lett. **14**, 250–252 (1975).

38. F. W. Hehl and W.-T. Ni, *Inertial effects of a Dirac particle*, Phys. Rev. D **42**, 2045–2048 (1990).

39. E. M. Corson, *Introduction to Tensors, Spinors, and Relativistic Wave-Equations* (Blackie, London, 1953).

40. F. W. Hehl, *On energy-momentum and spin/helicity of quark and gluon fields,* Invited talk, in: *XV Advanced Research Workshop on High Energy Spin Physics (DSPIN-13),* Dubna, Russia, 08–12 October 2013, Proceedings edited by A. V. Efremov and S.V. Goloskokov, Dubna (2014), pp. 65–74 [http://arXiv.org/pdf/1402.0261].

41. D. W. Sciama, *On the analogy between charge and spin in general relativity,* in: *Recent Developments in General Relativity,* Festschrift for Infeld (Pergamon Press, Oxford, UK; PWN, Warsaw, 1962), pp. 415–439.

42. L. O'Raifeartaigh, *The Dawning of Gauge Theory* (Princeton University Press, Princeton, NJ, 1997).

43. M. Blagojević, *Gravitation and Gauge Symmetries* (IoP Publishing, Bristol, UK, 2002).

44. L. Ryder, *Introduction to General Relativity* (Cambridge University Press, Cambridge, UK, 2009).

45. F. W. Hehl, P. von der Heyde, G. D. Kerlick, and J. M. Nester, *General relativity with spin and torsion: Foundations and prospects,* Rev. Mod. Phys. **48**, 393–416 (1976).
46. A. Trautman, *The Einstein–Cartan theory,* in: *Encyclopedia of Mathematical Physics,* vol. 2, J.-P. Françoise et al. (eds.) (Elsevier, Oxford, UK, 2006) pp. 189–195 [http://arXiv.org/pdf/gr-qc/0606062].
47. Y. Itin and F. W. Hehl, *Maxwell's field coupled nonminimally to quadratic torsion: axion and birefringence,* Phys. Rev. D **68**, 127701 (2003) [4 pages] [http://arXiv.org/pdf/gr-qc/0307063].
48. Y. Mao, M. Tegmark, A. Guth and S. Cabi, *Constraining torsion with Gravity Probe B,* Phys. Rev. **D76**, 104029 (2007) [26 pages] [http://arXiv.org/pdf/gr-qc/0608121].
49. D. Puetzfeld and Y. N. Obukhov, *Probing non-Riemannian spacetime geometry,* Phys. Lett. A **372**, 6711–6716 (2008) [http://arXiv.org/pdf/0708.1926].
50. F. W. Hehl, Y. N. Obukhov, and D. Puetzfeld, *On Poincaré gauge theory of gravity, its equations of motion, and Gravity Probe B,* Phys. Lett. A **377**, 1775–1781 (2013) [http://arXiv.org/pdf/1304.2769].
51. Y. Itin, *Energy-momentum current for coframe gravity,* Class. Quantum Grav. **19**, 173–189 (2002) [http://arXiv.org/pdf/gr-qc/0111036].
52. T. Sauer, *Field equations in teleparallel spacetime: Einstein's Fernparallelismus approach towards unified field theory,* Historia Math. **33**, 399–439 (2006) [http://arXiv.org/pdf/physics/0405142].
53. C. N. Yang, *Integral formalism for gauge fields,* Phys. Rev. Lett. **33**, 445–447 (1974). Reprint 19.1 in Ref. [11].
54. K. F. Shie, J. M. Nester, and H. J. Yo, *Torsion cosmology and the accelerating universe,* Phys. Rev. D **78**, 023522 (2008) [16 pages] [http://arXiv.org/pdf/0805.3834].
55. E. W. Mielke, *Spontaneously broken topological SL(5,R) gauge theory with standard gravity emerging,* Phys. Rev. D **83**, 044004 (2011) [9 pages].

Paving the Way for Transitions—A Case for Weyl Geometry

Erhard Scholz

Abstract This paper presents three aspects by which the Weyl geometric generalization of Riemannian geometry, and of Einstein gravity, sheds light on actual questions of physics and its philosophical reflection. After introducing the theory's principles, it explains how Weyl geometric gravity relates to Jordan–Brans–Dicke theory. We then discuss the link between gravity and the electroweak sector of elementary particle physics, as it looks from the Weyl geometric perspective. Weyl's hypothesis of a preferred scale gauge, setting Weyl scalar curvature to a constant, may get new support from an interplay between the gravitational scalar field and the electroweak one (the Higgs field). This has surprising consequences for cosmological models and adds to the motivation for putting central features of the present cosmological model into a wider perspective.

1 Introduction

When *Johann Friedrich Herbart* discussed the 'philosophical study' of science he demanded that the sciences should organize their specialized knowledge about core concepts (*Hauptbegriffe*). Philosophy should then strive "*... to pave the way for adequate transitions between the concepts ...*" in order to establish an integrated system of knowledge.[1] In this way philosophy and the specialized sciences were conceived as a common enterprise. Only together they would be able to generate a connected system of knowledge and contribute to the 'many-sidedness of education' Herbart had in mind.

This is not exactly what is usually understood by 'metatheory'; but the concept of the workshop which gave rise to this volume was to go beyond the consideration of working theories in themselves and to reflect on possible mutual connections between

---

[1]"... und gilt uns [im philosophischen Studium, E.S.], dem gemäß, *alle Bemühung, zwischen den Begriffen die gehörigen Uebergänge zu bahnen ...* " [73, 275, emphasis in original].

E. Scholz (✉)
LS Didaktik Mathematik, Bergische Universität Wuppertal FB
C—Mathematik/Naturwissenschaften, Wuppertal, Germany
e-mail: scholz@math.uni-wuppertal.de

© Springer Science+Business Media, LLC 2017   171
D. Lehmkuhl et al. (eds.), *Towards a Theory of Spacetime Theories*,
Einstein Studies 13, DOI 10.1007/978-1-4939-3210-8_6

different spacetime theories, and perhaps beyond. This task comes quite close to what Herbart demanded from 'speculation' as he understood it. In this contribution, I want to use the chance offered by the goal of the workshop to discuss how *Weyl geometry* may help to 'pave the way for transitions' between certain segments of physical knowledge. We deal here with connections between theories some of which came into existence long after the invention of Weyl geometry and are far beyond Weyl's original intentions during the years 1918 to 1923.

Mass generation of elementary particle fields is one of the topics. In general relativity mass serves as the active and passive charge of the gravitational field; high energy physics has made huge progress in analyzing the basic dynamical structures which determine the energy content, and thus the gravitational charge, of field constellations. The connection between high energy physics and gravity is still wide open for further research. Most experts expect the crucial link between the two fields to be situated close to the Planck scale, viz shortly after the big bang, with the Higgs 'mechanism' indicating a phase transition in the early universe. This need not be so. The Weyl geometric generalization of gravity considered here indicates a more structural connection between gravitation and the electroweak scalar field, independent of cosmological time. The dilationally invariant Lagrangians of (special relativistic) standard model fields translate to scale invariant fields on curved spaces in an (integrable) Weyl geometry. The latter offers a well-adapted arena for studying the transition between gravity and standard model fields. Scalar fields play a crucial role on both sides, the question will be to what extent they are interrelated mathematically and physically.

Similar, although still more general, questions with regard to the transition from conformal structures to gravity theory have already been studied by Weyl. In his 1921 article on the relationship between conformal and projective differential geometry, [151] he argued that his new geometry establishes a peculiar bridge between the two basic geometrical structures underlying general relativity, conformal and projective. The first one was and still is the mathematical expression of the causal structure (light cones) and the second one represents the most abstract mathematization of inertial structure (free fall trajectories under abstraction from proper time parametrization). Weyl indicated a kind of 'transition' to a fully metric gravity theory into which other dynamical fields, in his case essentially the electromagnetic one, could be integrated. He showed that a Weylian metric is uniquely determined if its conformal and its projective structures are known. In principle, such a metric can be determined by physically grounded structural observations without any readings of clocks or measurements with rods; i.e. Weyl geometry allows to establish a connection between causal structure, free fall and metrical geometry in an impressingly basic way.

To make the present contribution essentially self-contained, we start with a short description of Weyl geometry, already with physical meaningful interpretations in mind, exemplified by the well-known work of Ehlers/Pirani/Schild (section 2). In a first transition we see how Jordan–Brans–Dicke (JBD) theory with its scalar field, 'non-minimally' coupled to gravity, fits neatly into a Weyl geometric framework (section 3). The different 'frames' of JBD theory correspond to different choices of scale gauges of the Weylian approach. Usually this remains unnoticed in the literature, although the basic structural ingredients of Weyl geometry are presupposed and dealt with in a non-explicit way.

The link is made explicit in a Weyl geometric version of generalized Einstein theory with a non-minimally coupled scalar field, due to Omote, Utiyama, Dirac e.a. (WOD gravity), introduced in section 4. Strong reasons speak in favour of its integrable version (iWOD gravity) close to, but not identical with, (pseudo-) Riemannian geometry. An intriguing parallel between the Higgs field of electroweak theory and the scalar field of iWOD gravity comes into sight if one includes the gravitational coupling into the potential of the scalar field. This suggests to consider a common biquadratic potential for the two scalar fields (section 5). In its minimum, the ground state of the scalar field specifies a (non-Riemannian) scale choice of the Weyl geometry which establishes units for measuring mass, length, time, etc., and gives rise to the vacuum expectation value and mass of the Higgs field.

In his correspondence with Einstein on the physical acceptability of his generalized geometry Weyl conjectured, or postulated, an adaptation of atomic clocks to (Weylian) scalar curvature. In this way, according to Weyl, measuring devices would indicate a scaling in which (Weylian) scalar curvature becomes constant (Weyl gauge). This conjecture is supported, in a surprising way, by evaluating the potential condition of the gravitational scalar field. If, moreover, the gravitational scalar field 'communicates' with the electroweak Higgs field, clock adaptation to the ground state of the scalar field gets a field theoretic foundation in electroweak theory (section 5.3, 5.4). The question is now open, whether such a transition between iWOD gravity and electroweak theory indicates a physical connection or whether it is only an accidental feature of the two theories.

Reconsidering Weyl's scale gauge condition (constant Weylian scalar curvature) necessitates another look at cosmological models (section 6). The warping of Friedmann–Robertson–Walker geometries can no longer immediately be interpreted as an actual expansion of space (although that is not excluded). Cosmological redshift becomes, at least partially, due to a field theoretic effect (Weylian scale connection). From such a point of view, much of the cosmological observational evidence, among it the cosmological microwave background and quasar distribution over redshift, ought to be reconsidered. The enlarged perspective of integrable Weyl geometry and of iWOD gravity elucidate, by contrast, how strongly some realistic claims of present precision cosmology are dependent on specific facets of the geometrico-gravitational paradigm of Einstein–Riemann type. Many empirically sounding statements are insolvably intertwined with the data evaluation on this basis. Transition to a wider framework may be helpful to reflect these features—perhaps not only as a metatheoretical exercise (section 7).

2   On Weyl Geometry and the Analysis of EPS

Weyl geometry is a generalization of Riemannian geometry, based on two insights:
(i) The automorphisms of both, of Euclidean geometry and of special relativity, are the *similarities* (of Euclidean, or respectively of Lorentz signature) rather than the

congruences. No unit of length is naturally given in Euclidean geometry, and likewise the basic structures of special relativity (inertial motion and causal structure) can be given without the use of clocks and rods. (ii) The development of field theory and general relativity demands a conceptual implementation of this insight in a consequently *localized mode* (physics terminology).[2]

Based on these insights, Weyl developed what he called *reine Infinitesimalgeometrie* (purely infinitesimal geometry) [149, 150]. Its basic ingredients are a conformal generalization of a (pseudo-) Riemannian metric $g = (g_{\mu\nu})$ by allowing point-dependent rescaling $\tilde{g}(x) = \Omega(x)^2 g(x)$ with a nowhere vanishing (positive) function Ω, and a scale ('length') connection given by a differential form $\varphi = \varphi_\mu dx^\mu$, which has to be gauge transformed $\tilde{\varphi} = \varphi - d \log \Omega$ when rescaling $(g_{\mu\nu})$. The scale connection (φ_μ) expresses how to compare lengths of vectors (or other metrical quantities) at two infinitesimally close points, both measured in terms of a scale, i.e. a representative $(g_{\mu\nu})$ of the conformal class. The typical symmetry of the geometry, in the infinitesimal, is thus the scale-extended Poincaré group, sometimes called the *Weyl group* (although the same name is used in Lie group theory in a completely different sense).[3]

Of course, Weyl's generalization of Riemannian geometry may be embedded in E. Cartan's even wider program of geometries with infinitesimal symmetries. In the case of the scale extended Poincaré group one then arrives at a Cartan–Weyl geometry with a translational Cartan connection and *torsion* as the typical extension of the structure. Here we restrict to the case without torsion and, as we shall see in a moment, to the most simple case of an integrable scale connection.[4] That even in this most simple case (no torsion and an integrable scale connection) the Cartan geometric approach to Weyl geometry can be conceptually and physically illuminating, because of the link between infinitesimal translational symmetries to the energy–momentum current, may be inferred from F. Hehl's contribution to this book. But the interesting question of a 'teleparallel' version of integrable Weyl geometric gravity will not be a topic in the present chapter.

2.1   Scale Connection, Covariant Derivative, Curvature

Metrical quantities in Weyl geometry are directly comparable only if they are measured at the same point p of the manifold. Quantities measured at different points $p \neq q$ of finite, i.e. non-infinitesimal distance can be metrically compared only after

---

[2]In mathematical terminology, the implementation of a similarity structure happens at the *infinitesimal*, rather than at the local, level. For a concrete ('passive') description of (i) and (ii) in a more physical language, see Dicke's postulate cited in section 3.1.

[3]For more historical and philosophical details see, among others, [120, 127, 147], from the point of view of physics [2, 12, 114, 116, 131], and for the view of differential geometres [56, 64, 76] (as a short selection in all three categories).

[4]For a modern presentation of Cartan geometry, including the Cartan–Weyl case, see, e.g. [138, chap. 7]; for the physical aspects of the extension studied since the 1970s [13, chap. 8].

an integration of the scale connection along a path from p to q. Weyl realized that this structure is compatible with a uniquely determined affine connection $\Gamma = (\Gamma^\mu_{\nu\lambda})$ (the affine connection of Weylian geometry). If $_g\Gamma^\mu_{\nu\lambda}$ denotes the Levi-Civita connection of the Riemannian part g only, the Weylian affine connection is given by

$$\Gamma^\mu_{\nu\lambda} = {}_g\Gamma^\mu_{\nu\lambda} + \delta^\mu_\nu \varphi_\lambda + \delta^\mu_\lambda \varphi_\nu - g_{\nu\lambda}\varphi^\mu. \tag{1}$$

The *covariant derivative* with regard to Γ, will be denoted by $\nabla = \nabla_\Gamma$. A change of scale neither changes the connection (the left-hand side of (1)) nor the covariant derivative; only the composition from the underlying Riemannian part and the corresponding scale connection (right-hand side) is shifted.

Curvature concepts known from 'ordinary' (Riemannian) differential geometry follow, as every connection defines a unique curvature tensor. The Riemann and Ricci tensors, *Riem*, *Ric*, are scale invariant by construction, although their expressions contain terms in φ. On the other hand, the scalar curvature involves 'lifting' of indices by the inverse metric and is thus scale covariant of weight -2 (see below).

Field theory gets slightly more involved in Weyl geometry, because for vector and tensor fields (of 'dimensional' quantities) the appropriate scaling behaviour under change of the metrical scale has to be taken into account. If a field, expressed by X (leaving out indices) with regard to the metrical scale $g(x) = (g_{\mu\nu}(x))$ transforms to $\tilde{X} = \Omega^k X$ with regard to the scale choice $\tilde{g}(x)$ as above, X is called a *scale covariant field* of *scale* or *Weyl weight* $w(X) := k$ (usually an integer or a fraction). Generally the covariant derivative, ∇X, of a scale covariant quantity X is not scale covariant. However, scale covariance can be reobtained by adding a weight dependent term. Then the *scale covariant derivative* D of a scale covariant field X is defined by

$$DX := \nabla X + w(X)\varphi \otimes X . \tag{2}$$

For example, ∇g is not scale covariant, but Dg is. Moreover, one finds that $Dg = \nabla g + 2\varphi \otimes g = 0$; i.e. in Weyl geometry g *appears* no longer *constant* with regard to the derivative ∇ but *with regard to the scale covariant derivative D*.

In physics literature, an affine connection Γ with $\nabla_\Gamma g \neq 0$ is usually regarded as 'non-metric', and $\nabla_\Gamma g$ is considered its non-metricity.[5] These concepts hold in the Riemannian approach. In Weyl geometry, in contrast,

$$\nabla g = -2\varphi \otimes g \iff Dg = 0 \tag{3}$$

expresses the *compatibility* of the affine connection Γ with the *Weylian metric* represented by the pair (g, φ).

Geodesics can be invariantly defined as auto-parallels by the Weyl geometric affine connection (so did Weyl himself). But one can just as well, in our context even better, consider scale covariant geodesics of weight -1 (see section 6.1).

---

[5] See the contribution by F. Hehl, this volume.

Under a change of scale $g \mapsto \tilde{g} = \Omega^2 g$ and the accompanying gauge transformation for the scale connection $\varphi \mapsto \tilde{\varphi} = \varphi - d \log \Omega$, the compatibility condition transforms consistently, $\nabla_\Gamma \tilde{g} = -2\tilde{\varphi} \otimes \tilde{g}$. Equ. (3) ensures, in particular, that geodesics (i.e. auto-parallels) with initial direction along a nullcone of the conformal metric remain directed along the nullcones. This is the most important geometric feature of metric compatibility in Weyl geometry.[6]

2.2 Weyl Structures and Integrable Weyl Geometry (IWG)

In the more recent mathematical literature a *Weyl structure* on a manifold is defined by a pair $(\mathscr{C}, \nabla)$ consisting of a *conformal structure* $\mathscr{C} = [g]$ (an equivalence class of pseudo-Riemannian metrics) and the covariant derivative of a *torsion-free linear connection* ∇, constrained by the condition

$$\nabla g + 2\varphi_g \otimes g = 0 ,$$

with a differential 1-form φ_g depending on $g \in \mathscr{C}$.[7] The change of the conformal representative $g \mapsto \tilde{g} = \Omega^2 g$ is accompanied by a change of the 1-form

$$\varphi_{\tilde{g}} = \varphi_g - d \log \Omega , \tag{4}$$

i.e. by a 'gauge transformation' as introduced by Weyl in [149]. Formally, a *Weyl metric* consists of an equivalence class of pairs (g, φ_g) with scale and gauge transformations defining the equivalences. Given the scale choice $g \in \mathscr{C}$, φ_g represents the scale connection.

In Weyl's view of a strictly 'localized' (better: infinitesimalized) metric, metrical quantities at different points p and q can be compared only by a 'transport of length standards' along a path γ from p to q, i.e. by multiplication with a factor

$$l(\gamma) = e^{\int_0^1 \varphi(\gamma')} . \tag{5}$$

$l(\gamma)$ will be called the *length* or *scale transfer* function (depending on p, q and γ). The *curvature* of the *scale connection* is simply the exterior differential, $f = d\varphi$ with components, $f_{\mu\nu} = \partial_\mu \varphi_\nu - \partial_\nu \varphi_\mu$, where $\partial_\mu := \frac{\partial}{\partial x^\mu}$.

For vanishing scale curvature, $f = 0$, the scale transfer function can be integrated away, i.e. there exist local choices of the scale, $\tilde{g}$, with vanishing scale connection, $\varphi_{\tilde{g}} = 0$. In this case one deals with *integrable Weyl geometry* (IWG). Then the Weyl

---

[6]Weyl understood the compatibility of the scale connection with the metric in the sense that parallel transport of a vector $X(p)$ by the affine connection along a path γ from p to q to $X(q)$ leads to consistency with length transfer along the same path. Compare the compatibility condition given, in a different mathematical framework, by [44].

[7][20, 64, 76, 106].

metric may be locally represented by a Riemannian metric[8]; we call this the *Riemann gauge* (equivalently *Riemannian scale choice*) of an integrable Weyl metric. In this gauge the Weylian curvature tensor does not contain terms in φ. For integrable Weyl geometry vanishing of the Riemann tensor, $Riem = 0$ is of course equivalent to local flatness.

Whether a reduction to Riemannian geometry makes sense physically, depends on the field theoretic content of the theory. If a scalar field plays a part in determining the scale—physically speaking, if scale symmetry is 'spontaneously' broken by a scale covariant scalar field—the result may well be different from Riemannian geometry (see below, sections. 4ff.).

2.3   *From Ehlers/Pirani/Schild to Audretsch/Gähler/Straumann*

Weyl originally hoped to represent the potential of the electromagnetic field by a scale connection and to achieve a geometrical unification of gravity and electromagnetism by his 'purely infinitesimal' geometry. The physical difficulties of this approach, usually presented as outright inconsistencies with observational evidence, have been discussed in the literature [65, 147]. But, of course, there is no need to bind the usage of Weyl geometry to this specific, and outdated, interpretation. Since the early 1970s a whole, although minoritarian and heterogeneous, literature of Weyl geometric investigations in the foundations of gravity has emerged. In this contribution I want to take up, and pursue a little further, an approach going back to M. Omote, R. Utiyama, and P.A.M. Dirac, which was later extended in different directions (section 6, below).[9] But before we follow these more specific lines we have to briefly review the foundational aspects of Weyl geometry for gravity theory analyzed in the seminal paper of Ehlers et al. [44] (EPS).

Like Weyl in 1921, these three authors based their investigation on the insight that the causal structure of general relativity is mathematically characterized by a conformal (cone) structure, and the inertial structure of point particles by a projective path structure. They investigated the interrelation of the two structures from a foundational point of view in a methodology sometimes called a 'constructive axiomatic' approach. Their axioms postulated rather general properties for these two structures and demanded their compatibility. EPS concluded that these properties suffice for

---

[8]Here 'local" is used in the sense of differential geometry, i.e. in (finite) neighbourhoods. Physicists usage of 'local', in contrast, refers in most cases to point-dependence or 'infinitesimal' neighbourhoods. In the following, both language codes are used, not always with further specification. The respective meaning will be clear from the context.

[9]The interpretation of the quantum potential in Weyl geometric terms proposed by [122, 123] and others indicate a completely different route of attempted 'transitions' than reviewed here. It is not further considered in the following.

specifying a unique Weylian metric [44].[10] The axioms of Ehlers, Pirani and Schild were motivated by the physical intuition of inertial paths (of classical particles) and the causal structure. Other authors investigated connections to quantum physics. J. Audretsch, F. Gähler, N. Straumann (AGS) found that wave functions (Klein–Gordon and Dirac fields) on a Weylian manifold behave acceptable only in the integrable case. As a criterion of acceptability they studied the streamlines of wavefront developments in an WKB approximation (WKB: Wentzel–Kramers–Brioullin) and found that, for $\hbar \to 0$, the streamlines converge to geodesics if and only if $d\varphi = 0$, i.e. in the case of an *integrable* Weyl metric [4]. Therefore the integrability of the Weyl structure seems necessary for consistency between the geodesic principle of classical particles and the decoherence view of the quantum to classical transition.

The gap between the structural result of EPS (Weyl geometry in general) and the pseudo-Riemannian structure of ordinary (Einstein) relativity was considerably reduced in the sense of integrability, but still it was not clear that the Riemannian scale choice of IWG had to be chosen. The selection of Riemannian geometry remained ad hoc and was not based on deeper insights. It had to be stipulated by an additional postulate involving clocks and rods. The transition from the EPS axiomatics to Einstein gravity still contained a methodological jump and relied on reference to observational instruments external to the theory, which Weyl wanted to exclude from the foundations of general relativity.[11] So even after the work of EPS and their successors the question remained whether the transition to Riemannian geometry and Einstein gravity is the only one possible. Alternatives were sought for by a different group of authors who started more or less simultaneous to EPS, investigating alternatives based on a scale invariant Lagrangian (section 4) similar to the one studied by Jordan, Brans, and Dicke in the Riemannian context. It was not noticed at the time that even the latter can be analyzed quite naturally in the framework of Weyl geometry.

3   Jordan–Brans–Dicke Theory in Weyl Geometric Perspective

In the early 1950s and 1960s P. Jordan, later R. Dicke and C. Brans (JBD) proposed a widely discussed modification of Einstein gravity.[12] Essential for their approach was a (real valued) scalar field χ, coupled to the traditional Hilbert action with Lagrangian density

---

[10]For the compatibility see fn. 6. A recent commentary of the paper is given in [144]. How $f(R)$ theories of gravity may lead back to the EPS paper is discussed in [23].

[11]Although in his 1918 debate with Weyl, Einstein insisted on the necessity of clock and rod measurements in general relativity as the empirical basis for the physical metric, he admitted that rods and clocks should not be accepted as fundamental. He reiterated this view until late in his life [45, 555f.], cf. [91].

[12][17, 35, 77]; for surveys on the actual state of JBD theory and its applications to cosmology see [50, 62], for a participant's recollection of its history [16]; a broad contextual history is presented in [87].

$$\mathscr{L}_{\mathscr{J}\mathscr{B}\mathscr{D}} = (\chi R - \frac{\omega}{\chi}\partial^\mu \chi \, \partial_\mu \chi)\sqrt{|det\, g|}\,, \tag{6}$$

where ω is a free parameter of the theory. For $\omega \to \infty$ the theory has Einstein gravity as limiting case. All three authors allowed for conformal transformations, $\tilde{g} = \Omega^2 g$, under which their scalar field χ transformed with weight -2 (matter fields and energy tensors T of weight $w(T) = -2$ etc.).[13] Jordan took up the discussion of conformal transformations only in the second edition of his book [77], after Pauli had made him aware of such a possibility. Pauli knew Weyl geometry very well, he was one of its experts already as early as 1919 but neither he nor Jordan or the US-American authors looked at JBD theory from that point of view.

3.1 Conformal Rescaling in JBD Theory

For introducing conformal rescaling Dicke argued as follows:

> It is evident that the particular values of the units of mass, length, and time employed are arbitrary and that the *laws of physics must be invariant under a general coordinate dependent change of units* [35, 2163][emph. ES].

By 'coordinate dependent change of units' Dicke indicated a point dependent rescaling of basic units. In the light of the relations established by the fundamental constants (velocity of light c, (reduced) Planck constant $\hbar$, elementary charge e and Boltzmann constant k) all units can be expressed in terms of one independent fundamental unit, e.g. time, and the fundamental constants (which, in principle can be given any constant numerical value, which then fixes the system).[14] Thus only one essential scaling degree of units remains and Dicke's principle of an arbitrary, point dependent unit choice came down to a 'passive' formulation of Weyl's localized similarities in the framework of his scale gauge geometry.[15] It was not so clear, however, how Dicke's postulate that the 'laws of physics must be invariant' under point-dependent rescaling ought to be understood in JBD theory. Its modified Hilbert term was, and is, not scale invariant and assumes correction terms under conformal rescaling (vanishing only for $\omega = -\frac{3}{2}$).

---

[13]Weights rewritten in adaptation to our convention.

[14]The present revision of the international standard system SI is heading towards implementing measurement definitions with time as only fundamental unit, $u_T = 1\,s$ such that "the ground state hyperfine splitting frequency of the caesium 133 atom $\Delta v(^{133}Cs)_{\text{hfs}}$ is exactly 9 192 631 770 hertz" [19, 24f.]. In the 'New SI', four of the SI base units, namely the kilogram, the ampere, the kelvin and the mole, will be redefined in terms of invariants of nature; the new definitions will be based on fixed numerical values of the Planck constant, the elementary charge, the Boltzmann constant, and the Avogadro constant (www.bipm.org/en/si/new_si/). The redefinition of the meter in terms of the basic time unit by means of the fundamental constant c was implemented already in 1983. Point dependence of the time unit because of locally varying gravitational potential was inbuilt in this system. For practical purposes it can be outlevelled by reference to the *SI second on the geoid* (standardized by the International Earth Rotation and Reference Systems Service IERS).

[15]Compare principles (i) and (ii) at the beginning of section 2.

On the other hand, the principles of JBD gravity were moved even closer to Weyl geometry by all three proponents of this approach considering it as self-evident that the *Levi-Civita connection* $\Gamma := {}_g\Gamma$ of the Riemannian metric g in (6) remains *unchanged* under conformal transformation of the metric. Probably the protagonists considered that as a natural outcome of assuming invariance of the 'laws of nature' under conformal rescaling.[16] In any case, they kept the affine connection Γ fixed and rewrote it in terms of the Levi-Civita connection ${}_{\tilde{g}}\Gamma$ of the rescaled metric, $\tilde{g} = \Omega^2 g$, with additional terms in partial derivatives of Ω. Let us summarily denote these additional terms by $\Delta(\partial\Omega)$,[17] then

$$\Gamma = {}_{\tilde{g}}\Gamma + \Delta(\partial\Omega) \,.$$

Conformal rescaling, in addition to a fixed affine connection, have become *basic tools of JBD theory*.

3.2   IWG as Implicit Framework of JBD Gravity

The variational principle (6) of JBD gravity determines a connection with covariant derivative $\nabla = {}_g\nabla$ and a scalar field χ. The theory allows for conformal rescalings of g and χ without changing ∇. That is, JBD theory *specifies a Weyl structure* $(\mathscr{C}, \nabla)$ with $\mathscr{C} = [g]$. Transformation between different frames happen in this framework, even though this remains unreflected by most of its authors.

In the JBD tradition, a choice of units is called a *frame*. In terms of Weyl geometry, such a frame corresponds to the selection of a scale gauge. Two frames play a major role:

- *Jordan frame*: the one in which $\nabla = {}_g\nabla$ (g the metric of (6)), i.e. the affine connection is the Levi-Civita one of the Riemannian metric,
- *Einstein frame*: the one in which $\tilde{\chi} = const$; then the affine connection is different from the Levi-Civita one of the reference metric.

The Jordan frame is such that, by definition, the dynamical affine connection is identical to the Levi-Civita connection of g. Expressed in Weyl geometric terms, this implies vanishing of the scale connection, $\varphi = 0$. Thus this frame corresponds to what we have called the *Riemann gauge* of the underlying integrable Weylian metric (section 2). In Einstein frame the scalar field ($\neq 0$ everywhere) is scaled to a constant; we may call this the *scalar field gauge*. Another terminology for it is

---

[16]If the trajectories of bodies are governed by the gravito-inertial 'laws of physics' they should not be subject to change under transformation of units. The same should hold for the affine connection which can be considered a mathematical concentrate of these laws.

[17]For our purpose the explicit form of $\Delta(\partial\Omega)$ is not important. R. Penrose noticed that the additional terms of the (Riemannian) scalar curvature are exactly cancelled by the partial derivative terms of the kinematical term of χ if and only if $\omega = -\frac{3}{2}$. In this case the Lagrangian (6) is conformally invariant [109].

Einstein gauge. In this gauge, the gravitational 'constant' appears as a true constant, contrary to Jordan's motivation. By obvious reasons, Jordan tended to prefer the other frame; thus its name.

Clearly in the Einstein frame JBD gravity does not reduce to Einstein gravity, as the affine connection is deformed with regard to the metrical component of the gauge. Scalar curvature in Einstein frame can easily be expressed in terms of Weyl geometrical quantities, but usually it is not. Practitioners of JBD theory prefer to write everything in terms of $\tilde{g}$, take its Levi-Civita connection $_{\tilde{g}}\Gamma$ as representative for the gravito-inertial field and consider the modification terms as arising from the transformation from Riemann gauge to scalar field gauge. Sometimes they appear as additional ('fifth') force.[18]

From our point of view, we observe

• Structurally, JBD theory presupposes and works in an *integrable Weyl structure*, although its practitioners usually do not notice.[19]
• *Scale covariance*, not scale invariance, is often the game of JBD theoreticians. That lead to a debate (sometimes confused), which frame should be considered as 'physical' and which not. Jordan frame used to be the preferred one. In the recent literature of JBD some, maybe most, authors argue in favour of Einstein frame as 'physical' [51].
• Some authors studied the conformally invariant version of the JBD Lagrangian, corresponding to $\omega = -\frac{3}{2}$, and investigated the hypothesis of a conformally invariant theory of gravity at high energies, which gets 'spontaneously broken' by the scalar field taking on a specific value [34, 48]. That was achieved by adding additional polynomial terms in χ with coefficients usually of 'cosmological' order of magnitude. Problems arose in the conformal JBD approach from the sign of ω; a negative sign indicated a 'ghost field' with negative energy [62, 5].[20]
• Empirical high precision tests of gravity in the solar system concentrated on the Jordan frame and found increasingly high bounds for the parameter ω. To the disillusionment of JBD practitioners, ω was found to be $> 3.6 \cdot 10^3$ at the turn of the millenium [157]; today these values are even higher. So the leeway for JBD theory *in Jordan frame* deviating from Einstein gravity became increasingly reduced. That does not hinder authors in cosmology to assume Jordan frame models for the expansion of universe shortly after the big bang.[21] Shortly after the big bang, the world of mainstream cosmology seems to be Feyerabendian.

From the Weyl geometric perspective, a criterion of scale invariance for observable quantities supports preference of the Einstein frame. In any case, Weyl geometry is a conceptually better adapted framework for JBD gravity than Riemannian geometry.

---

[18]For a critical discussion see [115].

[19]A discussion from a slightly different view can be found in [3, 115, 117].

[20]Some authors choose to switch the sign of the 'gravitational constant', e.g. [34, 250]. This strategy indicated that there is a basic problem for the conformal JBD approach ($\omega = -\frac{3}{2}$) in spite of its attractive basic idea.

[21]E.g. [10, 67, 78, 81].

Perhaps that was felt by some physicists at the time. Be that as it may, about a decade after the rise of JBD theory two groups of authors in Japan and in Europe, independently of each other, started to study a similar type of coupling between scalar field and gravity in a Weyl geometric theory of gravitation.

4   Weyl–Omote–Dirac Gravity and Its Integrable Version (iWOD)

In 1971 M. Omote proposed a Lagrangian field theory of gravity with a scale covariant scalar field coupling to the Hilbert term like in JBD theory, but now explicitly formulated in the framework of Weyl geometry. A little later R. Utiyama and others took up the approach for investigations aiming at an overarching theory of strongly interacting fields and gravity.[22] Independently P.A.M. Dirac initiated a, formally, similar line of research with a look at possible connections between fields of high energy physics, gravity, cosmology and geophysics [36]. In particular, he was fond of the idea of a 'time dependent' gravitational constant $G(t)$.[23] It did not take long until the idea of a spontaneously broken conformal gauge theory of gravitation was also considered in the framework of Weyl geometry and brought into first contact with the rising standard model of elementary particle physics [25, 72, 97, 140]. Important for this move seemed to be that the obstacle of a negative energy ('ghost') scalar field or wrong sign of the gravitational constant, arising in the strictly conformal version of JBD theory, could be avoided in this framework.[24] Here we are not interested in historical details, but aim at sketching the potential of the approach from a more or less philosophical point of view.[25]

4.1   The Lagrangian of WOD Gravity

The affine connection of Weyl geometry is scale invariant; the same holds for its Riemannian curvature $Riem = (R^\kappa_{\lambda\mu\nu})$ and the Ricci tensor $Ric = (R_{\mu\nu})$ as its contraction.[26] Scalar curvature $R = g^{\mu\nu} R_{\mu\nu}$ is scale covariant of weight

---

[22] [70, 102, 103, 145, 146]—thanks to F. Hehl to whom I owe the hint to Omote's works.

[23] For a detailed study of the connection to geophysics see [87].

[24] Cf. fn. 20.

[25] For a first rough outline of the history see [132]. For a commented source collection of much wider scope [13].

[26] We use abbreviated symbols of geometrical objects, $Riem$, Ric, φ, ∇, etc., together with their indexed coordinate description. The whole collection of indexed quantities will be denoted by round brackets like in matrix notation, e.g. $Ric = (R_{\mu\nu})$ or $\varphi = (\varphi_1, \ldots, \varphi_n)$, in short $\varphi = (\varphi_\mu)$. The latter is somehow analogous to φ_μ in 'abstract index notation', often to be found in the literature. In our

$w(R) = w(g^{\mu\nu}) = -2$. Coupling of a norm squared real or complex scalar field[27] ϕ of weight -1 to the scalar curvature of Weyl geometry gives, for the Lagrangian density of the modified Hilbert term

$$\mathscr{L}_{HW} = L_{HW}\sqrt{|det\, g|} = -\frac{1}{2}\xi^2|\phi|^2 R\sqrt{|det\, g|}\,, \tag{7}$$

a total weight $-2 - 2 + 4 = 0$ and thus scale invariance.[28] If R denotes just that of Riemannian geometry and if one adds the kinematical term of the scalar field, Penrose's criterion for conformal invariance only holds for $\alpha = -\frac{1}{6}$. It is crucial to realize that in the Weyl geometric framework local scale invariance holds for *any* *coefficient*.

Conformal rescaling leads to different ways of decomposing covariant or invariant terms into contributions from the Riemannian component g and the scale connection φ of a representative (a 'scale gauge') (g, φ) of the Weylian metric. We characterize these components by subscripts put in front; e.g. for scalar curvature the decomposition is summarily written as $R =_g R +_\varphi R$, with $_g R$ the scalar curvature of the Riemannian part g of the metric alone and $_\varphi R$ the term due to the respective scale connection. For dimension $n = 4$ of spacetime one obtains (independently of the signature)

$$_\varphi R = -(n-1)(n-2)\varphi_\lambda\varphi^\lambda - 2(n-1)_g\nabla_\lambda\varphi^\lambda = -6\varphi_\lambda\varphi^\lambda - 6_g\nabla_\lambda\varphi^\lambda\,, \tag{8}$$

where $_g\nabla$ denotes the covariant derivative (Levi-Civita connection) of the Riemannian part g of the metric. Of course, the merging of scale dependent terms to scale invariant aggregates is of primary conceptual import, besides being calculationally advantageous.[29]

The gradient term of the scalar field in Omote–Dirac gravity is modelled after the kinematical term of a Klein–Gordon field:

$$L_\phi = \varepsilon_{sig}\frac{1}{2}D_\nu\phi^* D^\nu\phi\,, \qquad \mathscr{L}_\phi = L_\phi\sqrt{|det\, g|} \tag{9}$$

with scale covariant derivative $D_\nu\phi = (\partial_\nu - \varphi_\nu)\phi$, according to equ. (2), is scale invariant, as $w(L_\phi) = -4$. Here ε_{sig} specifies a signature dependent sign: $\varepsilon_{sig} = 1$ for $sig = (1, 3)$ i.e. $(+ - - -)$ and $\varepsilon_{sig} = -1$ for $sig = (3, 1) \sim (- + + +)$. In this paper we shall work with this kinematical term. In other contexts, e.g. in a Weyl geometric adaptation of the AQUAL approach to relativistic MOND dynamics, one

(Footnote 26 continued)

notation the bracketed symbol stands for the whole collection of indexed quantities, the unbracketed symbol for a single indexed quantity $\varphi_\mu \in \{\varphi_1, \ldots, \varphi_n\}$.

[27]Later the scalar field is allowed to take values in an isospin $\frac{1}{2}$ representation of the electroweak group, section 4.5.

[28]$w(\sqrt{|det\, g|}) = \frac{1}{2}4 \cdot 2 = 4$, $w(L_{HW}) = -2 - 2 = -4$.

[29]The authors of the 1970s usually did not use the aggregate notation.

has to allow for other forms of L_ϕ. In particular a cubic gradient terms may lead to new insights in modified gravity on the galactic and the cluster level.[30]

A polynomial potential for the scalar field $V(\phi)$ leads to a scale invariant Lagrange term if and only if the degree of V is four, i.e. for a quartic monomial

$$L_V = -\frac{\lambda}{4}|\phi|^4, \qquad \mathscr{L}_V = L_V\sqrt{|det\ g|}\ . \tag{10}$$

Considering the scale connection φ as a dynamical field, the 'Weyl field' with its quantum excitation, called 'Weyl boson' or even 'Weylon' by [25], demands to add a Yang–Mills action for the scale curvature $f = (f_{\mu\nu})$:

$$L_{YM\varphi} = -\frac{\beta}{4}f_{\mu\nu}f^{\mu\nu} \tag{11}$$

So did Omote, Dirac and later authors.[31]

The whole scale invariant Lagrangian of Weyl–Omote–Dirac gravity including the scalar field, neglecting for the moment further couplings to matter and interactions fields, is given by

$$L_{WOD} = L_{R^2} + L_{HW} + L_\phi + L_V + L_{YM} + L_m\ ,$$

where L_{R^2} contains all second-order curvature contributions. They seem to be necessary if one wants to study (perturbative) quantization, starting from this classical template. L_m denotes matter and interaction terms, for example the adapted standard model fields, $L_m = L_{SM}$ (lifted to curved Weyl space).[32]

$$L_{WOD} = L_{R^2} - \varepsilon_{sig}\frac{1}{2}\xi^2|\phi|^2 R - \frac{\lambda}{4}|\phi|^4 + \varepsilon_{sig}\frac{1}{2}D_\nu\phi^* D^\nu\phi - \frac{\beta}{4}f_{\mu\nu}f^{\mu\nu} + L_m \tag{12}$$

Formally it contains a Brans–Dicke like modified Hilbert action, a 'cosmological' term, quartic in ϕ, and dynamical terms for the scalar field and the scale connection. The Weyl geometric expressions for scalar curvature and scale covariant derivative ensure scale invariance of the Lagrangian density $\mathscr{L}_{WOD} = L_{WOD}\sqrt{|det\ g|}$. Scale invariance forces the polynomial part of the potential with constant coeffi-

---

[30]AQUAL stands for the 'aquadratic Lagrangian' approach, which was the first attempt at a relativistic version of MOND dynamics [9]. An adaptation to Weyl geometric gravity is investigated in [134, 135].

[31]Dirac, curiously, continued even in the 1970s to stick to the interpretation of the scale connection as electromagnetic potential. No wonder that this proposal was not accepted even in the selective reception of his work cf. [87].

[32]Signs are chosen such that ϕ has positive energy density (no ghost field) [62, 5]. In [13, equ. (8.5)] the coefficient α has to be assumed negative—compare with their source paper 8.3 [97], eqs. (2) and (7). For the role of L_{R^2} in quantum gravity see [22, 18ff., 62ff.] and, historically, [124]. For steps towards adapting the standard model Lagrangian to Weyl geometry (basically by writing it locally scale invariant) see, among others, [7, 41, 96, 98, 116].

cients to be exclusively quartic. Later we shall see that the assimilation of the standard model Lagrangian L_{SM} to gravity makes it necessary to modify the potential term $L_V = V(\phi) = -\frac{\lambda}{4}$ by introducing a combined quartic potential $V(\phi, \Phi)$ for the gravitational scalar field and the Higgs field Φ.

4.2 From WOD to iWOD Gravity

A closer look at the WOD-Hilbert term shows that, because of equ. (8), it contains a mass-like term for the scale connection (the 'Weyl field'):

$$\frac{1}{2}m_\varphi^2 \varphi_\lambda \varphi^\lambda = \frac{1}{2}6\xi^2|\phi|^2\varphi_\lambda\varphi^\lambda \tag{13}$$

If WOD describes a realistic modification of Einstein gravity, its Hilbert term has to approximate the latter very well under the limiting conditions $|\phi| \to const$, $\varphi \to 0$. Then $\xi^2|\phi|^2$ must be comparable to the inverse of the gravitational constant $\xi^2|\phi|^2 \approx [\hbar c](8\pi G)^{-1} = \frac{m_{pl}^2}{8\pi} = M_{pl}^2$ with reduced Planck mass M_{pl}.[33] Then the 'Weylon' (Cheng, Nishino/Rajpoot e.a.) turns out to be sitting a little above the reduced Planck mass (but below the unreduced one):

$$m_\varphi \approx 2.5 M_{pl} \approx 0.5\, m_{pl} \tag{14}$$

Variation of the Lagrangian shows that it satisfies a Proca equation with this tremendously high mass [25, 140]. Because of the scaling behaviour of ϕ the Proca-like mass term does not destroy scale invariance of the Lagrangian.[34]

If one assumes a physical role for the Weyl field, its (immediate) range, in the sense of its Compton wave length, would be restricted to Planck scale physics. On all scales accessible to experiments and to direct observation the *curvature of the Weyl field vanishes effectively*. This result agrees with the integrability result of Audretsch, Gähler and Straumann on the compatibility of Weyl geometry with quasi-classical relativistic quantum fields (section 2).[35] Although the scale curvature field (the Weylon) stays in the background it may become important for stabilizing (quantum) fluctuations of the scalar field, if one starts to investigate such problems more closely. Here we can, for most of our purposes, *pass to integrable Weyl geometry*.[36]

---

[33] $m_{pl}^2 = \frac{\hbar c}{G}$, with 'reduced' $M_{pl} := \sqrt{\frac{\hbar c}{8\pi G}}$.

[34] Therefore the, otherwise interesting, discussion of the gravitational scalar field as a kind of 'Stückelberg compensator' by [100] seems a bit artificial.

[35] This does not preclude the possibility for the Weylian scale connection (the 'Weyl vector field') to play a proper dynamical role in a high energy regime (at present far beyond accelerator energies). For a recent exploration of such a perspective see [101].

[36] In four space-time dimensions the collection of quadratic curvature terms then reduces to $L_{R^2} = -\alpha_1 R^2 - \alpha_2 R^{\lambda\nu}R_{\lambda\nu}$ [90]. The reduced form is assumed in [97, 389], [140, 260], [41, 1028]. It also

At many occasions also L_{R^2} may be neglected[37]; then the Lagrangian of *integrable Weyl–Omote–Dirac* (iWOD) gravity reduces effectively to

$$L_{iWOD} = -\varepsilon_{sig}\frac{\xi^2}{2}|\phi|^2 R + \varepsilon_{sig}\frac{1}{2}D_\nu\phi^* D^\nu\phi - \frac{\lambda}{4}|\phi|^4 + L_m \,. \tag{15}$$

That is very close to the Lagrangian used in recent publications on Jordan–Brans–Dicke theory, e.g. [62]. In Riemann gauge it is nearly identical with the 'modernized' JBD Lagrangian of the Jordan frame, the only difference being the $|\phi|^4$-term and the *explicit* scale invariance of the Lagrangian. In other gauges (frames) the derivative terms of the rescaling function are 'hidden' in the Weyl geometric terms.[38]

4.3   The Dynamical Equations of iWOD

Variation of the Lagrangian with regard to the Riemannian component of the metric leads to an Einstein equation very close to the 'classical' case; but now the curvature terms appear in *Weyl geometric* form.[39] For $\mathscr{L}_{iWOD}$ without further matter terms the modified *Einstein equation* becomes

$$Ric - \frac{R}{2}g = \Theta^{(\phi)} = \Theta^{(I)} + \Theta^{(II)} \,, \tag{16}$$

where the right-hand side is basically the energy–momentum $\Theta^{(\phi)}$ of the scalar field (multiplied by $(\xi|\phi|)^{-2}$). It decomposes into a term proportional to the metric, $\Theta^{(I)}$, therefore of the character of vaccum energy or 'dark energy', and another one which behaves matter-like (compare the special case studied in section 6.2), $\Theta^{(II)}$:

$$\Theta^{(I)} = |\phi|^{-2}\left(-D^\lambda D_\lambda|\phi|^2 + \varepsilon_{sig}\xi^{-2}\frac{\lambda}{4}|\phi|^4 - \frac{\xi^{-2}}{2}D_\lambda\phi^* D^\lambda\phi\right) g$$

$$\Theta^{(II)}_{\mu\nu} = |\phi|^{-2}\left(D_{(\mu}D_{\nu)}|\phi|^2 + \xi^{-2}D_{(\mu}\phi^* D_{\nu)}\phi\right) \tag{17}$$

(Footnote 36 continued)
covers the simplified expression of the gravitational Lagrangian in Mannheim's conformal gravity built on $L_{conf} = C_{\lambda\mu\nu\kappa}C^{\lambda\mu\nu\kappa}$, with C the Weyl tensor [93].

[37] P. Mannheim indicates that this may be acceptable only in the medium gravity regime; he considers the conformal contribution to extremely weak gravity as crucial [93].

[38] Cf. [3]. The old version of the JBD parameter corresponds to $\omega = \frac{1}{2}\xi^{-2}$. Contrary to what one might think at first glance, (15) *does not stand in contradiction* to high precision solar system observations, because the 'scale breaking' condition for the scalar field by the quartic potential prefers scalar field gauge ('Einstein frame')—see below.

[39] If one varies the Riemannian part of the metric g and the affine connection Γ separately (Palatini approach), the variation of the connection leads to the *compatibility condition* (3) of Weyl geometry [3, 112]. That gives additional (dynamical) support to the Weyl geometric structure. Further indications of its fundamental role comes from a completely different side, a $f(R)$ approach enriched by an EPS-like property [23].

The ('ordinary') summands with factor ξ^{-2} are derived from the kinematical ϕ-term of the Lagrangian; the other summands arise from a boundary term while varying the modified Hilbert action. Because of the variable factor $|\phi|^2$, the boundary term no longer vanishes like in the classical case.[40] The additional term is often considered as an 'improvement' of the energy momentum tensor of the scalar field [21].[41]

All terms of the modified Einstein equation of iWOD gravity (16) are *scale invariant*,[42] although the geometrical structure is richer than conformal geometry. Of course there arises the question whether such a geometrical framework may be good for physics, without specifying a preferred scale; i.e. before 'breaking' of scale symmetry. We shall see in the next section that there is a natural mechanism for such 'breaking', which is not mandatory (at the classical level) on purely theoretical grounds.

Constraining the variation to *integrable* Weylian metrics leaves no dynamical freedom for the scale connection; thus no dynamical equation arises for φ.[43] Varying with regard to a real scalar field ϕ, on the other hand, gives a Klein–Gordon type equation with a 'funny' mass-like term:

$$D_v D^v \phi + 2(\xi^2 R + \varepsilon_{sig} \lambda |\phi|^2)\phi + \frac{\delta L_m}{\delta \phi} = 0 \tag{18}$$

In a way, the scale connection φ and the scalar field ϕ are closely related. It is possible to scale ϕ to a constant, then in general $\varphi \neq 0$; on the other hand one can scale $\varphi = 0$, then in general $\phi \neq const$. The'kinematical' (descriptive) freedom of φ is essentially governed by the dynamics of ϕ. The *scalar field ϕ*, not the scale connection φ encodes the additional *dynamical degree of freedom* in the integrable (iWOD) case, far below Planck scale.

4.4   Ground State of the Scalar Field

There are no reasons to assume that ϕ represents an elementary field. Like many other scalar fields of known physical relevance it may characterize an aggregate state. From our context we may guess that it could represent an *order parameter of a collective quantum state*, perhaps a condensate, of the Weyl field. Such a conjecture

---

[40] [143, 64ff.], [12, 96ff.], [62, 40ff.].

[41] Callan, Coleman, and Jackiw postulated these terms while studying perturbative scattering theory in a weak gravitational field. They noticed that the ordinary energy momentum tensor of a scalar field does not lead to finite matrix elements 'even to the lowest order in λ'. The 'improved' terms lead to finite matrix terms to all orders in λ [21].

[42] Sometimes the scale transformations are called 'Weyl transformations' in this context, e.g. in [12].

[43] The variation of the Riemannian component of the metric can be restricted to Riemann gauge $(g, 0)$. Note the analogy to the variation in JBD gravity of the Riemannian metric with regard to the Jordan frame.

has been stated in [71, 263], [72, 1096], and similarly already in [97, 140]. Here we are not interested in details of the dynamics given by its variational Klein–Gordon equation, but mainly in the ground state which may be indicative for the transition to Einstein gravity.

Transition to integrable Weyl geometry is not yet sufficient to get rid of rescaling freedom. A *full breaking of scale symmetry*—like that of any other gauge group—contains *two* ingredients:

(a) effective vanishing of the curvature (field strength) at a certain scale,
(b) physical selection of a specific gauge.[44]

Up to now only step (a) has been taken. (b) involves a ground state of the scalar field with respect to the biquadratic potential given by its gravitational coupling if the scalar field has the chance to govern the behaviour of physical systems serving as 'clocks' or as mass units (see section 5).

For field theoretic investigations signature $sig(g) = (1, 3)$ is best suited, so that $\varepsilon_{sig} = +1$. Abbreviating the gravitational terms we get $L_{iWOD} = \frac{1}{2} D_\nu \phi^* D^\nu \phi - V_{grav}(\phi)$ with

$$V_{grav}(\phi) = \frac{1}{2} \xi^2 |\phi|^2 R + \frac{\lambda}{4} |\phi|^4 . \tag{19}$$

In most important cases, scalar curvature R of cosmological models is negative.[45] Thus the effective gravitational potential of the scalar field is biquadratic and of 'Mexican hat' type with two minima symmetric to zero, like in electroweak theory. Here, however, the coefficient of the quadratic term $\frac{\xi^2}{2} R$ is a point-dependent function, but may be scaled to a constant.

The scalar field assumes the gravitational potential minimum for

$$|\phi_o|^2 = -\frac{\xi^2 R}{\lambda} \quad \text{(in reciprocal length units)}, \tag{20}$$

and the 'funny' mass term of the Klein–Gordon equation (18) vanishes in the undisturbed ground state. For the moment we have to leave it open, which kind of disturbances might shift the scalar field away from its potential minimum of (20).

Of course, there is a scale gauge in which $|\phi_o|$ assumes constant values. We call it the *scalar field gauge* (of Weyl geometric gravity). Starting from any gauge (g, φ) of the Weylian metric, just rescale by $\Omega := C^{-1} |\phi_o|$ with any constant C. Because of it having scale weight -1, the norm of the scalar field then becomes $|\phi_o(x)| \doteq C$ in inverse length units; equivalently in energy units

---

[44] 'Physical' means a selection with observational consequences. Mathematically, the selection of a gauge corresponds to the choice of a section (not necessarily flat) in the corresponding principle fibre bundle, at least locally (in the sense of differential geometry).

[45] The higly symmetric Robertson–Walker models of Riemannian geometry, with warp (expansion) function $f(\tau)$ and constant sectional curvature κ of spatial folia, have scalar curvature $_gR = -6 \left((\frac{f'}{f})^2 + \frac{f''}{f} + \frac{\kappa}{f^2} \right)$ in signature $(1, 3) \sim (+ - - -)$. For $\kappa \geq 0$, or at best moderately negative sectional curvature, and accelerating or 'moderately contracting' expansion, $_gR < 0$.

$$|\phi_o(x)|[\hbar c] \doteq C\hbar c =: |\phi_c| \tag{21}$$

with some constant energy value $|\phi_c|$. The *dotted equality* $\doteq$ expresses that the relation is no longer scale invariant but holds in a specific gauge only, here in the scalar field gauge.

With C such that $\xi^2 C^2 = (8\pi G)^{-1}[\frac{c^4}{\hbar c}]$ (G gravitational constant) the coefficient of the iWOD-Hilbert term (15) goes over into the one of Einstein gravity. Then

$$|\phi_c| = \xi^{-1}\left(\frac{\hbar c^5}{8\pi G}\right)^{\frac{1}{2}} = c^2 M_{pl} = E_{pl} \quad \text{(reduced Planck energy)}, \tag{22}$$

and the coupling constant ξ^2 turns out to be basically a squared hierarchy factor between the scalar field ground state in energy units and Planck energy E_{pl}.

4.5 Scale Invariant Observables and a New Look at 'dark Energy'

It is easy to extract a *scale invariant observable magnitude* $\hat{X}$ from a scale covariant field X of weight $w(X) = k$. One only has to form the proportion with regard to the appropriate power of the scalar field's norm

$$\hat{X} := X/|\phi|^{-k} = X|\phi|^k; \tag{23}$$

then clearly $w(\hat{X}) = 0$.

Scale invariant magnitudes $\hat{X}$ are directly indicated, up to a globally constant factor in scalar field gauge, i.e. the gauge in which $|\phi_o| \doteq const.$[46] Conceptually the problem of scale invariant magnitudes is solvable, even with full scaling freedom, but there are physical effects which lead to actually breaking scale symmetry. Atomic 'clocks' and 'rods' (atomic distances) express a preferred metrical scale. They stand in good agreement with other periodic motions of physics on different levels of magnitude.

The ordinary energy–momentum terms with scale covariant derivatives of ϕ in (17) get suppressed by the inverse squared hierarchy factor $\xi^{-2} < 10^{-32}$ (see section 5.3). Only the λ-term corresponding to the old cosmological term survives because it is of fourth order in $|\phi|$ and $|\phi|$ is sufficiently large. In the ground state $|\phi|^2$ can be expressed in terms of the scalar curvature, (20). Then the energy–momentum of the scalar field simplifies to (remember: $g = (g_{\mu\nu})$ stands for the whole metric):

---

[46]In [145] ϕ is therefore called a 'measuring field'; cf. [131].

$$\Theta^{(I)} \approx \left(-\frac{R}{4} - |\phi|^{-2} D^\lambda D_\lambda |\phi|^2 \right) g =: \Lambda g \tag{24}$$

$$\approx \left(-\frac{R}{4} - R^{-1} D^\lambda D_\lambda R \right) g$$

$$\Theta^{(II)}_{\mu\nu} \approx |\phi|^{-2} D_{(\mu} D_{\nu)} |\phi|^2 = R^{-1} D_\mu D_\nu R \tag{25}$$

This expresses a peculiar back-reaction of curvature (gravity) on itself via the scalar field, which is not present in Einstein gravity. Of course it also complicates the dynamical equations.[47]

Taking traces on both sides of the (iWOD) Einstein equation shows that in the *matter free* case, $L_m = 0$,

$$|\phi|^{-2} D^\lambda D_\lambda |\phi|^2 \approx 0 \,, \tag{26}$$

and the vacuum Einstein equation can be written in a trace free form

$$Ric - \frac{R}{4} g \approx \Theta^{(II)} \tag{27}$$

These identities signal a remarkable change in comparison with Einstein gravity and its problems with the cosmological constant. $\Theta^{(I)}$ represents a functional equivalent to the traditional 'vacuum energy' term, but here it is due to the scalar field. The coefficient Λ in (24) *depends on the geometry of iWOD gravity and thus, indirectly, on the matter distribution.* Moreover, $\Theta^{(II)}$ is an additional contribution to the energy momentum of the scalar field (25). Perhaps we can expect that some of the effects ascribed to *dark matter* may be due to it.

4.6 A First Try of Connecting to Electroweak Theory

It seems tempting to consider the electroweak energy scale v as a candidate for the value of the gravitational scalar field in scalar field gauge,

$$|\phi_c| = v \approx 246 \, GeV \,.$$

In this case, the value of the hierarchy factor would be $\xi = \frac{E_{pl}}{v} \sim 10^{16}$.
With

$$\lambda \sim 10^{-56}, \tag{28}$$

---

[47]In general, the order of the Einstein equation is raised to four, although in Weyl gauge it remains of second order!

the value of the scalar field's ground state is located, by (20), at the electroweak scale[48]:

$$[\hbar c]|\phi_o| = \hbar c \frac{\xi\sqrt{|R|}}{\sqrt{2\lambda}} \stackrel{\sim}{\sim} 10^{16-33+28} \, eV \sim 10^{11} \, eV \, , \quad |\phi_o| \stackrel{\sim}{\sim} 10^{16} \, cm^{-1} \quad (29)$$

This observation indicates a logically possible connection between Weyl gravity (iWOD) and electroweak theory, although the order of magnitude of λ looks quite suspicious. In the next section we explore a related, but more convincing transition which gives up the idea that the gravitational scalar field might be immediately identified with the Higgs field. Our goal is to find out whether there is a chance for the scalar field to determine the rate of clock ticking and to influence the units of mass by some relation to the electroweak theory.

5   A Bridge Between Weyl Geometric Gravity and ew Theory

Let us try to explore whether the Weyl geometric setting may contribute to conceptualizing the 'generation of mass' problem of elementary particle physics. Mass is the charge of matter fields with regard to the inertio-gravitational field, the affine connection of spacetime. In flat space, and thus in special relativity, that may fall into oblivion because there the affine connection is hidden under the pragmatic form of partial derivatives. The exercise of importing standard model fields to 'curved spaces', i.e. Lorentzian or Weyl–Lorentzian manifolds is conceptually helpful even if it is done on a classical level as a first step. Using Weyl geometry seems all the more appropriate, as nearly all of the Lagrange terms of the standard model of elementary particle physics (SM) are already conformally invariant. The only exception is the quadratic term of the Higgs field, $\frac{\mu^2}{2}|\Phi|^2$, with the dimensional factor μ^2. By means of the gravitational scalar field it can easily brought into a scale covariant form of the correct weight.[49]

5.1   Importing Standard Model Fields to IWG

Most contributions to the special relativistic Lagrange density $L_{SM}(\psi)dx$ of the standard model of elementary particles (SM) are invariant under dilations in Minkowski space. Dilational invariance is closely related to unit rescaling, but not identical. Assigning Weyl weight $w = -d$ to a field ψ of dilational weight d (often called

---

[48]Here $|R|\stackrel{\sim}{\sim}H^2$ with $H = H_1 \approx 7.6 \cdot 10^{-29} \, cm^{-1}$, respectively $\hbar c \, H \approx 1.5 \cdot 10^{-33} \, eV \sim 10^{-32} \, eV$. In section 6 we find good reasons to consider $R \doteq 24H^2$ (59).

[49][7, 96, 98, 100].

'dimension') gives an invariant Lagrangian density under global unit rescaling in special relativity.[50] Unit rescaling can be made point dependent, if the fields can be generalized to the Weyl geometric framework.

An energy/mass scale is introduced into the SM by the Lagrangian of the Higgs-e.a. mechanism.[51] One usually assumes that the Higgs field is an *elementary* scalar field with values in an isospin-hypercharge representation $(I, Y) = (\frac{1}{2}, 1)$ of the electroweak group $G_{ew} = SU(2) \times U(1)$.[52] At least two generations of particle physicists have been working in the expectation that this scalar field is carried by a massive boson of rest mass at the electroweak level ($\sim 100\ GeV$). Experimenters at the LHC have finally found striking evidence for such a boson with mass $m_H \approx 125$–$126\ GeV$ [28, 29].

Without going too much into detail, it can be stated that all the fields and differential operators of the standard model Lagrangian can be imported into Weyl geometry. The most subtle question is the representation of the Weylian covariant derivative for fermionic fields.[53]

The kinetic term of the special relativistic Dirac action $\frac{i}{2}(\psi^* \gamma^o \gamma^\mu \partial_\mu \psi - (\gamma^\mu \partial_\mu \psi)^* \gamma^o \psi)$ is conformally invariant if ψ is given the scaling weight $w(\psi) = -\frac{3}{2}$. After orthogonalizing the Levi-Civita connection by introducing tetrad coordinates (in the tangent bundle) it is locally given by a 1-form ω with values in $so(1, 3)$. Using the appropriate spin representation it can be 'lifted' to spinor fields.[54] In this way the Dirac action on 'curved' Lorentzian spaces acquires the form

$$\frac{i}{2}(\psi^* \gamma^o \gamma^j \nabla_j \psi - (\gamma^j \nabla_j \psi)^* \gamma^o \psi)\,, \qquad (30)$$

---

[50]Under the active dilation of Minkowski space $x \mapsto \tilde{x} = \Omega x$ ($\Omega > 0$ constant) a field ψ of dilational weight d transforms by $\psi(x) \mapsto \Omega^d \psi(\Omega^{-1}x)$ [111, 682ff.]. Invariance of the action $S = \int L(\psi)dx$ holds if $\int L(\psi(x))dx = \int \tilde{L}(x)\Omega^{-4}dx$. That is the case if and only if $\tilde{L} = \Omega^4 L$, thus $d(L) = 4$ and $w(L) = -4$ for Lagrangians invariant under dilations. Rescaling $\eta = diag(1, -1, -1, -1)$ by $\eta \mapsto \tilde{\eta} = \Omega^2 \eta$ leads to $L\sqrt{|det\ \eta|} = \tilde{L}\sqrt{|det\ \tilde{\eta}|}$ and thus to a scale invariant Lagrange density.

[51]Spelt out, Brout–Englert–Guralnik–Hagen–Higgs–Kibble 'mechanism'.

[52]With the ordinary Gellmann–Nishijima relation $Q = I_3 + \frac{1}{2}Y$ usually assumed in the literature. Drechsler uses a convention for Y, such that $Q = I_3 + Y$.

[53]Here we are mainly concerned with the Higgs sector, so we do not need to consider all details of the Weyl geometric version of $\mathscr{L}_{SM}$. For a complete formulation see [98, 100], similarly, from a purely conformal view [96]; for the ew sector see [37, 41, 131]. The scalar field and scale connection (Weylon) sector is introduced in [25]. A short discussion of the local bundle construction in Weyl geometry is given by [39]; for the Riemannian case see, e.g. [57, chap. 19].

[54]In 1929, Weyl and Fock noticed independently that in this construction a point-dependent phase can be chosen freely without affecting observable quantities. That implied an additional $U(1)$ gauge freedom and gave the possibility to implement a $U(1)$-connection [129]. Their original proposal to identify the latter with the electromagnetic potential was not accepted because all fermions would seem to couple non-trivially to the electromagnetic field. [108] gives the interesting argument that in electroweak theory the hypercharge field can be read as operating on the spinor phase, exactly like Weyl and Fock had proposed for the electromagnetic field [55, 154].

where the latin indices $i, j, k \ldots$ indicate tetrad coordinates, γ^j constant, standard Dirac matrices and ∇_j (here) the covariant spinor derivative. Notation here: $\psi^* = {}^t\overline{\psi}$, $\overline{\psi}$ complex conjugate, ^{t} transposition. All this can be done globally if the underlying spacetime manifold M is assumed to be *spin*, otherwise only locally.[55] The action is conformal invariant and is used in conformal approaches to gravity and SM fields [11, 85].[56]

For different choices of the representative of the metric, the conformal approach refers to *different* affine connections, but uses scale invariant Lagrangians and equations. In the Weyl geometric approach, on the other hand, rescaling does *not change* the affine connection and covariant derivative (see sect. 2.1, eq. (1)). Therefore the 'orthogonalized' Weyl geometric connection $\omega = (\omega^i{}_j)$, written as 1-form with values in $so(1, 3)$, contains a contribution of the scale connection φ, $\omega = {}_g\omega + {}_\varphi\omega$ (${}_g\omega$ the orthogonalized Levi-Civita connection of g).[57] This contribution is a *specific attribute* of the Weyl geometric coupling of the scale connection to spinor fields, while the usual gauge interaction vanishes (see below). ${}_\varphi\omega$ takes care for the spin connection being *unchanged* under rescaling. Without it the Audretsch/Gähler/Straumann consideration on streamlines of the WKB approximation could not hold indepedendently of the scale gauge (section 2.3).

Finally, the scale covariant derivative for Dirac spinors becomes

$$D_\mu\psi = \partial_\mu\psi + \frac{1}{4}[\gamma^i, \gamma^j]\,\omega_{ij\mu}\,\psi - \frac{3}{2}\varphi_\mu\psi$$

$$\mathcal{D}\psi = [\hbar c]\,\gamma^\mu D_\mu\psi\,, \tag{31}$$

with $w(\gamma^\mu) = -1$ and $w(\mathcal{D}\psi) = -\frac{5}{2}$.[58] The kinetic term of the action is formed analogous to (30). In simplified form (passing over the chiral decomposition of the spinor fields) the massless Dirac action and the corresponding Yukawa mass term can be written as

$$L_\psi = \frac{i}{2}(\psi^*\gamma^o\,\mathcal{D}\psi - (\mathcal{D}\psi)^*\gamma^o\psi) \tag{32}$$

$$L_Y = -\mu_\psi|\phi|\,\psi^*\gamma^o\psi$$

L_ψ and L_Y are of weight -4. Thus in the Weyl geometric theory not only the massless Dirac field but also the *massive* one has a *scale invariant* Lagrangian density. Due to

---

[55] M is *spin*, iff it admits a global $SL(2, \mathbb{C})$ bundle; then the Dirac operator can be defined globally, otherwise only locally (in the sense of differential geometry). A sufficient criterion is $H_2(M, \mathbb{Z}_2) = 0$.

[56] Thanks to P. Mannheim for insisting on this point; cf. [93, fns. 20, 21]. It is important for clarifying the specific difference between the conformal Dirac action and its Weyl geometric twin.

[57] ${}_\varphi\omega$ is of the form $(\varphi_i\eta_{jk} - \varphi_j\eta_{ik})\vartheta^k = \omega_{ijk}\vartheta^k$, where $\{\vartheta^i\}$ denotes the selected coframe basis, η the Minkowski metric, and Latin indices i, j of φ indicate its coframe coordinates [39, eq. (2.16)], [12, eq. (4.39b)].

[58] $\{\gamma^\mu, \gamma^\nu\} = 2g^{\mu\nu}$ implies $w(\gamma^\mu) = -1$, while by the same reason $w(\gamma_j) = w(\gamma^j) = 0$.

hermitian symmetrization the real-valued gauge couplings $-\frac{3}{2}\varphi_k\psi$ from (31) cancel in (32).[59]

We rebuild crucial aspects of the Higgs field in our framework by extending the scalar field of iWOD gravity to an electroweak bundle of appropriate maximal weight for G_{ew}, $(I, Y) = (\frac{1}{2}, 1)$. The scalar field turns into a field Φ with values in a point dependent representation space isomorphic to $\mathbb{C}^2$,

$$\Phi(x) = (\phi_1(x), \phi_2(x)) \,. \tag{33}$$

5.2 Two Steps in the Geometry of Symmetry Breaking

The usual 'mechanism' for electroweak symmetry breaking on the classical level consists of two components.

(I) By a proper choice of $SU(2)$ gauge $\Phi(x)$ is transformed into a 'down' state at every point; $\Phi(x) = (0, h(x))$, with complex valued $h(x)$.

(II) In the ground state of Φ, its (squared) norm, physically spoken the expectation value $< \Phi^*\Phi >$, is assumed to lie in a minimum of a quartic ('Mexican hat') potential. We write $\Phi_o = (0, h_o(x))$. In the classical Higgs theory its norm is a constant, $|h_o(x)| = const = v$.

In the physics literature (I) is considered as a *spontaneous breaking* of the $SU(2)$ symmetry. This happens without reducing the symmetry of the Lagrangian. For step (II) in the usual understanding of the Higgs procedure, a mass scale is introduced into the otherwise (globally) scale invariant Lagrangian of the standard model ; i.e. scale symmetry is *explicitly broken*. In our context, we have to reconsider the last point. But before we do so, we shall have a short look at the features of the spontaneous breaking in step (I). This will help us in transforming step (II) into a breaking of the spontaneous type, which we want to address in section 5.3.

The first step presupposes the ability to specify 'up' and 'down' states with regard to which the 'diagonal' subgroup of $SU(2)$ with generator $\sigma_3 = \frac{i}{2} diag(1, -1)$ is defined. Otherwise the $U(1)$ subgroup could be any of infinitely many conjugate ones.[60] Stated in more physical terms: How do we know in which 'direction' (inside $\mathbb{C}^2$) the 3-component of isospin has to be considered? This question, already important in special relativistic field theory, becomes pressing in a consequently 'localized' (in the physical sense) version of the theory; i.e. in passing to general relativity.

---

[59]Even for the non-integrable case this cancelling takes place [12, 81, ex.1], [94], already noted by [69, 440]. Although there is no gauge coupling of the Yang–Mills type, the scale covariance term $-\frac{3}{2}\varphi_k\psi$ has to be retained for consistency reasons (in the Lagrangian and the resulting Dirac equation). Dynamical effects of the scale connection result only from $_\varphi\omega$ ($\rightarrow$ fn. 57).

[60]There are infinitely many *maximal tori* subgroups, all of them can serve with equal right as 'diagonal' (Cartan) subgroup. The 'localization' (in the sense of physics) allows to make the selection point dependent.

In the following we shall consider the Weyl geometrically extended Higgs field Φ and investigate whether the (complex valued) down state component $h(x)$ of the Higgs field may be related to the gravitational scalar field $\phi(x)$.

It seems natural to assume that the *ground state of the electroweak vacuum field* $\Phi(x)$ *defines the down state of the vacuum representation* of the electroweak group, $(I, Y) = (\frac{1}{2}, 1)$, at every point x. Thus a subgroup $U(1)_o \subset SU(2)$ is specified as the isotropy group (fix group) of the complex ray generated by $\Phi(x)$ at each point. It singles out the I_3 and charge eigenstates in all associated representations of G_{ew}, and thus for the elementary fields. In consequence, an *adapted basis in each of the representation spaces* can be chosen at every point, such that wave functions of the up/down states get their usual form. The scalar field, e.g. goes over into the form of the preferred electroweak gauge (often called 'unitary gauge')

$$\Phi(x) = (0, h(x)),\qquad(34)$$

and the only degrees of freedom for Φ are those of h, a complex valued field.

In this way the Higgs field specifies, at each point $x \in M$, a subgroup $U(1)_o \subset SU(2)$, mathematically a maximal torus of $SU(2)$, in $G_{ew} = SU(2) \times U(1)$. The eigenspaces of $U(1)_o$ are the I_3 eigenstates of the corresponding isospin representation spaces with $I \in \{\frac{1}{2}k \mid k \in \mathbb{N}\}$. In physical terms, the ew dynamics is 'informed' by the Higgs field how the weak and the hypercharge group (or Liealgebra) are coordinated in the generation of electric charge, also for other (fermionic) representation spaces.[61] In this sense, the electroweak symmetry does not treat every maximal torus $(U(1))$ subgroup of the $SU(2) \subset G_{ew}$ equivalent to any other. The Higgs field, encoding an important part of the physical vacuum structure, seems to be crucial for the distinction.

In this way the Higgs-e.a. mechanism, can be imported to the general relativistic framework. The whole structure can still be transformed under point-dependent $SU(2)$ operations without being spoiled, i.e. it may be gauge transformed.[62] And even more importantly, if a $\mathfrak{su}(2)$ or $\mathfrak{g}_{ew}$ connection of non-vanishing curvature, i.e. an electroweak field, is present,[63] it is not reduced to one of vanishing curvature by

---

[61]Experiment has shown that for left-handed elementary fields (and for the 'vacuum') $I = \frac{1}{2}$. At any point of spacetime the charge eigenstates of left-handed elementary matter fields are specified by the dynamical structure of the vacuum as the eigenstates $(I_3 = \pm\frac{1}{2})$ of $U(1)_o$ and $Q = I_3 + \frac{1}{2}Y$. $(I, Y) = (\frac{1}{2}, -1)$ for (left-handed) leptons, $(I, Y) = (\frac{1}{2}, \frac{1}{3})$ for (left-handed) quarks, and $(I, Y) = (\frac{1}{2}, 1)$ for the 'vacuum'. For right-handed elementary fields the isospin representation is trivial, $(I, Y) = (0, 2Q)$.

[62]'Active' gauge transformations operate on the whole setting of $\Phi(x)$, $U(1)_o$ and the corresponding frame of up/down bases—similar to the diffeomorphisms of general relativity, considered as gauge transformations; they carry the metrical structure with them. The active transformations can be countered by 'passive' ones which, in mathematical terminology, are nothing but an adapted change of the trivialization of a principle fibre bundle and accompanying choices of standard bases (I_3 eigenvectors) in the associated representation spaces. After a joint pair of active and passive gauge transformations the wave functions expressed in 'coordinates' remain the same.

[63]Curly small letters like $\mathfrak{su}(2)$ and $\mathfrak{g}_{ew}$ denote the Liealgebra of the corresponding groups.

the pure presence of the scalar (Higgs) field. In that respect, *gauge symmetry remains intact* in the sense of both automorphism structure and dynamics.

The metaphor of 'breaking' gauge symmetries has been discussed broadly, often critically, in philosophy of science, cf. [60, 61]. It did not pass without objection among physicists either, e.g. [38]. For an enlightening historical survey of the rise of the electroweak symmetry breaking narrative and its important *heuristic and systematic role* see [15, 82]. From our point of view, it does not seem a particularly happy choice to speak of 'breaking' the $SU(2)$ symmetry at this stage. But it is true that the physical specification of the $U(1)_o$ subgroup (maximal torus) in $SU(2)$ by the scalar field allows to introduce standard sections (I_3 bases) and preferred trivializations of the representation bundles, corresponding to step (b) in the characterization of section 4 (above footnote 44). In this sense, the otherwise free choice of a trivialization is 'broken', and one can say that a full reduction of the electroweak symmetry, which presupposes vanishing of the curvature (the field strength) is *foreshadowed* by the presence of the scalar field. In such a sense there is no problem with the language of 'spontaneous breaking' of symmetries.

A *full breaking* of the dynamical symmetry will be accomplished when, in addition to a preferred gauge choice (trivialization), the physical conditions for an effective vanishing of the $SU(2)$ curvature component are given (step (a) in section 4.4). That is the *result of the gauge bosons acquiring mass*, rather than the origin and explanation of mass generation, although the mass splitting of the fermions is 'foreshadowed' by the physical choice of $U(1)_o$ subgroup (the 'I_3 direction' in more physical terms). This agrees well with S. Friederich's convincing discussion, including quantum field aspects, of the Higgs mechanism in [61]. We come back to this point in a moment.

The second aspect of the usual ew symmetry breaking scenario, (II) in the characterization above, consists of reducing the underdetermination of the (squared) norm of Φ, respectively the vacuum expectation value of $\Phi^*\Phi = |h|^2$. In the ordinary Higgs-e.a. mechanism that is achieved by ad hoc postulating a quartic potential of 'Mexican hat' type for the Higgs field. In the iWOD approach, a similar potential for the gravitational ϕ is *naturally* given by (19), with ground state in (20). It remains to be seen whether the Higgs potential can be related to it in a mathematically and physically convincing way.

Crucial for the Higgs-e.a. mechanism is the fact that covariant derivative terms of the scalar field in ew theory (the ew bundle) lead to *mass-carrying Lagrange terms* for the gauge fields, which are nevertheless *consistent with the full gauge symmetry*. This is, of course, just so in the ew-extended iWOD model. The kinematical term of the scalar field becomes now

$$L_\Phi = \frac{1}{2}\tilde{D}_\nu\Phi^*\tilde{D}^\nu\Phi,\tag{35}$$

$$\tilde{D}_\mu\Phi := (\partial_\mu - \varphi_\mu + \frac{1}{2}gW_\mu + \frac{1}{2}g'B_\mu)\Phi,$$

where the W_μ and B_μ denote the connections in the $\mathfrak{su}(2)$ and $\mathfrak{u}(1)$ component of the electroweak group respectively. The ew covariant derivative terms of (35) lead to scale covariant formal mass terms for the ew bosons.[64] After the settling of Φ in a ground state, $\Phi_o = (0, h)$, we hope to find an appropriate scale in which $h \doteq const = v$ ('*Higgs gauge*'). Then, after a change of basis (Glashow–Weinberg rotation), the formal mass terms turn into explicit ones

$$m_W^2 = \frac{g^2}{4} v^2 , \qquad m_Z^2 = \frac{g^2}{4 \cos \Theta^2} v^2 , \qquad (36)$$

with $\cos \theta = g \, (g^2 + g'^2)^{-\frac{1}{2}}$ like in special relativistic field theory.

Already in the special relativistic case it is much more difficult to establish G_{ew} invariant and scale invariant Lagrangian densities for the fermionic fields, in particular with regard to the mass terms.[65] The transfer to the Weyl geometric context is a smaller problem, once that has been achieved.[66] Basically one has to adapt the Dirac operator (31) to the Weyl geometrical context. In simplified form, the resulting Lagrangian for electrons can be written as

$$L_e = \frac{i}{2} (\psi_e^* \gamma^o \not{D} \psi_e - (\not{D} \psi_e)^* \gamma^o \psi) - \mu_e |\phi| \, \psi_e^* \gamma^o \psi_e , \qquad (37)$$

with μ_e the coupling coefficient for the interaction of ϕ and the electron field.

The fermions and the weak gauge bosons acquire their mass from their interactions with Higgs field in its ground state Φ_o. For the electron

$$m_e = \mu_e [hc] |\Phi_o| \doteq \mu_e v . \qquad (38)$$

Once the weak bosons have acquired mass m_w, the range of the exchange forces mediated by them is limited to the order of $l_w = \frac{\hbar c}{mc^2} \sim 10^{-16} \, cm$. At distances $d \gg l_w$ the curvature of the weak component in the group $G_{ew} = SU(2) \times U(1)$ vanishes effectively, the weak gauge connection can be 'integrated away', and the symmetry can be effectively reduced to $U(1)$. As a result, *electroweak symmetry is broken down to the electromagnetic subgroup.* That happens *because of the mass acquirement of the weak bosons—not the other way round.* In this respect the physical interpretation of our stepwise reduction deviates slightly from the standard account, although the basic structure of the Higgs-e.a. mechanism has been taken over in most respects.

We still have to face the fact that in the Weyl geometric setting even the ground state Φ_o has to be scale covariant of weight $w(\Phi) = -1$, just like the gravitational scalar field ϕ. We therefore have to look for a modification of the classical Higgs potential, adapting it to Weyl geometry and forging a bridge, a 'transition', between the electroweak (Higgs) scalar field and the gravitational one.

---

[64] $\frac{1}{4} g^2 |\Phi|^2 W_\mu W^\mu$ and $\frac{1}{4} g'^2 |\Phi|^2 B_\mu B^\mu$.

[65] Decomposition in chiral (left and right) states and the transformation on mass eigenstates for quarks (Cabibbo–Kobayashi–Maskawa (CKM) matrix) and leptons (Maki–Nakagawa–Sakate (MNS) matrix) have to be taken into account. The Yukawa Lagrangian for the fermions are simplest, if written in unitary gauge (34), but are gauge invariant, cf. fn 62.

[66] [38, 100, 131], cf. [96].

5.3 Intertwinement of the Higgs Field with Gravity's Scalar Field

The usual Higgs 'mechanism' works with a Lagrangian of the form

$$\mathscr{L}_\Phi = (\frac{\mu^2}{2}|\Phi|^2 - \frac{\lambda}{4}|\Phi|^4 + \frac{1}{2}D_\nu\Phi^*D^\nu\Phi + \ldots)\sqrt{|det\,\eta|}, \tag{39}$$

with η the Minkowski metric, $|\Phi|^2 = \Phi^*\Phi$, and μ^2, λ the effective values for the quadratic and quartic coefficients of the SM Lagrangian at the ew energy level. The coefficient of the quadratic term $\frac{\mu^2}{2}$ is dimensionful and of type energy/mass squared. In our convention it would correspond to a quantity of scale weight $w(\mu^2) = -2$. Formally this Lagrangian bares a close resemblance to the one of the Weyl geometric gravitational scalar field

$$\mathscr{L}_\phi = (-\frac{1}{2}\xi^2|\phi|^2R - \frac{\lambda}{4}|\phi|^4 + \frac{1}{2}D_\nu\phi^*D^\nu\phi + \ldots)\sqrt{|det\,g|}. \tag{40}$$

But we have seen that a direct identification is impossible because of the empirical constraints for the coupling coefficients.[67]

In order to make (39) locally scale invariant, we first replace the definite mass value μ by a scale covariant quantity which, for the sake of local scale covariance, has to be a scale covariant scalar function with real or complex values. Nevertheless a preferred scale indicating the level of electroweak energy $v = 246\,GeV$ for the expectation value of Φ,[68] clearly a constant relative to the definitions of measurement units has to arise naturally. For that we need some kind of 'spontaneous breaking' of the scale symmetry.

In section 4.4 we have observed that the gravitational scalar field ϕ shares features of such a spontaneous breaking by its coupling to the Weyl geometric scalar curvature (20), analogous to criterion (I) in section 5.2. In our context it seems very natural to consider the hypothesis that the Higgs field acquires it preferred ('broken') expectation value, and thus its mass, by its coupling to the gravitational scalar field.

The simplest form to achieve this is to assume a biquadratic potential

$$V_{bi}(\phi, \Phi) = \frac{\lambda}{4}(|\Phi|^2 - \alpha^2\phi^2)^2 + \frac{\lambda'}{4}\phi^4, \tag{41}$$

in addition to the modified Hilbert term from (40). Simplifying $|\Phi|^2 = h^2$ according to (34), the full gravitational potential becomes

---

[67]Moreover, a closer look at galactic and cluster dynamics may speak in favour of introducing another form of the gradient term L_ϕ, e.g. the cubic one of a weylianized AQUAL theory.

[68]More precisely we deal here with the square root of the expectation value $< \Phi^*, \Phi >$ abbreviated by $|\Phi|^2$.

$$V(\phi, h) = \frac{1}{2}\xi^2|\phi|^2 R + \frac{\lambda}{4}(h^2 - \alpha^2\phi^2)^2 + \frac{\lambda'}{4}\phi^4. \tag{42}$$

And the scalar field part of the Lagrangian $\mathscr{L}_{\phi\Phi} = L_{\phi\Phi}\sqrt{|det\ g|}$ is

$$L_{\phi\Phi} = -\frac{\xi^2}{2}|\phi|^2 R + \frac{1}{2}D_\nu\phi^* D^\nu\phi + \frac{1}{2}D_\nu\Phi^* D^\nu\Phi - V_{bi}(\phi, \Phi). \tag{43}$$

Similar locally scale invariant Lagrangians of the scalar field sector have been introduced and studied by several author groups during the last few years.[69] Of course, the Lagrangians studied in these papers differ among each other.[70] Here we consider (43) as a paradigmatic example and concentrate on the role of the gravitational coupling of ϕ for the emergence of a fixed (constant) value for $h(x)$ and, in this sense, for the 'spontaneous breaking' of scale symmetry.

For that we have to investigate, whether a common ground state of the two scalar fields exist, that is, we have to ask for a (local) minimum of $V(\phi, h)$ in both variables. An easy calculation shows that the gradient $grad\ V = (\partial_\phi V, \partial_h V)$ vanishes for $h_o = \alpha|\phi|_o$, $\phi_o^2 = -\frac{\xi^2}{\lambda'}R$, and that $V(\phi_o, h_o)$ is, indeed, a local minimum.[71] That shows that the ground state of the gravitational scalar field ϕ_o, compared with (20), is not affected by its coupling to h. The common ground state (ϕ_o, h_o) of the two fields is

$$\phi_o^2 = -\frac{\xi^2}{\lambda'}R, \quad h_o^2 = \alpha^2\phi_o^2 = -\frac{\alpha^2\xi^2}{\lambda'}R. \tag{44}$$

Like in section 4.4 we see that the *scalar field gauge* agrees with the *Weyl gauge*, if ϕ is in its ground state. Moreover, (44) shows that the 'Higgs field gauge' (i.e. the gauge in which the Higgs field is scaled to a constant vacuum expectation value) is identical to scalar field gauge and to Weyl gauge: There is one gauge in which all three, gravitational scalar field, Higgs field, and Weyl geometric scalar curvature are scaled to a constant norm. By obvious reasons we call it *Einstein–Weyl gauge* and denote the respective values by ϕ_c, h_c, R_c (lower c for 'constant').

---

[69] [7, 96, 98, 116] and others. A similar form of the scalar Lagrangian with *global scale invariance* is considered in [63]. The last-mentioned authors introduce their Lagrangian as the 'minimal scale invariant extension' of the SM and GR.

[70] All of the mentioned papers include a direct coupling of the Higgs field to the scalar curvature, but conclude that the effects can be neglected. Some are fascinated by the perspective to study the role of the Higgs field for cosmological 'inflation'. Meissner/Nicolai and Bars/Steinhardt/Turok do not use a Weyl geometric framework but consider conformal or 'Weyl scaling'. The last-mentioned group of authors study the effects of considering ϕ as a ghost field (inverse signs of gravitational couplings of ϕ and Φ and inverse signs of kinematical terms) on geodesic completability of cosmological models. Although all investigations deserve attention in themselves we need not, and cannot, go into more details here.

[71] $\partial_\phi V = R\xi^2\phi - \alpha^2\lambda\phi(h^2 - \alpha^2\phi^2) + \phi^3\lambda'$, $\partial_h V = \lambda h(h^2 - \alpha^2\phi^2)$, and for ϕ_o, h_o as above $Hessian\ (V)_{|(\phi_o, h_o)} > 0$ (positive definite).

From empirical observation we get constraints

$$\xi^2 \phi_c^2 = M_{pl}^2 \approx (2.4 \cdot 10^{18} \, GeV)^2, \qquad \alpha^2 \phi_c^2 = v^2 \approx (246 \, Gev)^2. \qquad (45)$$

Therefore

$$\frac{\xi}{\alpha} \sim 10^{16}, \quad \text{the ew-Planck hierarchy factor.} \qquad (46)$$

If we assume $\lambda' \sim 1$, we read off from (44) and (45) that ϕ_c lies 'logarithmically in the middle' between R (in Einstein–Weyl gauge) and M_{pl}; i.e. ϕ_c is the geometrical mean between the two[72]:

$$|R|^{\frac{1}{2}} \xrightarrow{\ \xi/\sqrt{\lambda'}\ } |\phi_c| \xrightarrow{\ \xi\ } M_{Pl}$$

For $R \overset{\sim}{\sim} 10 \, H^2$ with Hubble constant H and $\hbar c H \sim 10^{-33}$ and $\lambda' \sim 1$, we find[73] that the order of magnitude of the second hierachy factor (between the energy level of the scalar field's ground state and Planck energy) is $\xi \sim 10^{30}$. The ew scale lies close to the geometrical mean between ϕ_c and M_{pl}:

$$H \xrightarrow{\ \xi\ } |\phi_c| \xrightarrow{\ \alpha\ } v \xrightarrow{\ \alpha'\ } M_{pl},$$

where $\xi = \alpha\alpha'$, $\alpha \sim 10^{14}$, $\alpha' \sim 10^{16}$. The effective (classical) value of λ is constrained by the observational values of the Higgs mass $m_h \approx 126 \, GeV$ and $v \approx 256 \, GeV$ to $\lambda \approx 0.24$.[74]

At first glance one might expect that λ' is constrained by dark energy considerations. But this is not the case, as one can check by inspecting the changes in the energy tensor (17) of the scalar sector after introducing the Higgs field. In the ground state of the scalar fields the only changes arise from the contribution of the kinematical terms of h. They are suppressed like the ones of ϕ in the effective approximation (24, 25).[75] Thus the *energy momentum tensor of the combined scalar fields* (ϕ, Φ) *in their ground state is still given by* (24), (25) like in the single gravitational scalar field case of section 4.4.

---

[72]This relation may lie at the bottom of some of the 'large number coincidences' which fascinated Eddington, to a lesser degree Weyl, and others.

[73]For the estimates of R and H see footnote 48.

[74]The tachyonic mass term of the Higgs field $\frac{1}{4}\alpha^2|\phi_c|^2 = \frac{1}{2}v^2$ turns into a real mass term for the Higgs excitation, $m_h^2 = \lambda v^2$, thus the value for λ.

[75]In (42, 43) the Higgs field Φ is not coupled to R; therefore no boundary term of the variation of the Hilbert term appears. (This is similar for the direct Higgs coupling to R considered by the authors mentioned in fn. 69, because in all cases the Higgs coupling to scalar curvature is by far outweighed by the dominating ϕ term ($\sim M_{pl}^2$)). The quartic term of h does not deliver a contribution to the energy tensor because in the ground state it is cancelled by the contribution from $\alpha^2\phi^2$. The additional kinematical terms for Φ, (those with factor ξ^2 in (17) are suppressed as indicated in the main text.

The often discussed question why the quartic term of the Higgs field does not dominate gravitational *vacuum energy* in the cosmological term finds a completely convincing *explanation on the classical level*. Moreover, while in Einstein gravity the cosmological constant term results in the anomalous feature of vacuum energy of being able to influence the dynamics of matter and geometry without backreacting to them, this problematic feature is dissolved here (like in other JBD-like approaches).

These are pleasing results of our investigation of the intertwinement between the Higgs field and the gravitational scalar field. Let us resume the most important qualitative (structural) results:

– The Higgs coupling to gravity considered here does *not affect the energy–momentum tensor* of the scalar sector.
– In its ground state the intertwined two gravitational scalar fields adapt to the Weyl geometric scalar curvature like in the case of 'pure' gravitational scalar field (section 4.4).
– Therefore the vacuum energy not only influences matter and geometrical dynamics, but also backreacts to the latter.
– Different to what one finds in the respective literature,[76] there is *no complete decoupling* of the electroweak sector from gravity in the 'low' energy regime ...,
– ... because the dimensional parameter μ^2 of the ordinary Higgs mechanism is derived from the scale covariant coupling with the gravitational scalar field.
– The ground state of the latter *is determined by the coupling to gravity* ($\xi^2\phi^2 R$ term).
– In this sense, the *two scalar fields* are gravitationally combined like *twins*.[77] Only taken together they induce a kind of 'spontaneous breaking' of (local) scale symmetry.

The last point deserves to be discussed in more detail in the next section.

5.4 A Weylian Hypothesis Reconsidered

The proportionality between the squared scalar field's value with R has the most important consequences for our understanding of measurement processes. Quantum mechanics teaches us how atomic spectra depend on the mass of the electron. The energy eigenvalues of the Balmer series in the hydrogen atom are governed by the Rydberg constant R_{ryd},

$$E_n = -R_{ryd}\frac{1}{n^2}, \qquad n \in \mathbb{N}. \tag{47}$$

---

[76]Cf. fn 69.

[77]We may hope that a deeper understanding of the emergence of the scalar field sector can lead to a common quantum field theoretical origin of the two related classical fields.

The latter (expressed in electrostatic units) depends on the fine structure constant α_f and on the electron mass, thus finally on the norm of Higgs field[78]:

$$R_{ryd} = \frac{e^4 m_e}{2\hbar^2} = \frac{\alpha_f^2}{2} m_e c^2 = \frac{\alpha_f^2}{2} \sqrt{\mu_e} \, v c^2 \qquad (48)$$

This equation is a classical idealization; with field quantization the fine structure constant α_f, and with it R_{ryd}, become scale dependent.[79]

In our scale covariant approach the masses of elementary fermions depend on indirect coupling to gravity as argued in 5.3. The Rydberg 'constant' turns into a scale covariant quantity of weight -1 and scales with ϕ, while the electron charge is considered as a 'true' (nonscaling) constant. In scalar field gauge (in other words, in Einstein gauge) the Rydberg factor is also scaled to constant (on the classical level) together with ϕ and h. In terms of (44) it is

$$R_{ryd} = \frac{\alpha_f^2}{2} \mu_e \, h_o \, c^2 = \frac{\alpha_f^2}{2} \alpha \, \mu_e \, |\phi_0| \, c^2 \doteq \frac{\alpha_f^2}{2} \alpha \, \mu_e \, |\phi_c| \, c^2 \, . \qquad (49)$$

Similarly, the usual atomic unit of length for a nucleus of charge number Z is the Bohr radius $l_{Bohr} = \frac{\hbar}{Z e^2 m_e}$ and gets rescaled just as well, like $|\phi|^{-1}$.

That is, typical *atomic time intervals* ('clocks') and *atomic distances* ('rods') are *regulated by the ew scalar field's ground state* $|h_o|$. If the discussion of section 5.3 hits the point, it is linked to the ground state of the gravitational scalar field and thus to Weyl geometric scalar curvature. Under the assumptions of section 5.3, a definition of units for central physical magnitudes like in the new SI rules establishes a measurement system in which *the value of* $|h|$ *is set to a constant by convention*, If it is evaluated in the framework of iWOD gravity (and presupposing the correctness of the laws linking measurement procedures to natural constants on which the SI regulations are based) that corresponds to fixing Einstein gauge for actual measurements.[80]

In the end, the scaling condition of Einstein gauge (= Weyl gauge) and (49) gives a surprising justification for an ad hoc assumption introduced by Weyl during his 1918 discussion with Einstein. Weyl conjectured that atomic spectra, and with them rods and clocks, adjust to the 'radius of the curvature of the world' [152, 309]. In his view, natural length units are chosen in such a way that scalar curvature is scaled to a constant, the defining condition of what we call *Weyl gauge*. In the fourth edition of *Raum–Zeit–Materie* (translated into English by H.L. Brose) he wrote:

> In the same way, obviously, the length of a measuring rod is determined by adjustment; for it would be impossible to give to *this* rod at *this* point of the field any length, say two or three times as great as the one that it now has, in the way that I can prescribe its direction arbitrarily. The world-curvature makes it theoretically possible to determine a length by adjustment. In

---

[78]Vacuum permissivity $\varepsilon_o = (4\pi)^{-1}$; then $e^2 = 2\alpha_f \varepsilon_o hc = \alpha_f \hbar c$.

[79]C. Hölbling and R. Harlander made me aware of this problem.

[80]Cf. fn. (14). Although the calculation of the spectral lines of ^{133}Caesium is more involved, the dependence on electron mass remains.

consequence of this constitution the rod assumes a length which has such and such a value in relation to the radius of curvature of the world. [152, 308f.]

The electroweak link explored in section 5.3 thus underpins a feature of Weyl geometric gravity which was introduced Weyl in a kind of 'a priori' speculative move. In the fifth (German) edition of *Raum–Zeit–Materie* Weyl already called upon Bohr's atom model as a first step towards justifying his scaling conjecture:

> Bohr's theory of the atom shows that the radii of the circular orbits of the electrons in the atom and the frequencies of the emitted light are determined by the constitution of the atom, by charge and mass of electron and the atomic nucleus, and Planck's action quantum.[81]

At the time when this was written, Bohr had already derived (47) and (48) for the Balmer series of the hydrogen atom and for the Rydberg constant [107, 201]. Weyl saw, at first, no reason to give up his scale gauge geometry. He rather continued:

> The most recent development in atomic physics has made it likely that the electron and the hydrogen nucleus are the fundamental constituents of all matter; all electrons have the same charge and mass, and the same is true for all hydrogen nuclei. From this it follows with all evidence that *the masses of atoms, periods of clocks and lengths of measuring rods are not preserved by some tendency of persistence; it rather is a result of some equilibrium state determined by the constitution of the structure (Gebilde), onto which it adjusts so to speak at every moment anew* (emphasis in original).[82]

The claim that 'it follows with all evidence' was, of course, an overstatement. It is well known how Weyl himself shifted his gauge concept from scale to phase only a few years later (in the years 1928–1929). After this shift he reinterpreted the Bohr frequency condition. In later discussions he referred to it as an argument *against* the physicality of his scale gauge idea.[83]

This shift gives evidence to a paradoxical double face of Weyl's remarks with regard to the Bohr frequency condition. For Weyl it may have contained a germ for the later distantiation from his first gauge theory, hidden behind an all too strong rhetoric of 'evidence'. But now it appears again in a completely new light. Read in a systematical perspective, Weyl's remarks from 1922/23 can now even appear as foreshadowing *a halfway marker on the road towards a bridge between gravity and atomic physics*. Whether this bridge resists depends, of course, on the answer to the question whether or not the link discussed here between the scalar fields of gravity and of ew theory is realistic ('physical'). This question is open for further research.

---

[81] "Die Bohrsche Atomtheorie zeigt, daß die Radien der Kreisbahnen, welche die Elektronen im Atom beschreiben und die Frequenzen des ausgesendeten Lichts sich unter Berücksichtigung der Konstitution des Atoms bestimmen aus dem Planckschen Wirkungsquantum, aus Ladung und Masse von Elektron und Atomkern ..." [153, 298].

[82] "Die neueste Entwicklung der Atomphysik hat es wahrscheinlich gemacht, daß die Urbestandteile aller Materie das Elektron und der Wasserstoffkern sind; alle Elektronen haben die gleiche Ladung und Masse, ebenso alle Wasserstoffkerne. Daraus geht mit aller Evidenz hervor, daß *sich die Atommassen, Uhrperioden und Maßstablängen nicht durch irgendeine Beharrungstendenz erhalten; sondern es handelt sich da um einen durch die Konstitution des Gebildes bestimmten Gleichgewichtszustand, auf den es sich sozusagen in jedem Augenblick neu einstellt.*" (loc. cit., emph. in or., 298).

[83] Compare, for example, Weyl's remarks in [155, 83].

At the end of the 1920s there was no chance for anticipating the electroweak pillar of the bridge. Historically, Weyl was completely right in considering the Bohr frequency condition as an indicator that his early scale gauge geometry could not be upheld as a physical theory in its original form. Weyl's original interpretation of the scale connection as the electromagnetic potential became obsolete in the 1920s, but his ad hoc hypothesis that *Weyl gauge* indicates measurements by material clocks and 'rods' most directly may now get new support.[84]

6 Another Look at Cosmology

It is of interest to see how cosmology looks from the vantage point of scale covariant gravity, not only in order to test the latter's formal potentialities on this level of theory building, but also because certain features of recent observational evidence of cosmology are quite surprising: dark matter and dark energy, distribution and dynamics of dwarf galaxies, lacking correlation of metallicity with redshift of galaxies and in quasars (i.e. no or, at best, highly doubtful indications of evolution), too high metallicity in some deep redshift quasars and the intriguing, but as yet unexplained, distribution of quasar numbers over redshift.[85] It would not be surprising if some of these develop into veritable anomalies for the present standard model of cosmology. At least they seem to indicate that some basic changes in the conceptual framework for cosmological model building may be due.

At the moment we neither can claim that these (potential) anomalies will be resolved by Weyl geometric gravity, nor are cosmological investigations in the framework of Weyl geometry per se bound to go beyond the present picture of an expanding universe plus 'inflation'. Often these studies are still committed to the standard picture.[86] But the above-mentioned problems may be taken as a reason for reflecting the status of present cosmology and to compare it with alternative approaches.

Weyl geometric gravity is not the only alternative 'on the market'; many others are being explored.[87] The number of publications which accept the present standard cosmology in the observable part but develop alternatives to the 'big bang' singularity seems to be rising.[88] Some of them may be worth considering in philosophical 'meta'-reflections on cosmology, complementary to philosophical investigations centred on more mainstream lines of investigation in cosmology.[89]

---

[84]This idea is discussed in more detail in [133].

[85][31, 68, 88, 89, 121, 125, 142].

[86]E.g. [100, 116].

[87]Some of them have been reviewed from a contemporary history view in [84–86] and the (quasi) steady state approach in [92]. Less discussed are different kinds of static or neo-static approaches [30, 95, 128, 130], or explorations of unconventional views on vacuum energy like in [49].

[88]Among them [7, 14, 110, 141].

[89]Very selectively, [8, 119, 139] and the recent volume *46* of *Studies in History and Philosophy of Science, Part A*.

6.1 Friedmann–Lemaitre Models in iWOD Gravity

One often uses approximate descriptions of cosmological spacetime by models with maximal symmetric spacelike folia, i.e. *Friedmann–Lemaitre–Robertson–Walker (FLRW)* manifolds with metric of the form

$$\tilde{g}: \quad d\tilde{s}^2 = d\tau^2 - a(\tau)^2 d\sigma_\kappa^2, \tag{50}$$

$$d\sigma_\kappa^2 = \frac{dr^2}{1 - \kappa r^2} + r^2(d\Theta^2 + \sin^2 \Theta \, d\phi^2).$$

The underlying manifold is $M \approx I \times S^{(3)}$, with $I \subset \mathbb{R}$ and $S^{(3)}$ three-dimensional. $S^{(3)}$ is endowed with a Riemannian structure of constant sectional curvature κ, locally parametrized by spherical coordinates (r, Θ, ϕ).[90]

For Weyl geometric FLRW models the behaviour and calculation of cosmological redshift is very close to what is known from the standard approach. The energy of a photon describing a null-geodesic $\gamma(\tau)$ considered by cosmological observers along trajectories of a cosmological time flow unit field $X(p)$, $p \in M$, $X = x'(\tau)$, is given by $E(\tau) = g(\gamma'(\tau), X(\gamma(\tau)))$.[91] Cosmological redshift is expressed by the ratio

$$z + 1 = \frac{E(\tau_o)}{E(\tau_1)} = \frac{g(\gamma'(\tau_o), X(\gamma(\tau_0)))}{g(\gamma'(\tau_1), X(\gamma(\tau_1)))}. \tag{51}$$

As we are working with geodesics of weight -1, $w(X) = -1$, and $w(g) = 2$, energy expressions for photons with regard to cosmological observers are *independent* of scale gauge; so is *cosmological redshift*.

In the standard view the warp function $a(\tau)$ is considered as an expansion of space with the cosmological time parameter τ. After an embedding of Einstein gravity into iWOD this view is no longer mandatory.[92] Even more, it is no longer convincing. If electroweak coupling—or any other mechanism leading to an analogous scale gauge behaviour—is realistic, Friedmann–Robertson–Walker geometries are better considered in Weyl gauge, i.e. scaled to constant scalar curvature in the Weylian generalization, than in Riemann gauge. In consequence, a large part of what appears as 'space expansion' $a(\tau)$ in present cosmology, perhaps even all of it, is encoded by the scale connection φ after rescaling to Weyl gauge.

In the result, the *cosmological redshift need not (exclusively) be due to expansion; it can just as well be a result of field theoretic effects expressed, in scalar field gauge,*

---

[90]Here ϕ is the usual designation of an angle coordinate. Contextual reading disentangles the dual meaning for ϕ we allow here. —For a survey of models with less symmetry constraints see [47], but consider the argumentation in [8].

[91]Cf. [24, 110, 116], for Weyl geometric generalizations, e.g. [112, 117, 130].

[92]Every Riemannian model (M, g) with Lorentzian spacetime M and metric g can easily be considered as an integrable Weyl geometric model with Weyl metric $[(g, 0)]$. If the dynamics is enhanced by a scalar field and scalar curvature of the model is $\neq 0$ the extension which is dynamically non-trivial. For a discussion of consequences for the view of gravitational effects see [118].

by both the warping of the Riemannian component of the metric and by the scale connection.[93] A similar argument that redshift may result from 'varying particle masses' was recently given in the framework of JBD gravity by [148].[94]

The counter argument that a quantum mechanical explanation is lacking and a necessary prerequisite for accepting the explanation is self-defeating, as the explanation by space expansion does not provide one either. Expansion or scale connection, both are essentially (gravitational) field theoretic effects and, in a scale covariant theory, even mutually interchangable.

6.2 A Static Toy Model: Einstein–Weyl Universes

If we extend our view from the classical cosmological models built upon Einstein's theory to scale invariant gravity, the picture of the 'universe' may change considerably. Models come into sight without any expansion at all, where the *whole cosmological redshift* is due to the scale connection φ. Toy models of such a type have been studied in [130].[95] The constraint for the scalar field, established here by the potential condition (20), facilitates the analysis considerably and allows to derive a surprising result with regard to dynamic equilibrium.

In *Riemann gauge*, these models can be represented as particularly simple Friedmann–Robertson–Walker spacetimes with a varying scalar field (a 'varying gravitational constant') and a linear warp ('expansion') function $a(\tau) = H\tau$.[96] In *Weyl gauge*, on the other hand, they exhibit a non-expanding spacetime, of course now with a non-vanishing scale curvature which contains all the information of the former warp function. After reparametrization of the timelike parameter $\tau = H^{-1}e^{Ht}$, the Weylian metric is given by

$$ds^2 = dt^2 - \left(\frac{dr^2}{1 - \kappa r^2} + r^2(d\Theta^2 + \sin^2\Theta \, d\phi^2) \right) = dt^2 - d\sigma_\kappa^2 \quad (52)$$
$$\varphi = (H, 0, 0, 0),$$

---

[93] The scale connection in scalar field gauge corresponds, in Riemann gauge, to a 'varying cosmological constant' and 'varying' particle masses and measuring units, regulated by the scalar field. See [112, 128, 130].

[94] Wetterich's reputation in the physics community helped to bring his argument into the *Nature* online journal http://www.nature.com/news/cosmologist-claims-universe-may-not-be-expanding-1.13379.

[95] The balancing condition between matter and the scalar field assumed there did not yet take the link to ew theory into account; therefore the dynamical assumptions of [130] differ from those discussed here and lead only to provisional results.

[96] Reparametrization of the time coordinate in Riemann gauge gives the picture of a 'scale expanding cosmos' [95] with exponential scale growth $ds^2 = e^{2HT}(ds^2 - d\sigma_\kappa^2)$. H the Hubble parameter observed today, cf. fn (97).

$(d\sigma_\kappa^2$ the metrik on the spacelike folia of constant curvature). These models have been called *Weyl universes*, in particular *Einstein–Weyl* universes for $\kappa > 0$ [130]. They are time homogeneous in a Weyl geometric sense.

The cosmological time flow remains static $x(\tau) = (\tau, \tilde{x})$ with $\tilde{x} \in S^{(3)}$. Coefficients of the Weylian affine connection are easily derived from the classical case, in particular $\Gamma^0_{00} = H$ and $\Gamma^i_{oi} = H$ $(i = 1, 2, 3)$, while all Γ^k_{ij}, for $i, j, k = 1, 2, 3$, are those of the spacetime folia (3-spaces of constant curvature). The parameter

$$\zeta := \frac{\kappa}{H^2} \tag{53}$$

characterizes Weyl universes up to isomorphism (Weyl geometric isometries).

The increment in cosmological redshift in Weyl universes is constant, and thus

$$z + 1 = e^{Ht} \tag{54}$$

or $z + 1 = e^{Hc^{-1}d}$ for signals from a point of distance d on $S^{(3)}$ from the observer (depending on 'which' H is meant, H_o or H_1).[97] In Weyl gauge it is described by the time component of the scale connection, $\varphi_o = H$.

Ricci curvature (independent of scale gauge) and scalar curvature in Weyl gauge are[98]

$$Ric = 2(\kappa + H^2)d\sigma_\kappa^2, \tag{55}$$
$$R = -6(\kappa + H^2). \tag{56}$$

In Weyl gauge the left-hand side of the generalized Einstein equation (16) has timelike component $3(\kappa + H^2)$ and spacelike entries $(\kappa + H^2)g_{ii}$, i.e. $-(\kappa + H^2)d\sigma_\kappa^2$, $(i = 1, \ldots, 3)$. That is familiar from classical static universe models. The absolute value of negative pressure $p\, g_{ii}$ is here $|p| = \kappa + H^2$, i.e. one third of the energy density $3(\kappa + H^2)$. The only difference to classical Einstein universes is marked by the H^2 terms.

In Einstein gravity, static universes are stricken by problems, even inconsistencies, with regard to their dynamics. It turns out impossible to stabilize them by a cosmological vacuum energy term or by substitutes. A natural question seems to ask, whether this may change by taking the energy momentum of the scalar field into account. Calculation of the scale covariant derivatives of $|\phi|^2$ for Weyl universes leads to[99]

---

[97]More precisely, one could distinguish between the time dimensional Hubble constant $H_o \approx 2.27\,10^{-18}\,s^{-1}$ and its length dimensional version $H_1 = H_o c^{-1} \approx 7.57\,10^{-29}\,cm^{-1}$ with its inverse, the *Hubble distance* $H_1^{-1} \approx 4.28\,Mpc$.

[98]Cf., e.g. [104], or any other textbook about Robertson–Walker spacetimes.

[99]Note that the scale covariant derivative of a function f of weight $w(f) = -2$ need not be zero, even if f is gauged to a constant. For Weyl universes $D_0 f = -2Hf$ and $D_0 D_0 f = D_0 D^0 f = 6H^2 f$, because of $\Gamma^o_{00} = H$. Moreover, $D_1 D^1 f = -2H^2 f$, similarly for $j = 2, 3$ because of $\Gamma^i_{oi} = H$ for $i, j = 1, 2, 3$; thus $D_\nu D^\nu |\phi|^2 = 0$ (wrong calculation in [130], corrected in [131, 64]).

$$\Theta^{(I)} = \frac{3}{2}(\kappa + H^2)g \tag{57}$$

$$\Theta^{(II)} = \text{diag}\,(6H^2 g_{00}, -2H^2 g_{11}, -2H^2 g_{22}, -2H^2 g_{33}) \tag{58}$$

(24), (25), (26). Comparison with (55, 56) shows that the Einstein equation holds for exactly one value of the spatial curvature,[100]

$$\kappa_o = 3\,H^2, \quad \text{i.e.} \quad \zeta_o = 3, \quad \text{then} \quad R_o = -24\,H^2. \tag{59}$$

We may call this special case the *balanced Einstein–Weyl universe*.

An inspection of the Friedmann equations in the Weyl geometric case leads to the sobering result that this balance is instable in the class of Robertson–Walker–Weyl models including the scalar field equation.[101] That seems to indicate that the Einstein–Weyl model with $\zeta = 3$ suffers from instability problems similar to those of the classical Einstein universe, and is thus relegated to the status of a toy model. But even then, its characteristic features, in particular cosmological redshift without expansion, are an interesting hint on how the perspective may change under a moderate change of the geometric framework of our gravity theory. Moreover, the picture changes if we bring the potential condition (20) into play. If the latter is of physical relevance on the cosmological level, the balanced Einstein–Weyl model seems to be the only vacuum solution of the overdetermined system constituted by the Friedmann equations, the scalar field equation and the potential condition (for our choice (15) of the coupling constant of L_ϕ).

6.3 Theory-Ladenness of Cosmological Observations

Positive curvature for spatial folia and static geometry stand in harsh contrast to many features of the present standard model of cosmology. Moreover, observational evidence of the cosmic microwave background CMB and from supernovae magnitude–luminosity characteristics, measured with such impressing precision during the last decades, seem to outrule not only balanced Einstein–Weyl universes, but also the whole attempt at extending cosmological modelling to the framework of Weyl geometric gravity.

But we should be careful. If we want to judge the empirical reliability of a new theoretical approach we have to avoid rash claims of refutation on the basis of empirical results which have been evaluated and interpreted in a theoretical framework differing in basic respects from the new one. *Theory-ladenness of the interpretation*

---

[100]$\kappa = 3\,H^2$ corresponds to $\Lambda = 6H^2$ with relative value $\Omega_\Lambda = 2$. Note that the 'dark matter' term $\Theta^{(II)}$ has positive pressure, characterized by $\frac{p}{\rho} = \frac{1}{3}$, and contributes $\Omega_{\Theta^{(II)}} = 2$ to the relative energy density.

[101]Contrary to hopes, based on too naive heuristic considerations, expressed in earlier preprint versions of this paper.

of empirical data is *particularly strong in the realm of cosmology*. Enlarging the symmetry of the Lagrangian by scale invariance comes down to a *drastic shift in the constitutive framework* for the formulation of physical laws. Judgement of such a shift demands careful comparative considerations. That has to be kept in mind in particular for the evaluation and conclusions drawn from the high-precision studies of the cosmic microwave background (Planck and WMAP data).

In the Weyl geometric approach, cosmological redshift looks like a field theoretic effect on the classical level; in scalar field gauge it is modelled by the combination of the (integrable) scale connection and the remaining warp factor of the spatial component of the Riemannian metric, rather than by a realistically interpreted 'space expansion'. The CMB may be explainable be a quantum physical background equilibrium state of the Maxwell field excited by stellar and quasar radiation, as was argued by I.E. Segal.[102] The correlation of the tiny inhomogeneities in the temperature distribution with large scale matter structures would be independent of the causal evolution postulated in the present structure formation theory. It has to be checked whether the flatness conclusion from CMB data is stable against a corresponding paradigm change.

Supernovae data have to be reconsidered in the new framework, in particular with view on possible observation selection effects.[103] Galaxy evolution would look completely different, as no big bang origin would shape the overall picture. In particular Seyfert galaxies and quasars can be understood as *late* developmental stages of mass accretion in massive galactic cores. Jets emitted from them seem to redistribute matter recycled after high energy cracking inside galactic cores. Structure formation would have to be reconsidered.[104] Nuclear synthesis would no longer appear as 'primordial' but could take place in stars on a much larger time scale than in the recieved view, and in galactic cores, respectively, quasars.[105]

Regenerative cycles of matter mediated by galactic cores, quasars and their jets are excluded as long as cosmology is based on Einstein gravity by the extraordinary role of its singularity structures ('black holes'). But these have to be reconsidered in the Weyl gravity approach.

Because of the Weyl gauge condition, local clocks tick slower in regions of strong gravity (large $_gR$) also in comparison with Riemann gauge. The resulting conformal rescaling demanded by the potential condition (20), Weyl gauge as Einstein gauge, and their influence on the rate of spectral clocks (47) changes the picture of the spacetime metric near singularities of the Riemann gauge (and also in comparison to Einstein gravity). We cannot be sure that the singularity structure is upheld. Conformal rescaling may change the whole geometry, similar to the effect that an initial

---

[102] According to Segal, the quantized Maxwell field on an Einstein universe will, under very general assumptions, build up an equilibrium radiation of perfect Planck characteristic [136].

[103] For a detailed argument that strong observation selection effects may come into the play in the selection procedures of the SNIa data see [30, sec. 4.6], for a first glance at supernovae data from the point of view of Einstein–Weyl universes [130].

[104] For a sketch of such a picture see [30] or [52].

[105] Then even the *Lithium 6/7 riddle* might dissolve quite unspectacularly.

singularity may be due to a 'wrong' (Riemannian) scaling of Friedmann–Robertson–Walker geometry in the case of Einstein–Weyl universes. Such investigations have started for Weyl geometric gravity by [113] and in a different perspective by [7].

But why should one head towards such an enterprise of basic reconsideration of the cosmological overall picture? Only a few astronomers or astrophysicists like to tackle such a complex task at the moment. Among them, David Crawford has been investigating for some time, how well different classes of observational evidence fits into the picture of a comological model with static spherical spatial folia. The outcome is not disappointing for this assumption [30]. The choice between an expanding space model or a (neo-)static one seems to be essentially determined by underlying (explicit or implicit) principles of gravity theory.[106]

Certain basic problems of the standard picture are being discussed in the present discourse on cosmology. There are different strategies to overcome them. The most widely known approaches for explaining the unexpected outer galaxy dynamics ascribe these effects to *dark matter* [121]. On larger scales the evolution and distribution of quasars deliver plenty empirical evidence, not always in full agreement with the 'old' picture. Quasar data of the Sloan Digital Sky Survey (SDSS), the 2dF group, and others outweigh the supernovae observations in number, precision and redshift range [125, 142]. A striking feature is that there is *no indication of evolution of metallicity* in quasars or galaxies along the cosmological timeline, i.e. in correlation to redshift.[107] Less well known, but perhaps even more important, are recent observations of distribution and dynamics of dwarf galaxies. They seem to indicate a fundamental inconsistency with the structure formation theory of the standard approach [88].

Such irritating observations, combined with diverging research strategies, are a worthwhile object for metatheoretical investigations in a pragmatic sense. The concentration on new classes of observational evidence is often crucial for the process of clarifying mutual vices and virtues of competing theories. That is the reason why we want to have a short glance at quasar distribution before we finish.

---

[106]Crawford assumes a peculiar dynamics of 'curvature cosmology' which claims to remain in the framework of Einstein gravity. It seems doubtful that this conception can be defended. But here we are mainly interested in the detailed investigation of observational evidence in parts I, II of [30].

[107]Another, at the moment isolated, inconsistency with the received picture of metallicity development is a quasar with redshift $z \approx 3.91$ and of extremely high metallicity (Fe/O ratio about 3) observed by [68]. Still it is considered as irritating only for the standard picture of star, galaxy, and quasar evolution [31]. But it could foreshadow more.

6.4 A Geometrical Explanation of Quasar Distribution?

The *distribution of quasars* in dependence of redshift shows a distinctive asymmetric bell shape with a soft peak between $z \approx 0.9$ and 1.6 and at first a rapid, then slackening, decrease after $z \approx 2$ shown in figure 1.[108]

In standard cosmology the regular distribution curve is a riddle which calls for ad hoc explanations of quasar formative factors. In the toy model of the balanced Einstein–Weyl universe, the distribution pattern would be easy to explain: Here it turns out to be *close to the volume increments of the backward lightcone* with rising redshift in the balanced Einstein–Weyl universe (fig. 2).[109]

The deviation of the SDSS number counts from the calculated curve of the balanced Einstein–Weyl universe consists of fluctuations and some remaining, rather plausible, observational selection effects: a moderate excess of counts below $Z = 1$ and a suppression of observed quasars above $z \approx 2$. All in all, *the curves agree surprisingly well* with the assumption of an *equal volume distribution of quasars in large averages* in the balanced Weyl universe. But there arises a new question: The conjugate point on the spatial sphere is reached at $z = e^{H\pi/c} - 1 = e^{\frac{\pi}{\sqrt{3}}} - 1 \approx 5.13$ ($r = \frac{1}{\sqrt{\kappa}}$ radius of the sphere). Interpreted in this model, quasars and galaxies with higher redshift than 5.13 ought to be images of objects 'behind' the conjugate point and should have counterparts with lower redshift on 'this' side. For terrestrial observers the two images are antipodal, up to the influence of gravitational deflection of the sight rays. In principle, it should be possible to check the 'prediction' of the Einstein–Weyl model of *paired antipodal objects* for the highest redshift quasars and galaxies with present observational techniques.[110]

At the moment such consequences have not yet been studied in sufficient detail. Maybe they never will, unless some curiosity of experts in gravity theory and in cosmology, both theoretical and observational, are directed towards studying some of the more technical properties of the iWOD approach.

For the 'metatheoretical' point of view, it becomes apparent already here and now, that important features of our present standard model of cosmology are not as firmly anchored in empirical evidence as is often claimed. They are highly dependent on the interpretive framework of Riemannian geometry which plays a constitutive role for Einstein gravity. Although we have very good reasons to trust this framework on closer, surveyable astronomical scales—at least on the solar system level—it

---

[108]Best data come from the 2dF collaboration and the Sloan Digital Sky Survey [125, 142]. Here we take the data of SDSS 5th data release; total number of objects 77 429 (fig. 1 upper curve), SDSS corrections for selection effects reduces the total number by half [125]; the total number of the corrected collective is 35 892. The maximum of the corrected distribution is manifestly a little above $z \approx 1$; the authors give $z = 1.48$ as the median of the collection.

[109]The maximum is reached around the equator of the spatial sphere. For $\kappa = 3H^2$ the equator corresponds to redshift $z_{eq} = e^{H\frac{\pi}{2}(\sqrt{3}H)^{-1}} - 1 \approx 1.47$ (54).

[110]The pairing of redshift and magnitudes are easy to calculate. But gravitational deflection of light disturbs the direction and local deviation from spherical symmetry close to the conjugate point blurs the focussing of light rays and affects magnitudes and redshift. Therefore an effective decision of this question could be a true challenge for observational cosmology.

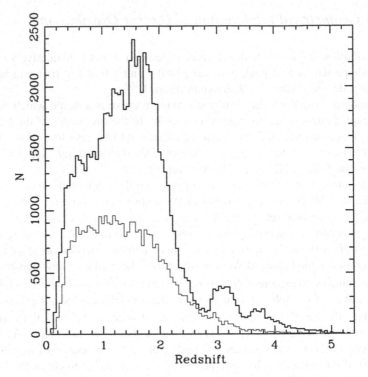

Fig. 1 Redshift distribution of quasars from SDSS, 5th data release, width of redshift bins 0.05; upper curve raw data, lower curve corrected for selection effects; source [125, Fig. 3].

is not clear at all whether we can trust its extrapolation to the gigantic scales far above cluster level. The proposal of modified Newtonian dynamics (MOND) for explaining galaxy rotation curves may be understood as a sign that we cannot be sure that Einstein gravity describes gravity with the necessary precision already at outer galaxy level.[111]

7 Review of 'transitions'

We have seen how Weyl geometry offers a well-structured intermediate step between the conformal structure and the projective path structure of physics and a fully metrical geometry (section 2). Riemannian geometry is only slightly generalized, if the Weyl geometric scale connection is integrable. Low energy quantum physics gives convincing arguments to accept this constraint for considerations far below the Planck scale (Audretsch/Gähler/Straumann, section 2.3, and mass of the 'Weyl

---

[111]For other anomalous evidence see fn. 85 and, in particular, the above mentioned study of dwarf galaxies in [88].

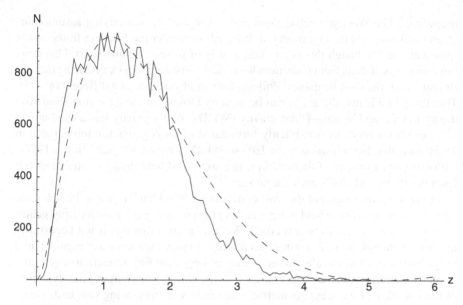

Fig. 2 Redshift distribution of quasars from SDSS, 5th data release, corrected for selection effects (zigzag curve), in comparison with equally distributed objects, volume increments over redshift bins of width 0.05, in Einstein–Weyl universe $\zeta = 3$ (dotted curve).

boson', section 4.2). As the standard model Lagrangian of elementary particle physics is (nearly) invariant under point-dependent rescaling, a scale invariant generalization of Einstein gravity is a natural, perhaps necessary, intermediate step for bridging the gap between gravitation theory and elementary particle fields. There are encouraging indications that integrable Weyl geometry may be helpful for the search of deeper interconnections between gravity and quantum structures. Recently, the authors of [26] have proposed a quantization procedure of a Weyl (scale) invariant classical Lagrangian, which preserves Weyl invariance for the effective (quantized) action. Some experts expect a resolution of the notorious fine tuning problem for the Higgs mass from such a move.[112]

In the 1980s, not in its beginnings,[113] Jordan–Brans–Dicke theory was explored for similar reasons, although in a different theoretical outlook and, up to now, without striking success [79, 80]. A conceptual look at Jordan–Brans–Dicke theory shows that the latter's basic assumptions presuppose, usually without being noticed, the basic structure of integrable Weyl geometry (section 3). From a metatheoretical standpoint it seems surprising that this has been acknowledged explicitly only very

---

[112]In a scale invariant Lagrangian the radiative corrections to the Higgs mass are expected to become only logarithmic rather than quadratic [7]; for global scale invariance see the similar argument in [137].

[113]The different motivations in the early phase are described by [87].

recently.[114] The Weyl geometric view makes some of the underlying assumptions clearer and supports the arguments of those who consider the Einstein frame as the 'physical' one (although this is an oblique way of posing the question). Physicists often seem to withhold from such metatheoretical considerations by declaring them as formal—and 'thus'—idle games. Philosophers of physics are of a different opinion. That this game is not idle at all, can be seen by looking at the transition from JBD theory to Omote–Utiyama–Dirac gravity (WOD). WOD gravity has a Lagrangian close to JBD theory, but is explicitly formulated in Weyl geometric terms (section 4). Historically, the transition from JBD to WOD gravity took place in the 1970s; but only a tiny minority of theoreticians in gravity and field theory contributed to it from the 1980s and 1990s until the present.[115]

Perhaps the mass factor of the scale connection ('Weyl field') close to Planck scale contributed to the widely held belief that Weyl geometric gravity is an empty generalization as far as physics is concerned. We have argued that this is not necessarily the case. Although the scale connection φ is able to play the role of a dynamical field only close to the Planck scale—where it may be important for a transition to quantum gravity structures—it is an *important geometric device* for studying the dynamics of the interplay of the Weyl geometric scalar field with measuring standards (scale gauges) on lower energy scales. It is therefore not negligible even in the integrable version of Weyl–Omote gravity and closely related to the scalar field ϕ which has to be considered as the *new dynamical entity* in the integrable case. The latter may represent a state function of a quantum collective close to the Planck scale.

By conceptual reasons iWOD does not need breaking of scale co- or invariance; it allows to introduce scale invariant observable magnitudes with reference to any scale gauge of the scalar field (section 4.5). There are physical reasons, however, to assume such 'breaking' of a spontaneous type, if one takes the potential condition for the scalar field's ground state into account. A quartic potential of Mexican hat type arises here from the gravitational coupling of the scalar field. Formally, it is so close to the potential condition of the Higgs mechanism in electroweak theory that it invites us to consider an extension of the Weyl geometric scalar field to the electroweak sector (section 5). We then recover basic features of the so-called Higgs mechanism of electroweak theory, but now without assuming an elementary field with an 'ordinary' mass factor in the classical Lagrangian. From a metatheoretical point of view this closeness allows to illucidate the usual narrative of 'symmetry breaking' in the electroweak regime. We have seen how the *mass acquirement of weak bosons and elementary fermions may come about by a bridge between the Higgs field to gravity* via a coupling of the two scalar fields (section 5.3). But of

---

[114]See the first preprint version of this paper, arXiv:1206.1559v1 and [115] which was first posted on arXiv in 2011.

[115]Of course other contributions could be mentioned. Perhaps most extensive, and not yet mentioned here, are the contributions of N. Rosen and M. Israelit, cf. the provisional survey in [132].

course we cannot judge, at the moment, whether such a link indicated by iWOD is more than a seducive song of the syrenes.[116]

From the point of view of the iWOD generalization of Einstein gravity we have reasons to reconsider our view of cosmology. The potential condition established by the electroweak link of the scalar field 'breaks' scale symmetry most naturally in such a way that Weyl geometric scalar curvature is set to a constant. That corresponds to an idea of Weyl formulated in 1918 (section 5.4). It may induce us to have a new look at the Friedmann–Lemaitre models of classical cosmology, readapted to the Weyl geometric context.

The consequences of such a shift cannot yet be spelled out in detail. Models of constant scalar curvature and time homogeneity (Weyl universes) show interesting unexpected features. The Einstein–Weyl universe with $\kappa = 3H^2$ is a *balanced vacuum solution* (section 6.2). Although it seems to be instable in the solution space of the Friedmann equations and the scalar field equation, it becomes distinguished if the potential condition (20) is added. Certain empirical data, in particular from quasar distribution and from metallicity, would even fit surprisingly well to the model (section 6.4).

On the conceptual level there is a fundamental argument in favour of the model. A (neo-) static universe of the Einstein–Weyl type *could bring back energy and momentum conservation to cosmology*. The 'expanding' universe with its permanent increase of energy in the observed part of the universe by the cosmological constant term ('dark energy') has very unpleasant consequences for the asymptotics of local field constellations. Einstein–Weyl universes have a group of automorphisms of type $SO(4) \times \mathbb{R}$, inside the larger group of ('gauge like') diffeomorphisms as in Einstein general relativity. For local inhomogeneities constellations with Einstein–Weyl asymptotics, we may therefore expect that asymptotic time homogeneity symmetry $(\mathbb{R}, +)$ and the 6 spacelike symmetry generators of the cosmological model lead to integral charge conservation for (on-shell) field constellations.[117] This difference to the expanding space view might invite physicists and philosophers alike to consider the chances and the advantages of a paradigm shift from the expanding view to the Einstein–Weyl framework.

In this framework, *dark energy* changes its character already at the classical level. It is generated by the metric proportional part of the energy momentum of the scalar field $\Theta^{(I)}$. Not only does it influence spacetime geometry, but it also reacts back to curvature. In addition, the question of *dark matter* might get a new face, if the respective gravitational effects can be explained by the part of the scalar field's energy momentum, $\Theta^{(II)}$, not proportional to the metric. At the moment this is only a

---

[116] A full-fledged (non-integrable) Weyl geometric framework in the strong energy regime of particle physics, in which the scale symmetry is broken down to Einstein gravity, not to integrable Weyl geometry, is studied in [101].

[117] Asymptotically 'conserved' (i.e. closed) $(n-2)$-forms derived from the superpotentials of Noether II currents have, in many similar cases, been shown to lead to conserved charges defined by the flux of the superpotential forms through the asymptotic (closed) $(n-2)$-dimensional boundaries of spacelike hypersurfaces [1, 6, 154]. See A. Sus' contribution to this volume.

speculation; an important open question would be to study the quantitative behaviour of inhomogeneities of $\Theta^{(II)}$ around galaxies and clusters in the iWOD approach.

In the end, the question is whether a MOND-like phenomenology can be recovered by Weyl geometric gravity. A chance for this may be opened by studying a modified kinetic term L_ϕ of the scalar field, similar to the one of Bekenstein/Milgrom's AQUAL theory. A first look at an adaptation of this approach to the Weyl geometric setting is encouraging, at least from a conceptual point of view. It would be interesting, if it works out for the gravitational dynamics of galaxies as well as MOND and, perhaps, even better for galaxy clusters.[118]

In such an approach we may have to give up the received view of cosmological redshift as an effect of 'space expansion' and to substitute it by an effect of the whole Weylian metric, including the scale connection (section 6.1). Rescaling of the metric, in particular in regions of strong gravity (high Riemannian component of scalar curvature), changes the effective measure of time and length so strongly that in this regime no immediate transfer of geometrical results derived in classical gravity to the new context is possible. It is no longer clear that cosmological geometry necessarily contains an initial singularity.[119]

Let us, at the end, come back to the philosopher quoted at the beginning of this article. Herbart—talking about metaphysics—described transitions between established theories, which he called the 'different formative stages' of knowledge, as *revolutions* which have to be traversed before research can generate concepts necessary for a 'distinguished enduring' state [74, 198, 199]. Also he spoke of the 'manifold delusions (mannigfaltige Täuschungen)' which our knowledge has to pass before such an enduring state can be reached. Riemann considered these remarks important enough for excerpting them.[120]

It may be that in present cosmology (although it is embellished by the attribute 'precision' at many occasions) we still have to leave behind 'manifold delusions', before we have a chance to arrive at a more enduring picture of how the universe in the large and the foundations of physics may go together.

Acknowledgements I thank the referees of the first version of this paper, in particular P. Mannheim. Their detailed remarks helped to improve content, structure and readability of the paper. I also want to thank G. Ellis for his interest in the Weyl geometric approach to gravity and for his comments. M. Krämer, R. Harlander and C. Hölbling gave advice with regard to section 5.3. F. Hehl was so kind to give helpful comments in spite of his general scepticism with regard to the integrable Weyl

---

[118]First steps in this direction are attempted in [134, 135].

[119]A similar question may be posed for localized singularities. Their external dynamics might be caused by finite matter concentrations which mimick structures of the black hole type if considered in Einstein gravity.

[120]"Wie die astronomische Betrachtung, die in die Tiefen des Weltbaues hinausgeht, so muß auch die metaphysische Forschung, welche in die Tiefen der Natur eindringt, mancherley Revolutionen durchlaufen, ehe sie so glücklich ist solche Begriffe zu *erzeugen*, welche der Erscheinung genugthun und mit sich selbst zusammenstimmen" [74, 198, emph. in original]. The section ends by the remark "Daraus folgt dann sogleich, *daß auch die Täuschungen, die in diesem Werden nach einander entstehen, sehr mannigfaltig, daß sie den verschiedenen Bildungsstufen angemessen sind, welche successiv erreicht werden;* ..." (ibid, 199). For Riemann's Herbart studies see [126].

geometric approach to gravity (for his critical remarks see his contribution to this volume). Recently H. Ohanian explained to me why, at least in his approach of including a fully dynamical Weylian scale connection to elementary particle physics at highest energies (close to the Planck scale), there is no chance for getting a low energy limit different from Einstein gravity. But most of all I express my gratitude to Dennis Lehmkuhl, the main editor of this book. Without his activity, his interest and the discussions in our interdisciplinary group on *Epistemology of the LHC*, based at Wuppertal, this essay would not have been written.

Postscript

The first version of this paper was written in summer 2012, some months before the Higgs detection was announced. Four years went by until publication. That gave plenty occasion for my rethinking basic questions of Weyl geometric gravity. Clear evidence of the author's 'manifold delusions' is documented in the successive versions of this paper in arXiv:1206.1559.
E.S., March 2016.

References

1. Abbott, L.F.; Deser, Stanley. 1982. "Charge definition in nonabelian gauge theories." *Physics Letters B* 195:76ff.
2. Adler, Ronald; Bazin, Maurice; Schiffer Menahem. 1975. *Introduction to General Relativity*. New York etc.: Mc-Graw-Hill. 2nd edition.
3. Almeida, T.S.; Formiga, J.B., Pucheu, M.L.; Romero, C. 2014. "From Brans-Dicke gravity to a geometrical scalar-tensor theory." *Physical Review D* 89:064047 (10pp.). arXiv:1311.5459.
4. Audretsch, Jürgen; Gähler, Franz; Straumann Norbert. 1984. "Wave fields in Weyl spaces and conditions for the existence of a preferred pseudo-riemannian structure." *Communications in Mathematical Physics* 95:41–51.
5. Bär, Christian; Fredenhagen, Klaus; eds. 2009. *Quantum Field Theory on Curved Spacetimes*. Vol. 786 of *Lecture Notes in Physics* Berlin etc.: Springer.
6. Barbich, Glenn; Brandt, Friedemann. 2002. "Covariant theory of asymptotic symmetries, conservation laws and central charges." *Nuclear Physics B* 633382. arXiv:hep-th/0111246.
7. Bars, Itzhak; Steinhardt, Paul; Turok, Neil. 2014. "Local conformal symmetry in physics and cosmology." *Physical Review D* 89:043515. arXiv:1307.1848.
8. Beisbart, Claus. 2009. "Can we Justifiably Assume the Cosmological Principle in order to Break Model Underdetermination in Cosmology?" *Journal for General Philosophy of Science* 40:175–205.
9. Bekenstein, Jacob; Milgrom, Mordechai. 1984. "Does the missing mass problem signal the breakdown of Newtonian gravity?" *Astrophysical Journal* 286:7–14.
10. Bezrukov, Fedor; Shapovnikov, Mikhail. 2007. "The standard model Higgs boson as the inflaton" *Physics Letters B* 659:703–706.
11. Birrel, N.D.; Davies, Paul C.W.. 1984. *Quantum Fields in Curved Spacetime*. Cambridge: University Press. 2nd edition (1st ed. 1982).
12. Blagojević, Milutin. 2002. *Gravitation and Gauge Symmetries*. Bristol/Philadelphia: Institute of Physics Publishing.
13. Blagojević, Milutin; Hehl, Friedrich W. 2013. *Gauge Theories of Gravitation. A Reader with Commentaries*. London: Imperial College Press.
14. Bojowald, Martin. 2009. *Zurück vor den Urknall*. Frankfurt/Main: Fischer.
15. Borrelli, Arianna. 2012. "The case of the composite Higgs: the model as a "Rosetta stone" in contemporary high-energy-physics." *Studies in the History and Philosophy of Modern Physics* 43:195–214.
16. Brans, Carl. 2005. "The roots of scalar-tensor theories: an approximate history." Contribution to Cuba workshop, 2004. arXiv:gr-qc/0506063.

17. Brans, Carl; Dicke, Robert H. 1961. "Mach's principle and a relativistic theory of gravitation." *Physical Review* 124:925–935.
18. Buchbinder, I.L.; Odintsov, S.D.; Shapiro, I.L. 1992. *Effective Action in Quantum Gravity*. Bristol/Philadelphia: Institute of Physics Publishing.
19. Bureau international des poids et mesures. 2011. "Resolutions adopted by the General Conference on Weights and Measures (24th meeting), Paris, 17–21 October 2011." www.bipm. org/en/si/new_si/.
20. Calderbank, David; Pedersen, Henrik. 2000. Einstein-Weyl geometry. In *Surveys in Differential Geometry. Essays on Einstein Manifolds*, ed. C. Le Brun; M. Wang. Boston: International Press pp. 387–423.
21. Callan, Curtis; Coleman, Sidney; Jackiw Roman. 1970. "A new improved energy-momentum tensor." *Annals of Physics* 59:42–73.
22. Capozziello, Salvatore; Faraoni, Valerio. 2011. *Beyond Einstein Gravity. A Survey of Gravitational Theories for Cosmology and Astrophysics*. Dordrecht etc.: Springer.
23. Capoziello, Salvatore; de Laurentis, M.; Fatibene, L.; Francaviglia, M. 2012. "The physical foundations for the geometric structure of relativistic theories of gravitation: from GR to extended theories of gravity through Ehlers-Pirani-Schild approach." *International Journal of Geometric Methods in Modern Physics* 9:12500727 (18pp.).
24. Carroll, Sean. 2004. *Spacetime and Geometry*. San Francisco: Addison Wesley.
25. Cheng, Hung. 1988. "Possible existence of Weyl's vector meson." *Physical Review Letters* 61:2182–2184.
26. Codello, Alessandro; D'Orodico, G.; Pagani, C.; Percacci, Roberto. 2013. "The renormalization group and Weyl invariance." *Classical and Quantum Gravity* 30:115015 (22 pp.). arXiv:1210.3284.
27. Coleman, Sidney; Weinberg, Erick. 1972. "Radiative corrections as the origin of sponteneous symmetry breaking." *Physical Review D* 7:1888–1910.
28. Collaboration, ATLAS. 2012. "Observation of a new particle in the search for the standard model Higgs boson with the ATLAS detector at the LHC." *Physics Letters B* 716:1–29. arXiv:1207.7214.
29. Collaboration, CMS. 2012. "Observation of a new boson at a mass of 125 GeV with the CMS experiment at the LHC." *Physics Letters B* 716:30–61. arXiv:1297.7235.
30. Crawford, David F. 2011. "Observational evidence favors a static universe." *Journal of Cosmology* 13. arxiv:1009.0953.
31. Cui, Jinglei; Zhang, Xin. 2011. "Cosmic age problem revisited in the holographic dark energy model." *Physical Letters B* 690:233–238. arXiv:1005.3587v3.
32. Dappiaggi, Claudio; Fredenhagen, Klaus; Pinamonti Nicola. 2008. "Stable cosmological models driven by a free quantum scalar field." *Physical Review D* 77:104015. arXiv:0801.2850.
33. Degrassi, Giuseppe; Di Vita, Stefano; Isidori, Gino e.a. 2012. "Higgs mass and vacuum stability in the Standard Model at NNLO." *CERN-PH-TH/2012-134*, arXiv:1205.6497.
34. Deser, Stanley. 1970. "Scale invariance and gravitational coupling." *Annals of Physics* 59:248–253.
35. Dicke, Robert H. 1962. "Mach's principle and invariance under transformations of units." *Physical Review* 125:2163–2167.
36. Dirac, Paul A.M. 1973. "Long range forces and broken symmetries." *Proceedings Royal Society London A* 333:403–418.
37. Drechsler, Wolfgang. 1991. "Geometric formulation of gauge theories." *Zeitschrift f. Naturforschung* 46a:645–654.
38. Drechsler, Wolfgang. 1999. "Mass generation by Weyl symmetry breaking." *Foundations of Physics* 29:1327–1369.
39. Drechsler, Wolfgang; Hartley, D.. 1994. "The role of the internal metric in generalized Kaluza-Klein theories." *Journal of Mathematical Physics*: 35, 3571–3586
40. Drechsler, Wolfgang; Mayer, Meinhard E. 1977. *Fibre Bundle Techniques in Gauge Theories. Lectures in Mathematical Physics at the University of Austin*. Vol. 67 of *Lecture Notes in Physics* Berlin etc.: Springer.

41. Drechsler, Wolfgang; Tann, Hanno. 1999. "Broken Weyl invariance and the origin of mass." *Foundations of Physics* 29(7):1023–1064. arXiv:gr-qc/98020
42. Dürr, Stefan.; Fodor, Zoltàn; Frison J. e.a. 2008. "Ab-initio determination of light hadron masses." *Science* 322(5095):1224–1227.
43. Dvali, Gia; Gomez, Cesar. 2013. "Quantum compositeness of gravity: black holes, AdS and inflation." arXiv:1312.4795.
44. Ehlers, Jürgen; Pirani, Felix; Schild Alfred. 1972. The geometry of free fall and light propagation. In *General Relativity, Papers in Honour of J.L. Synge*, ed. L. O'Raifertaigh. Oxford: Clarendon Press pp. 63–84.
45. Einstein, Albert. 1949. *Autobiographical notes*. Vol. 7 of *The Library of Living Philosophers* La Salle, Illinois: Open Court.
46. Eisenstaedt, Jean; Kox, Anne (eds.). 2005. *The Universe of General Relativity*. Vol. 11 of *Einstein Studies* Basel etc.: Birkhäuser.
47. Ellis, George; van Elst, Henk. 1998. "Cosmological models (Cargèse lectures 1998)." *NATO Advanced Study Institiute Series C Math. Phys. Sci.* 541:1–116. arXiv:gr-qc/9812046v5.
48. Englert, François; Gunzig, Edgar; Truffin C.; Windey P. 1975. "Conformal invariant relativity with dynamical symmetry breakdown." *Physics Letters* 57 B:73–76.
49. Fahr, Hans-Jörg; Heyl, Michael. 2007. "Cosmic vacuum energy decay and creation of cosmic matter." *Naturwissenschaften* 94:709–724.
50. Faraoni, Valerio. 2004. *Cosmology in Scalar-Tensor Gravity*. Dordrecht etc.: Kluwer.
51. Faraoni, Valerio; Gunzig, Edgard. 1999. "Einstein frame or Jordan frame." *International Journal of Theoretical Physics* 38:217–225.
52. Fischer, Ernst. 2007. "An equilibrium balance of the universe." Preprint. arXiv:astro-ph/0708.3577.
53. Flato, Moshé; Raçka, Ryszard. 1988. "A possible gravitational origin of the Higgs field in the standard model." *Physics Letters B* 208:110–114. Preprint, SISSA (Scuola Internazionale Superiore di Studi Avanzate), Trieste, 1987 107/87/EP.
54. Flato, Moshé; Simon, J. 1972. "Wightman formulation for the quantization of the gravitational field." *Physical Review D* 5:332–341.
55. Fock, Vladimir. 1929. "Geometrisierung der Diracschen Theorie des Elektrons." *Zeitschrift für Physik* 57:261–277.
56. Folland, George B. 1970. "Weyl manifolds." *Journal of Differential Geometry* 4:145–153.
57. Frankel, Theodore. 1997. *The Geometry of Physics*. Cambridge: University Press. ²2004.
58. Fredenhagen, Klaus. 2017. "Locally covariant quantum field theory". In *XIVth International Congress on Mathematical Physics*, ed. J.-C. Zambrini. Singapore: World Scientific. Lecture Notes in Physics 721:61–67. arXiv:hep-the/0403007.
59. Fredenhagen, Klaus; Rehren, Karl-Henning; Seiler Erhard. 2006. "Quantum field theory: Where we are." Preprint. arXiv:hep-th/0603155.
60. Friederich, Simon. 2011. "Gauge symmetry breaking in gauge theories – in search of clarification." *British Journal for the Philosophy of Science* 55:645–665. philsci-archive.pitt.edu/8854.
61. Friederich, Simon. 2014. "A philosophical look at the Higgs mechanism." *Journal for General Philosophy of Science* 42:335–350.
62. Fujii, Yasunori; Maeda, Kei-Chi. 2003. *The Scalar-Tensor Theory of Gravitation*. Cambridge: University Press.
63. Garcia-Bellido, Juan; Rubio, Javier, Shaposhnikov, Mikhail; Zenhäuser, Daniel. 1970. "Higgs-dilaton cosmology: From the early to the late universe." *Physical Review D* 84:123504, arXiv:1107.2163.
64. Gilkey, Peter; Nikcevic, Stana; Simon, Udo. 2011. "Geometric realizations, curvature decompositions, and Weyl manifolds." *Journal of Geometry and Physics* 61:270–275. arXiv:1002.5027.
65. Goenner, Hubert. 2004. "On the history of unified field theories." *Living Reviews in Relativity* 2004-2. http://relativity.livingreviews.org/Articles/lrr-2004-2.
66. Gray, Jeremy (ed.). 1999. *The Symbolic Universe: Geometry and Physics 1890–1930*. Oxford: University Press.

67. Guth, Alan; Kaiser, David. 1979. "Inflationary cosmology: Exploring the universe from the smallest to the largest scales." *Science* 307:884–890.
68. Hasinger, Günther; Komossa, Stefanie. 2002. "The X-ray evolving universe: (ionized) absorption and dust, from nearby Seyfert galaxies to high redshift quasars." In *Proceedings of the workshop "XEUS – studying the evolution of the hot universe"*, ed. G. Hasinger, T. Boller, A. Parmer, MPE Report 281 (2003):285ff. arXiv:astro-ph/0207321.
69. Hayashi, Kenji; Kasuya, Masahiro; Shirafuji, Takeshi. 1977. "Elementary particles and Weyl's gauge field." *Progress of Theoretical Physics* 57:431–440.
70. Hayashi, Kenji; Kugo, Taichiro. 1979. "Remarks on Weyl's gauge field." *Progress of Theoretical Physics* 61:334–346.
71. Hehl, Friedrich; Mielke, Eckehard; Tresguerres Romualdo. 1988. Weyl spacetimes, the dilation current, and creation of gravitating mass by symmetry breaking. In *Exact Sciences and their Philosophical Foundations; Exakte Wissenschaften und ihre philosophische Grundlegung*, ed. W. Deppert, K. Hübner e.a. Frankfurt/Main etc.: Peter Lang pp. 241–310.
72. Hehl, Friedrich W.; McCrea, J. Dermott; Mielke Eckehard; Ne'eman Yuval. 1995. "Progress in metric-affine theories of gravity with local scale invariance." *Foundations of Physics* 19:1075–1100.
73. Herbart, Johann F. 1807. *Über philosophisches Studium*. Göttingen: Heynrich Dietrich. In (75 Bd. 2, 227–296.).
74. Herbart, Johann F. 1825. *Psychologie als Wissenschaft. Zweiter analytischer Teil*. Königsberg. In (75 Bd. 2, 227–339).
75. Herbart, Johann F. 1850–1852. *Sämtliche Werke in chronologischer Reihenfolge: Hrsg. K. Kehrbach, O. Flügel*. Langensalza: Beyer. Reprint Aalen: Scientia.
76. Higa, Tatsuo. 1993. "Weyl manifolds and Einstein-Weyl manifolds." *Commentarii Mathematici Sancti Pauli* 42:143–160.
77. Jordan, Pascual. 1952. *Schwerkraft und Weltall*. Braunschweig: Vieweg. 2nd revised edtion 1955.
78. Kaiser, David. 1994. "Induced-gravity inflation and the density perturbation spectrum." *Physics Letters B* 349:23–28.
79. Kaiser, David. 2006. "Whose mass is it anyway? Particle cosmology and the objects of a theory." *Social Studies of Science* 36(4):533–564.
80. Kaiser, David. 2007. "When fields collide." *Scientific American* (June) pp. 62–69.
81. Kaiser, David. 2006. "Conformal transformations with multiple scalar fields." *Physical Review D* 81:084044 (8pp.).
82. Karaca, Koray. 2013. "The construction of the Higgs mechanism and the emergence of the electroweak theory." *Studies in History and Philosophy of Modern Physics* 44: 1–16.
83. Khoze, Valentin. 2013. "Inflation and dark matter in the Higgs portal of the classically scale invariant standard model." Journal of High Energy Physics 2013 (November): 215ff. arXiv:1308.6338.
84. Kragh, Helge. 2006. "Cosmologies with varying speed of light: A historical perspective." *Studies In History and Philosophy of Modern Physics* 37:726–737.
85. Kragh, Helge. 2009a. "Contemporary history of cosmology and the controversy over the multiverse." *Annals of Science* 66:529–551.
86. Kragh, Helge. 2009b. "Continual fascination: The oscillating universe in modern cosmology." *Science in Context* 22:587–612.
87. Kragh, Helge. 2016. *Varying Gravity. Dirac's Legacy in Cosmology and Geophysics*. Science Networks. Heidelberg etc: Birkhäuser/Springer.
88. Kroupa, Pavel; Famey, B.; de Boer K.S.; Dabringhausen J. e.a. 2010a. "Local-Group tests of dark-matter concordance cosmology . Towards a new paradigm for structure formation." *Astronomy and Astrophysics* 523 id.A32, 22p. arXiv:1006.1647.
89. Kroupa, Pavel; Pawlowski, Marcel. 2010b. "Das kosmologische Standardmodell auf dem Prüfstand." *Spektrum der Wissenschaft* pp. 22–31.
90. Lanczos, Cornelius. 1938. "A remarkable property of the Riemann-Christoffel tensor in four dimensions." *Annals of Mathematics* pp. 842–850.

91. Lehmkuhl, Dennis. 2014. "Why Einstein did not believe that general relativity geometrizes gravity." *Studies in History and Philosophy of Modern Physics* 46B:316–326. http://philsci-archive.pitt.edu/9825/.
92. Lepeltier, Thomas. 2005. "Nouveau dialogue sur les deux grands système du monde." *Revue des Questions Scientifiques* 176:163–186.
93. Mannheim, Philip. 2006. "Alternatives to dark matter and dark energy." *Progress in Particle and Nuclear Physics* 56:340f–445. arXiv:astro-ph/0505266.
94. Mannheim, Philip. 2014. "PT symmetry, conformal symmetry, and the metrication of electromagnetism." Preprint. arXiv:1407.1820.
95. Masreliez, John. 2004. "Scale expanding cosmos theory I — an introduction." *Apeiron* 11:99–133.
96. Meissner, Krzysztof; Nicolai, Hermann. 2009. "Conformal symmetry and the standard model." *Physics Letters B* 648:312–317. arXiv:hep-th/0612165.
97. Nieh, Hwa-Tung. 1982. "A spontaneously broken conformal gauge theory of gravitation." *Physics Letters A* 88:388–390.
98. Nishino, Hitoshi; Rajpoot, Subhash. 2004. "Broken scale invariance in the standard model." Report number CSULB-PA-04-2. arXiv:hep-th/0403039.
99. Nishino, Hitoshi; Rajpoot, Subhash. 2008. "Standard model and SU(5) GUT with local scale invariance and the Weylon." *AIP Conference Proceedings* 881:82–93. arXiv:0805.0613.
100. Nishino, Hitoshi; Rajpoot, Subhash. 2009. "Implication of compensator field and local scale invariance in the standard model" *Physical Review D* 79:125025. arXiv:0906.4778.
101. Ohanian, Hans. 2016. "Weyl gauge-vector and complex dilaton scalar for conformal symmetry and its breaking." *General Relativity and Gravity* 48(25):DOI:10.1007/s10714-016-2023-8. arXiv:0906.4778.
102. Omote, M. 1971. "Scale transformations of the second kind and the Weyl space-time." *Lettere al Nuovo Cimento* 2(2):58–60.
103. Omote, M. 1974. "Remarks on the local-scale-invariant gravitational theory." *Lettere al Nuovo Cimento* 10(2):33–37.
104. O'Neill, Barrett. 1983. *Semi-Riemannian Geometry with Applications to Relativity.* New York: Academic Press.
105. O'Raifeartaigh, Lochlainn. 1997. *The Dawning of Gauge Theory.* Princeton: University Press.
106. Ornea, Liviu. 2001. "Weyl structures on quaternionic manifolds. A state of the art." Preprint Bucharest. arXiv:math/0105041.
107. Pais, Abraham. 1986. *Inward Bound: Of Matter and Forces in the Physical World.* Oxford: Clarendon.
108. Pawłowski, Marek. 1999. "Gauge theory of phase and scale." *Turkish Journal of Physics* 23:895–902.
109. Penrose, Roger. 1965. "Zero rest-mass fields including gravitation: asymptotic behaviour." *Proceedings Royal Society London A* 284:159–203.
110. Penrose, Roger. 2010. *Cycles of Time. An Extraordinary New View of the Universe.* London: Bodley Head.
111. Peskin, Micheal; Schroeder, Daniel. 1995. *An Introduction to Quantum Field Theory.* New York: Westview Press.
112. Poulis, Felipe P.; Salim, J.M. 2011. "Weyl geometry as a characterization of space-time." *International Journal of Modern Physics: Conference Series* V 3:87–97. arXiv:1106.3031.
113. Prester, Pedrag. 2013. "Curing black hole singularities with local scale invariance". Preprint. arXiv:1309.1188.
114. Quiros, Israel. 2013. "Scale invariance and broken electroweak symmetry may coexist together." Preprint. arXiv:1312.1018.
115. Quiros, Israel; Garcìa-Salcedo, Ricardo; Madriz Aguilar, José E.; Mators, Tonatiuh. 2013. "The conformal transformations' controversy: what are we missing." *General Relativity and Gravitation* 45: 489–518 arXiv:1108.5857.
116. Quiros, Israel. 2013. "Scale invariant theory of gravity and the standard model of particles." Preprint. arXiv:1401.2643.

117. Romero, C., Fonsec-Neto J.B.; Pucheu M.L. 2011. "Conformally flat spacetimes and Weyl frames." *Foundations of Physics* 42:224–240. arXiv:1101.5333.
118. Romero, C., Fonsec-Neto J.B.; Pucheu M.L. 2012. "General Relativity and Weyl frames." *International Journal of Modern Physics A* 26(22). arXiv:1201.1469.
119. Rugh, Svend E. ; Zinkernagel, Henrik. 2009. "On the physical basis of cosmic time." *Studies in History and Philosophy of Modern Physics* 40:1–19.
120. Ryckman, Thomas. 2005. *The Reign of Relativity. Philosophy in Physics 1915–1925*. Oxford: University Press.
121. Sanders, Robert H. 2010. *The Dark Matter Problem. A Historical Perspective*. Cambridge: University Press.
122. Santamato, E. 1984. "Geometric derivation of the Schrödinger equation from classical mechanics in curved Weyl spaces." *Physical Review D* 29:216–222.
123. Santamato, E. 1985. "Gauge-invariant statistical mechanics and average action principle for the Klein-Gordon particle in geometric quantum mechanics." *Physical Review D* 32:2615 – 2621.
124. Schimming, Rainer; Schmidt, Hans-Jürgen. 1990. "On the history of fourth order metric theories of gravitation." *NTM Schriftenreihe für Geschichte der Naturwissenschaften, Technik und Medizin* 27(1):41–48.
125. Schneider, Donald P.; Hall, Patrick B.; Richards Gordon T. e.a. 2007. "The Sloan Digital Sky Survey Quasar Catalog IV. Fifth data release." *Astronomical Journal* 134:102–117. arXiv:0704.0806.
126. Scholz, Erhard. 1982. "Herbart's influence on Bernhard Riemann." *Historia Mathematica* 9:413–440.
127. Scholz, Erhard. 1999. Weyl and the theory of connections. In 66. pp. 260–284.
128. Scholz, Erhard. 2005a. Einstein-Weyl models of cosmology. In *Albert Einstein — Chief Engineer of the Universe. 100 Authors for Einstein. Essays*, ed. J. Renn. Weinheim: Wiley-VCH pp. 394–397.
129. Scholz, Erhard. 2005b. Local spinor structures in V. Fock's and H. Weyl's work on the Dirac equation (1929). In *Géométrie au vingtième siècle, 1930 – 2000*, ed. D.Flament, J. Kouneiher, P. Nabonnand, J.-J. Szczeciniarz. Paris: Hermann pp. 284–301.
130. Scholz, Erhard. 2009. "Cosmological spacetimes balanced by a Weyl geometric scale covariant scalar field." *Foundations of Physics* 39:45–72. arXiv:0805.2557.
131. Scholz, Erhard. 2011a. "Weyl geometric gravity and electroweak symmetry 'breaking'." *Annalen der Physik* 523:507–530. arXiv:1102.3478.
132. Scholz, Erhard. 2011b. Weyl's scale gauge geometry in late 20th century physics. arXiv:1111.3220.
133. Scholz, Erhard. 2015. "Higgs and gravitational scalar field together induce Weyl gauge." *General Relativity and Gravitation* 47:7pp. arXiv:1407.6811.
134. Scholz, Erhard. 2016a. "MOND-like acceleration in integrable Weyl geometric gravity." *Foundations of Physics* 46:176–208. arXiv:1412.0430.
135. Scholz, Erhard. 2016b. "Clusters of galaxies in a Weyl geometric approach to gravity." *Journal of Gravity* 46: (in print). arXiv::1506.09138.
136. Segal, Irving E. 1983. "Radiation in the Einstein universe and the cosmic background." *Physical Review D* 28:2393–2401.
137. Shaposhnikov, Mikhael; Zenhäusern, Daniel. 2009. "Quantum scale invariance, cosmological constant and hierarchy problem." *Physics Letters B* 671:162–166.
138. Sharpe, Richard W. 1997. *Differential Geometry: Cartan's generalization of Klein's Erlangen program*. Berlin etc.: Springer.
139. Smeenk, Chris. 2005. "False vacuum: Early universe cosmology and the development of inflation." In (46, 223–258).
140. Smolin, Lee. 1979. "Towards a theory of spacetime structure at very short distances." *Nuclear Physics B* 160:253–268.
141. Steinhardt, Paul; Turok, Neil. 2002. "A cyclic model of the universe." *Science* 296:1436–1439.

142. Tang, Su Min; Zhang, Shuang Nan. 2005. "Critical examinations of QSO redshift periodicities and associations with galaxies in Sloan Digital Sky Survey Data." *Astrophysical Journal* 633:41–51. arxiv.org/pdf/astro-ph/0506366.

143. Tann, Hanno. 1998. *Einbettung der Quantentheorie eines Skalarfeldes in eine Weyl Geometrie — Weyl Symmetrie und ihre Brechung*. München: Utz.

144. Trautman, Andrzej. 2012. "Editorial note to J. Ehlers, F.A.E. Pirani and A. Schild, The geometry of free fall and light propagation." *General Relativty and Gravity* 441:1581–1586.

145. Utiyama, Ryoyu. 1975a. "On Weyl's gauge field." *General Relativity and Graviation* 6:41–47.

146. Utiyama, Ryoyu. 1975b. "On Weyl's gauge field II." *Progress of Theoretical Physics* 53:565–574.

147. Vizgin, Vladimir. 1994. *Unified Field Theories in the First Third of the 20th Century. Translated from the Russian by J. B. Barbour*. Basel etc.: Birkhäuser.

148. Wetterich, Christian. 2013. "Universe without expansion." Preprint arXiv:1303.6878.

149. Weyl, Hermann. 1918a. "Gravitation und Elektrizität." *Sitzungsberichte der Königlich Preußischen Akademie der Wissenschaften zu Berlin* pp. 465–480. In (156, II, 29–42) [31], English in (150, 24–37).

150. Weyl, Hermann. 1918b. "Reine Infinitesimalgeometrie." *Mathematische Zeitschrift* 2:384–411. In (156, II, 1–28).

151. Weyl, Hermann. 1921. "Zur Infinitesimalgeometrie: Einordnung der projektiven und der konformen Auffassung." *Nachrichten Göttinger Gesellschaft der Wissenschaften* pp. 99–112. In (156, II, 195–207) [43].

152. Weyl, Hermann. 1922. *Space – Time – Matter. Translated from the 4th German edition by H. Brose*. London: Methuen. Reprint New York: Dover 1952.

153. Weyl, Hermann. 1923. *Raum - Zeit -Materie, 5. Auflage*. Berlin: Springer.

154. Weyl, Hermann. 1929. "Elektron und Gravitation." *Zeitschrift für Physik* 56:330–352. GA III, 245–267.

155. Weyl, Hermann. 1949. *Philosophy of Mathematics and Natural Science*. Princeton: University Press. ²1950, ³2009.

156. Weyl, Hermann. 1968. *Gesammelte Abhandlungen, 4 vols*. Ed. K. Chandrasekharan. Berlin etc.: Springer.

157. Will, Clifford. 2001. "The confrontation between general relativity and experiment." *Living Reviews in Relativity* 4:1–97. Update 2014-4.

A Model-Theoretic Analysis of Space-Time Theories

Claus Beisbart

Abstract This paper studies space-time theories from the perspective of the Semantic View of theories. Set-theoretic models are used to reconstruct several non-quantum space-time theories and to characterize their mutual relationships. Further, the Semantic View is adopted to discuss the question of what a space-time theory is to begin with. While the space-time theories incorporated in Newtonian theories, on the one hand, and in Einstein's General theory of relativity (GTR), on the other hand, are markedly different, GTR and many rival theories of gravitation do not differ on their space-time theory, but only on the way the structure of a space-time is explained.

1 Introduction

Some theories are space-time theories, others not. Which are? And why? And how are space-time theories related to each other and to other theories? If we are lucky, we can answer these questions in a systematic way and provide something like a theory of space-time theories.

One strategy to develop such a theory is to draw on a general theory of theories. This paper draws on the Semantic View of theories and uses a model-theoretic framework to answer some of the questions mentioned above. The basic idea behind the semantic approach is to account for theories using *systems* of which the axioms of the theory hold true. The systems and their evolutions are formalized using *models*, which are constructed using *sets*.[1] The Semantic View contrasts with the so-called Received View developed during logical positivism.[2]

---

[1] "Model" is a technical term in this context, the main idea being that models satisfy axioms. See [20] for a brief introduction to model theory.

[2] See [34] for a formulation and discussion of the Received View.

C. Beisbart (✉)
Institut für Philosophie, University of Bern, Bern, Switzerland
e-mail: Claus.beisbart@philo.unibe.ch

© Springer Science+Business Media, LLC 2017
D. Lehmkuhl et al. (eds.), *Towards a Theory of Spacetime Theories*,
Einstein Studies 13, DOI 10.1007/978-1-4939-3210-8_7

225

Why study space-time theories in the model-theoretic framework suggested by the Semantic View? First, regardless of whether the Semantic View provides the (whole) truth about theories, it is at least a useful framework to theorize about theories (see e.g., [17, 19]). The view suggests a couple of distinctions, and we can use them to put space-time theories into perspective. Second, to analyze a theory in the terms offered by the Semantic View is effectively to formalize or to axiomatize the theory in set-theoretic terms. Such an axiomatization is useful to clarify the concepts and claims associated with a theory. As [35], p. 244 puts it,

> "axiomatization is one constructive way of obtaining the sort of intellectual clarity and precision for which philosophers are always striving with respect to the foundations of the various sciences."

Axiomatizations of theories may even inspire new lines of research (cf. [31]). Third and finally, proponents of the Semantic View have extensively investigated relationships between theories, most notably varieties of reduction.[3] We can draw on this work to describe the relationships between different space-time theories.

As far as technical details are concerned, the ideas of the Semantic View have been elaborated in several ways.[4] In this paper, I try to be neutral between the various elaborations. Technical details will be glossed over whenever possible, and the focus is on the definitions of the models constitutive of a theory. The characterization of inter-theoretic relationships is more sketchy; I either put forward claims the proof of which is straightforward, or I sketch how one should proceed. In this sense, the present paper is only a first step toward a fully elaborated account of space-time theories.

To provide no more than this seems justifiable in view of the fact that no systematic study about space-time theories from the perspective of the Semantic View has yet been published. There is some work about the general theory of relativity though, on which I can draw [3, 30]. A further very interesting study is [11] who define set-theoretic structures and a topology to highlight assumptions that all classical space-time theories share. Their definition of space-time points in terms of underlying events may be used to avoid the sets, and models refer to primitive space-time points. My paper will not adopt their definition because my concern is not how space-time points may be constructed from underlying structures. But many claims in this paper may be elaborated further by using their definition.

The plan of my paper is as follows. Because I do not want to presume much familiarity with the semantic framework, I first recall some of its basic ideas in Sec. 2. The aim of Sec. 3 is to reconstruct the space-time theories part of Newtonian and general relativistic theories. Sec. 4 uses a few general distinctions suggested by the Semantic View to discuss space-time in physical theories. We can then distinguish different ways in which theories are space-time theories (Sec. 5). One possibility is that a theory is a space-time theory in that it can be used to explain why the spatiotemporal structure of the world is as it is. Sec. 6 turns to such theories, in

---

[3]See e.g., [1], Ch. VI and [29]/[30].
[4]See [1, 29, 30, 33]; cf. also [23].

particular to GTR and to some of its rivals. I explore the mutual inter-theoretic relationships between some space-time theories in Sec. 7. Conclusions are drawn in Sec. 8.

Before I begin, a few warnings and clarifications are in order. First, in this paper, quantum theories of space-time are bracketed. Their consideration would raise a number of issues that could not be dealt within a paper such as this. Second, for convenience, I will sometimes talk of a space-time even if I refer to theories that draw a sharp distinction between space and time. Third, I use the word "theory" in a very broad sense. In this paper, theories need not be more than small units of empirical research, e.g., representational models (theory elements in the terms of Balzer, Moulines and Sneed [1] = BMS, for short), although they may also be more comprehensive frameworks (e.g., theory nets in their terms; see their Ch. II and Sec. IV.2). Fourth, in the literature, set-theoretical models are often called structures. In this paper, I prefer talk of set-theoretical models; the term "structure" is only used in the context of space-time structure. Fifth and finally, in this paper, the term "classical" is employed to contrast broadly Newtonian theories with relativistic ones.

2   In the Space of Theories: The Semantic View of Theories

The semantic approach takes theories to be characterized by *models*. Models fulfill the theory. To a first approximation, the models are possible physical systems or ways in which such systems can behave according to the theory. They are faithfully described by solutions to the equations constitutive of the theory. We can thus say that certain systems are constitutive of a theory.[5]

Each theory is characterized by a *plurality* of models because i. most theories hold true of several distinct systems; ii. each particular system can behave in many different ways, depending on the initial conditions; and, maybe, iii. the same evolution of one system may be described in different ways. The last point applies only if the models of a theory are individuated in a fine-grained way that takes into account the way in which a system is described, e.g., the coordinates used.

Theories allow for a *systematic* account of a plurality of systems. To this end, they use theoretical concepts and mathematical objects such as numbers, functions, manifolds, tensor fields, and so on. But physical systems qua parts of the real world do not include mathematical objects, at least not obviously so. Consequently, we have to regiment and to formalize the systems when we use them to characterize a theory. The formalization is set-theoretic, i.e., the models are thought of as n-tuples

$$\langle S_1, ..., A_1, ..., s_1, ... \rangle \tag{1}$$

---

[5]It seems attractive of this view that it characterizes theories in terms of real stuff as it were. A different approach that nevertheless provides model-theoretic reconstructions of theories is advocated by G. Ludwig; see [23] for a recent outline.

with sets as components.[6] The *base sets* S_i are non-mathematical because their elements are (possible) physical objects in a wide sense.[7] The *auxiliary sets* A_1 are purely mathematical sets, e.g., the set of real numbers, $\mathbb{R}$, or $\mathbb{R}^3$. Since the auxiliary sets are most often well-known mathematical entities with standard names, we can omit them for the sake of brevity. Finally, the s_i are constructed on the basis of the S_i and the A_i. For instance, s_1 may be a function that assigns each particle from base set S_j a number from an auxiliary set to represent its mass in some units. Each s_i has a unique typification in terms of the base and auxiliary sets.[8]

The *literal claim* of a theory has it that certain target systems called the *intended applications* are within the class of systems constitutive of the theory. The claim makes sense only if the target systems are formalized in the same way as are the models of the theory. The claim is nontrivial because the range of systems associated with a theory is narrow, and intended applications may not be within this range.

But the literal claim of a theory is often too strong to be taken true. Many theories postulate entities and relationships that stretch beyond the realm of the observable, and scientists may use a theory without thinking that its theoretical posits exist. Proponents of the Semantic View account for this by associating an *idealized empirical claim* with a theory. According to the claim, the intended applications are not full models of the theory, but can be embedded in full models of the theory. There are two aspects of embedding. First, the models of the real-world systems may be poorer in that they do not contain certain theoretical components (more on theoreticity below). Second, the functions in the models of the real-world systems may only be restrictions of the corresponding functions in the theory. Intuitively, either the theories introduce functions that stretch beyond possible data, or the latter are sparser than the theories allows.[9] The name "idealized *empirical* claim" is justified because theoretical components do not matter for this claim.[10]

As a matter of fact, most theories hold only up to some approximation. This motivates the introduction of an *approximate empirical claim* of a theory.[11] To express

---

[6] For this and the following see [1], Ch. I and [29], Secs. II.1–II.2.

[7] For BMS, the sets S_i are merely not *necessarily* mathematical; they may contain physical objects, but need not. But in this way, BMS allow for many models that are irrelevant for the purposes of representing physical systems. I will not follow them in this regard. See their pp. 20–23 for a discussion.

[8] See BMS, p. 8 for a precise definition of typification.

[9] Data may of course also be richer than a theory allows. But this is just to say that a theory is restricted to certain aspects of a class of systems. See BMS, Sec. II.7 for details about the idealized empirical claim of a theory.

[10] Since theoreticity is theory-relative according to BMS (see below), the empirical claim is only empirical in a theory-relative sense. But below, we will often use the term "empirical" in the absolute sense of "subject to empirical scrutiny."

[11] See BMS, Sec. VII.2.3.

this claim, we need an approximation apparatus.[12] In this paper, I will bracket the approximation apparatus as far as possible.[13]

To study a particular theory in the semantic framework, we have to specify what its models are. As the models are set-theoretical constructions, we can pick them using a *set-theoretic predicate*.[14] Providing the set-theoretic predicate of a theory is the most important step to axiomatize or to reconstruct a theory. A full reconstruction of a theory includes further elements such as links to other theories and constraints (see below).

Some proponents of the Semantic View seem to think that a full reconstruction of this sort gives us what the theory *is*.[15] Such a claim would go too far. One and the same theory can be associated with different sets of models. Trivially, the base sets in the models may appear in different orders. Less trivially, the same systems may be formalized using slightly different mathematical constructions. All this casts doubts on the idea that a theory *is* at bottom a set of models because there seems no unique answer to the question as to which set the theory is. Consequently, I will here only use models to *represent* theories; questions that concern the ontology of theories are beyond the scope of this paper.[16]

In practice and regarding concrete examples, reconstructions of a theory in terms of models should take into account some desiderata. First, a reconstruction should be as free from "descriptive fluff" ([14], p. 27) as possible. The idea here is to abstract from representational devices and to concentrate on the systems the theory applies to. A second desideratum is motivated by the fact that each reconstruction of a theory suggests ontological commitments. If you take a theory to be literally true of a real-world system, you are committed to think that the elements of the base sets exist and provide the basic building blocks of the system.[17] Accordingly, theories should be reconstructed using base sets the elements of which we are prepared to take as existent. Third, the reconstructions of theories should make plain how the theories relate to experience. According to a very simple picture, some theories are basic in that they have only observable objects in their base sets and that their relations are subject to empirical scrutiny. Other theories may then build upon these theories

---

[12]See e.g., BMS, Sec. VII.2.

[13]There is an alternative way to associate claims with a theory. The idea is that a theory describes real-world systems not because the latter are among a theory's models, but rather because they are *isomorphic* to models constitutive of the theory (see e.g., [39], pp. 43–44; see [12] for a weakening). But the notion of isomorphism is a mathematical one, and to apply it to real-world systems, we have to formalize them in terms of models. We have to build up set-theoretical constructions of real-world stuff. If this is so, why not start with real-world models and then say that they can be embedded in models constitutive of a theory?

[14]See [36], Ch. XII or BMS, p. 15 for a definition.

[15]E.g., [2], p. 50.

[16]Cf. [17], pp. 278–279.

[17]This is a counterpart to Quine's claim that theories commit us to assume the existence of those things to which quantified variables refer [26]. Note though that proponents of the Semantic View need not assume that theories are, or should be, literally true. In fact, an influential proponent of the Semantic View [39], does not take serious parts of the models that stretch beyond the observable.

and introduce more theoretical layers. This would lead to a hierarchy of theories. But there are well-known difficulties with such a hierarchical picture (see [3]). The Semantic View does invite us to build up set-theoretic constructions from base sets, but that these sets are *base* sets does not mean that their elements are *basic* in the sense of empirical accessibility.

3 Making Space for Space-Time

Most physical theories describe how things (particles, fields, etc.) are distributed in space and how their positions change in time. Therefore, most physical theories use spatiotemporal notions and include or presuppose assumptions about space-time. If we want, we can assemble these assumptions into little space-time theories. The aim of this section is to analyze two space-time theories of this sort. We first consider the space-time theory assumed by classical particle mechanics (CPM, for short) and then turn to the space-time theory presumed in the general theory of relativity. We take it that CPM assumes what is called full Newtonian space-time even though other classical space-times would do as well.[18]

3.1 Space-Time in Classical Particle Mechanics

Let us first turn to classical particle mechanics. The theory traces the motion of particles in space. To simplify matters, I will only consider free particles; i.e., there are no forces. This does not make a difference for our purposes. Granted this simplification, the basic assumption of the theory is often put as

$$\frac{d^2}{dt^2}\mathbf{r}_i(t) = 0, \tag{2}$$

where $\mathbf{r}_i$ denotes the position of the ith particle for $i = 1, \ldots, N_{part}$, and t refers to time. The equation is a special case of Newton's Second Law.

Assuming the Semantic View, the theory is uniquely characterized by those possible systems to which the theory applies.[19] Let us thus characterize the theory by describing these systems in terms of models. An important relation in the models will be the entirety of functions $\mathbf{r}_i$. In Eq. (2), a derivative of $\mathbf{r}_i$ is taken, so it is natural to assume that each $\mathbf{r}_i$ maps real numbers t to vectors $\mathbf{r}_i(t)$ in $\mathbb{R}^3$.

$\mathbf{r}_i$ is not just any mathematical function of this sort, but rather has a certain empirical significance. To account for this significance, we take the real numbers t

---

[18]See [15], Chs. 2–3 for a discussion of classical space-times. Newtonian space-time is defined on his pp. 33–34.

[19]Set-theoretic axiomatizations of CPM have been provided by e.g., [33], Ch. VI and [1], Sec. I.7.

to label instances of time, and the vectors $\mathbf{r}_i$ to denote positions in space. To build up the models properly, we add S, i.e., the set of space points, and T, the set of instances of time, and coordinate charts that map elements from S and T to $\mathbb{R}$ and $\mathbb{R}^3$, respectively.[20] S and T are base sets because they are not built up using other sets.

Regarding coordinate charts, we could give each model two fixed charts (one for space, one for time), but this is not a good idea because coordinates are descriptive fluff and no proper part of physical systems. If we included particular coordinate charts in each model, the models would not be individuated in the same way as are systems.[21] We thus assign each model a *class* of coordinate charts.

The sets S and T are peculiar because their elements, space points and instances of time, are so, too. Do space points exist? Can they be identified through time? Whatever the answers are, it seems fair to say that CPM assumes there to be space points and instances of time, and every reconstruction of CPM should take this into account. Whatever the truth of the theory, the elements of S and T seem among its posits. Possibly more relationalist construals or varieties of CPM that avoid reference to points are beyond the scope of this paper. What is important though for our purposes is that the theory be empirical. Thus, space points and instances of time have to be connected to experience and to measurements.[22]

What we can measure about space and time are, e.g., distances between space points that are identified using physical objects, e.g., because they are occupied by particles, and temporal distances between physical events. Let us thus define distance functions and include them in the models. To this end, we need values of such functions:

- D_S, the set of possible spatial distances between space points;
- D_T, the set of possible temporal distances between instances of time.

These are again base sets because physical distances are not obviously set-theoretic constructions, nor are they just numbers.[23]

We can now define the physical distance functions:

- the physical spatial distance function: $d_S : S \times S \to D_S$; it returns the distance between two space points;
- the physical temporal distance function: $d_T : T \times T \to D_T$; it returns the temporal distance between two instances of time.

As a model is supposed to conceptualize a physical system, d_S and d_T cannot fully be characterized in terms of mathematics only; they have to return the real distances. Thus, d_S and d_T have to be such that they can be constrained empirically; they have

---

[20]The term "coordinate charts" is well-known from differential geometry and anticipates the terminology from differential geometry used later in this paper. In the present context, we have one coordinate chart for space points and one for instances of time.

[21]See [29], p. 222 for a discussion.

[22]Costa et. al. [11] construct points (in their case, space-time points) from underlying events, but we have then to posit the events.

[23]One could try to introduce distances though through equivalence classes of pairs of objects.

to return those distances that can be measured, e.g., using rigid rods or clocks. This requirement escapes a purely formal account, and we can only informally state that d_S and d_T are in fact distance functions.[24]

For a mathematical description of distances, the latter have to be labeled using numbers too. Ideally, the labels preserve certain features of distances. In particular, if one distance is a multiple of another in a physical sense, then the respective numbers behave in the same way. Suitable labelings are called "representations." Let us thus include the representations as typified components of the models:

- F_S, the set of representations $f_S : D_S \rightarrow \mathbb{R}_0^+$ of spatial distances through nonnegative real numbers;
- F_T, the set of representations $f_T : D_T \rightarrow \mathbb{R}_0^+$ of temporal distances through nonnegative real numbers.

So-called representation theorems ensure that the sets are not empty.[25] The representations are not unique; e.g., each mapping f_S fixes a conventional unit of length.

When we say that a function provides a representation, we do not just characterize the function mathematically, but rather relate it to physical systems. It would be desirable to spell out in set-theoretic terms what it means that the functions are representations. True, the relationships between physical systems and mathematical entities cannot be fully captured using mathematics (i.e., set theory) only, but in our example, we could at least use set theory and other relationships between physical objects to define what a representation is. Alternatively, we can say that the functions obtain their meaning via *links* to models of other theories. A link is a relation between models of different theories.[26] Models that fit each other, for instance because they contain the same function, are paired. In our example, we link the models of CPM to models of a theory that spells out how physical distances are represented numerically. The latter theory may be a theory of measurement, i.e., a theory that explains how distances are measured.[27]

Representation is also an issue concerning the space-time. The laws of CPM are most often stated assuming preferred charts. Most spatial coordinate charts are not useful because they do not match empirically significant spatial relationships. We can say that a coordinate c_S represents the distances between space points if, and only if, there is a representation f_S such that for all pairs of space points $x, y \in S$:

$$f_S(d_S(x, y)) = |c_S(x) - c_S(y)|, \tag{3}$$

where "$| \cdot |$" always denotes the usual square norm on $\mathbb{R}^n$ for a contextually fixed number n (here, $n = 3$).

---

[24]In differential geometry, distances are conceptualized using metric tensor fields (see below). Tensor fields are only defined on manifolds, and our space-time is not yet assumed to be a manifold. This is not a problem though because we are here not yet assuming the full mathematical formalism, but rather talking about physical distances.

[25]See e.g., [36], §12.3 for the notion of a representation theorem.

[26]See BMS, Def. DII-4 on p. 61 for a formal definition of (abstract) links.

[27]See [32] and [37] for measurement in a model-theoretic framework.

It is far from trivial that there are such representations. For instance, Eq. (3) implies the triangle inequality for distances between triples of space points x, y, z:

$$f_S(d_S(x, y)) + f_S(d_S(y, z)) \leq f_S(d_S(x, z)) . \tag{4}$$

The triangle inequality is trivial for a mathematical distance function, but not so for physical distances between space points.

In fact, to adopt Eq. (3) is to assume that space is Euclidean ([29], p. 50). And CPM does take space to be Euclidean. The components of our models have thus to obey Eq. (3). This is a significant assumption of the theory; it narrows down the range of models characteristic of the theory.[28]

CPM makes also assumptions about time. We can encapsulate them in the following requirement: For some f_T, there is a global chart c_T from T to $\mathbb{R}$ such that for all $t_1, t_2 \in T$

$$f_T(d_T(t_1, t_2)) = |c_T(t_1) - c_T(t_2)| . \tag{5}$$

This finishes the reconstruction of the space-time theory presumed by CPM. Let us call this theory CPM_{ST} and summarize our results in the following definition.

Definition 1 x is a model of CPM_{ST} if, and only if (iff)

$$x = \langle S, T, D_S, D_T, d_S, d_T, F_S, F_T, C_S, C_T \rangle, \tag{6}$$

where

1. the base sets S, T, D_S, and D_T comprise the space points, instances of time, spatial distances, and temporal distances, respectively;
2. $d_S : S \times S \to D_S$ and $d_T : T \times T \to D_T$ are spatial/temporal distance functions;
3. F_S, F_T are sets of representations f_S, f_T of spatial/temporal distances;
4. C_S and C_T are sets of coordinate charts c_S and c_T for space and time, respectively;
5. there are coordinate charts c_S, c_T and representations f_S and f_T such that

$$f_S(d_S(x, y)) = |c_S(x) - c_S(y)|, \tag{3}$$

$$f_T(d_T(t_1, t_2)) = |c_T(t_1) - c_T(t_2)|. \tag{5}$$

CPM_{ST} fixes the space-time structure uniquely. It thus only picks one model. The literal claim of the theory then is that, if regimented in the right way, the real world coincides with the only model of CPM_{ST}. As there are no theoretical components in the model of CPM_{ST}, the idealized empirical claim is that all intended applications can be embedded in the model of CPM_{ST}.

---

[28]Equation (3) fixes the Euclidean nature of space using coordinate charts. Alternatively, one could try to do without them. This would avoid some descriptive fluff. For instance, we could demand that the triangle inequality, Eq. (4) holds. Even this equation is not free of descriptive fluff, but at least it dispenses with coordinate charts. However, it is much easier to use coordinate charts to fix the properties of a space ([29], p. 50).

An important question though is how empirical this claim really is. If the claim is to be empirical, then it must be possible to determine empirically whether or not space points stand in the relationships they are supposed to stand in. The problem with this is that we can only pick space points using physical objects. For instance, the spatial location of a point particle defines a space point. Consequently, possible data for the theory have to refer to physical objects such as particles or fields, at least indirectly. But CPM$_{ST}$ itself does not contain any particles or fields. Thus, CPM$_{ST}$ has only an empirical claim if we understand it that the space points are defined or individuated using physical objects.[29]

Now CPM (rather than CPM$_{ST}$) does have particles in its ontology, and it thus makes a claim that is less elusive than that of CPM$_{ST}$. The models of CPM are of the type

$$\langle S, T, D_S, D_T, \mathscr{P}, d_S, d_T, F_S, F_T, C_S, C_T, r \rangle, \tag{7}$$

where we have added two new components, viz.

- $\mathscr{P}$, a set of particles

and

- $r : \mathscr{P} \times T \to S$ a function for the trajectories of the particles.

The models have to obey Newton's Second Law assuming zero forces. Thus, for all particles $p \in \mathscr{P}$ and for all charts c_S and c_T that are representations, we have: $c_S(r(p, c_T^{-1}(\cdot))) : \mathbb{R} \mapsto \mathbb{R}^3$ is twice differentiable and for all $a \in \mathbb{R}$

$$\frac{\mathrm{d}^2}{\mathrm{d}a^2} c_S(r(p, c_T^{-1}(a))) = 0 . \tag{8}$$

Here, a serves as a label for time. Newton's Second Law looks more complicated than usual (cf. Eq. 1) because the trajectories of the particles are described using a function that has non-mathematical sets as domain and range.[30]

CPM has additional implications over and above CPM$_{ST}$, and some of them are more clearly empirical. CPM entails how the distances between certain point particles evolve in time. Distances between their positions can be measured and are thus subject

---

[29]The status of other space points not picked using physical objects remains peculiar anyway. In GTR, this peculiarity is much discussed in debates about the hole argument. The latter is often taken to show that the assumption of space-time points in empty space has untenable consequences [13, 21]. But the hole argument does not raise any problem peculiar to our approach. First, most formulations of GTR quantify over all points of the manifold, regardless how they are identified empirically, so we do not have an objection specific to our approach. Second, what the hole argument shows is only that space-time points identified independently of any physical events are problematic. If we understand it that the elements of the space-time do not have identity apart from the roles that they take in the models, our formalization does not seem problematic.

[30]It can be shown that the equation holds for all representations if it holds for one pair of representations (c_T, c_S).

to empirical scrutiny. Related measurements, e.g., using rigid rods, can be described using a theory of measurement.

Further, physical measurements are themselves physical processes and we may hope that the latter can be described by the theory itself or, maybe, by an extension thereof. In the former case, we call the theory complete; in the latter, the theory can at least be completed.[31] In our reconstruction, CPM does not contain rigid rods nor standard clocks, so it may seem that CPM is incomplete. We could try to complete CPM by introducing rigid rods and standard clocks. But this would not be very elegant because the ontology of the theory then would display less unity. It would be more satisfactory to construct measurement devices on the basis of the ontology of CPM, i.e., only using particles. The question of whether this is possible is beyond the scope of this paper. But if a theory is complete, then we do not have to link its models to those of other theories to give their components meaning. The components of the models either have empirical significance in a very straightforward way or are theoretical.

We can reconstruct CPM, particularly the space-time theory it presupposes, using models in an alternative way. To this purpose, we use notions from differential geometry, notably that of a manifold. A manifold M is a set of points with additional mathematical structure, viz., with a topology and a set of coordinate charts. Effectively, several components of our models used thus far are condensed into one mathematical object, and the models become simpler because they have less components. Strictly speaking, a manifold is not a base set any more because it is a set-theoretic construction built up out of more basic sets, but to simplify matters, we will treat the manifold like a base set. A manifold allows for a number of other well-known constructions, notably of the tangent space $T_q(M)$ for each point q of the manifold and of the co-tangent space, i.e., the dual space to $T_q(M)$. In what follows, I will take these constructions for granted and not explicitly include them in the models.

We can now reconstruct the space-time using a four-dimensional manifold that is built up from space-time points (events) and describe distances using metrices. This is preferable in view of the General Theory of Relativity to which we turn soon. In what follows, all manifolds are assumed to be four-dimensional, differentiable, connected, and para-compact Hausdorff spaces.

To obtain full Newtonian space-time, we have to introduce absolute time, to ensure that space is flat and to identify space points throughout time. To this end, we need additional objects.[32] Here then are the models of the space-time theory behind CPM.

Definition 2 x is a model of CPM_{ST} iff

$$x = \langle M, D_S, D_T, d_S, d_T, F_S, F_T, g_S, g_T, \Gamma, v \rangle, \tag{9}$$

---

[31] See [41], pp. 18–19; for the completeness of space-time theories see [9] and [10].

[32] See [15], Ch. 2, particularly pp. 33–34; consult [18], Sec. III.1 for a slightly different alternative.

where

1. M is a manifold of space-time events;
2. D_S, D_T are sets of spatial/temporal distances as before;
3. d_S, d_T are physical distance functions as before;
4. F_S, F_T are sets of representation functions of spatial/temporal distances as before;
5. g_S is a symmetric, contravariant tensor field of signature $(0, +, +, +)$;
6. g_T is a symmetric, covariant tensor field of signature $(+, 0, 0, 0)$;
7. Γ is an affine connection;
8. v is a vector field;

and the following conditions hold:

a. There are representations f_S, f_T such that g_S and g_T "return" distances in units provided by f_S, f_T;
b. g_S and g_T are compatible with Γ, Γ is flat, v is constant with respect to the connection, the contraction of g_S and g_T equals zero, and $g_T(v^\star, v^\star) = 1$ everywhere, where $v^\star$ is the dual of v.[33]

It is desirable, but fairly complicated to spell out condition $a.$ in detail because g_S and g_T are defined locally. Here is nevertheless a brief sketch how g_S returns distances (cf. [18], p. 77). Consider a space-like curve, i.e., for every tangent vector to the curve, w, $g_T(w^\star, w^\star) = 0$. $g_S(w, w)$ equals the length of the tangent vector. An integration over the whole curve returns the length of the curve in some units. The distance between two events that can be connected via a space-like curve is the minimal length of the curves connecting the points.

Instead of relying on such complicated descriptions, we can hope for completeness, i.e., that the theory can be completed such that it can describe measurements within it.

3.2 Space-Time in General Relativistic Theories

The space-time theory incorporated in CPM has been overthrown. Let us thus consider a general relativistic counterpart to CPM, call it "general relativistic particle mechanics" or "GPM," and its space-time theory, GPM$_{ST}$. What are its models like? To simplify matters, we assume again free particles, i.e., that particles are freely falling; further, we neglect units and thus representations of distances. The corresponding set-theoretical constructions will be dropped in our models.

According to GTR, space and time cannot be separated in an observer-independent way. In particular, there are no space points that could be identified through time. Accordingly, the models of CPM$_{ST}$ cannot include sets of space points and of instances of time. S and T have to be replaced by a set of *space-time points* or

---

[33] See [40], pp. 35–36 for details about compatibility, consult ibid., pp. 20–21 for contraction and ibid., p. 19 for the dual.

events, M, and we need four-dimensional coordinate charts. M is assumed to be a manifold. Manifolds were already used for our second reconstruction of CPM_{ST} in Def. 2.

Following GTR, we can measure spatial and temporal distances, but they are not observer-independent any more. Accordingly, we keep D_S and D_T in the models, but we have to drop the typified relations d_S and d_T. Further, g_S and g_T are replaced by one metric tensor field g.

As before, g cannot be any old mathematical tensor field on the manifold, but has to be the metric. It must be possible to obtain physical distances using g, yet in a different manner than before because there are no observer-independent spatial and temporal distance functions any more.

Turn first to temporal distances as measurable by standard clocks. Let W be the typified set of all world lines of freely falling massive test particles on the manifold. These are the paths that such particles can take in the space-time. What they are is not fixed using other components in the models; rather, we take it that, for each specific space-time, there are matters of fact what these world lines are.[34] Each world line can be parameterized using a curve (i.e., a differentiable map from a real interval to the manifold). The following considerations do not depend on the way in which a world line is parameterized.

We associate with each world line in W a function that assigns a temporal distance to each pair of events on the world line. This distance is supposed to be the proper time that a clock traveling on the world line measures. Again, this function is not fixed a priori; there are matters of fact what these distances are.

Consider now a word-line in W with parameterization $r(\lambda)$, where λ is the parameter. For each event $r(\lambda)$ on the curve, the tangent to the word-line defines a vector in the tangent space at $r(\lambda)$, call it $V_{r(\lambda)}$.[35] We can thus form the following quantity:

$$\int_{\lambda_1}^{\lambda_2} d\lambda \sqrt{-g(V_{r(\lambda)}, V_{r(\lambda)})} \, . \tag{10}$$

We require this integral to equate the length of the time interval read off from a standard clock that follows the word-line. More precisely, for all possible pairs of events on all possible world lines of freely falling standard clocks, for all parametrizations, the integral has to return the proper time in the same units (i.e., using the same f_T).

Another restriction is needed to ensure that g measures spatial distances in the usual way. One option is to postulate that, in the models of the theory, in special coordinates, certain parts of g return spatial distances as measured with small rigid rods ([27], p. 143). To use rigid rods is problematic though since GTR does not allow for rigid rods. The problem is avoided by [38], Ch. III who uses proper temporal distances and certain configurations of events. Ehlers et al. [16] only draw on light

---

[34] Note though that we are here talking not about trajectories of real particles, but rather about hypothetical world lines.

[35] See [40], p. 17 for technical details.

rays and freely falling particles (instead of freely falling clocks). But such an approach is not needed for our purposes.[36]

These restrictions ensure that the tensor field g is the physical metric. The restrictions implicitly assume the geodesic hypothesis, i.e., that test particles follow geodesics.[37] This hypothesis is built into *metric* theories such as the GTR, but not into *every* theory of gravitation. The hypothesis follows from the strong principle of equivalence.[38]

All this leaves us with the following models (where we bracket entities needed to equip g with its meaning).

Definition 3 x is a model of GPM$_{ST}$ iff $x = \langle M, D_S, D_T, g, \Gamma \rangle$, where

1. M is a manifold as before;
2. D_S and D_T are sets of physical distances as before;
3. g is a symmetric covariant tensor field of signature $(-, +, +, +)$, which measures physical distances in the way sketched above;
4. Γ is an affine connection;

and Γ and g are compatible.

Here, the connection Γ fixes the affine structure of the space-time, i.e., its geodesics ([18], pp. 39–40). It is used to define a covariant derivative. Compatibility with the metric uniquely fixes the connection ([40], pp. 35–36).

The metric is not further restricted by axioms of the theory. Accordingly, unlike CPM$_{ST}$, GPM$_{ST}$ is characterized by a plurality of models that differ on the metric structure. In this sense, GPM$_{ST}$ has less content than CPM$_{ST}$. The content of GPM$_{ST}$ is only that physical distances can be determined using the metric in the way sketched. Part of this content is that hypothetical freely falling standard clocks minimize the integral in Eq. (10).

We have used hypothetical freely falling standard clocks to equip g with its meaning. Because we did not include standard clocks in the models, we have to link g to other theories. GPM$_{ST}$ then is incomplete because it does not describe how g is constrained empirically. In our reconstruction, GPM$_{ST}$ does not even contain any objects.

By contrast, GPM does have objects because it describes massive particles in the space-time. Consequently, the models of GPM are richer. They include a set of particles, $\mathscr{P}$, and a family of their world lines, r. A further axiom of the theory, viz., the geodesic hypothesis states that the world lines (ranges of differentiable curves on the manifold) are geodesics with respect to the metric. To add such particles is a step toward completeness because particles can be used to empirically constrain the metric. There is a price to pay though because, if we add further components,

---

[36]For a discussion on whether the unparameterized geodesics in a space-time uniquely fix the metric see [24].

[37]See [40], p. 67.

[38]See e.g., [41], Sec. 2.3.

we arguably move beyond the *space-time* theory within GPM. We had similar problems with CPM$_{ST}$ above. It thus seems that mere space-time parts of theories are abstractions that cannot really be isolated from more comprehensive theories.

Even if it should not prove sensible to isolate space-time theories from more comprehensive theories, we can still ask how space-time and assumptions about it enter physical theories. This is a question we have effectively answered for CPM and GPM in this section. From the perspective of the Semantic View, the models of the theories include a space-time manifold of events as a component. The space-time has a structure due to its metric. The events in the space-time manifold and the metric only obtain physical significance due to objects moving in the space-time. They latter define certain events, and it may be determined by measurement in which relations they stand.

4 Time for Distinctions

The Semantic View suggests a couple of issues that may be used to discuss the role of space-time in physical theories (more on *space-time* theories shortly).

Co-variance and symmetries.

The basic idea of the semantic approach taken in this paper is that theories are characterized using systems. It proves useful to use coordinate charts to characterize the models of a theory, and we have thus included sets of such charts in the models. Coordinate charts are descriptive fluff, but our account does at least not individuate models using coordinate charts. In this sense, our approach is friendly to coordinate–free representations of a theory. The axioms of GTR, in particular Einstein's field equations, are usually stated in a coordinate-independent way.[39]

Coordinate-free versions of a law or an axiom should be distinguished from symmetries of a theory. Naturally, our focus is here on symmetries with respect to the space-time. These are invariances of functions characterizing the space-time under automorphisms of the space-time manifold. For an example, consider CPM in our first formalization, but assume for simplicity that S is a manifold. Suppose now that ι is an automorphism of S onto itself and let d be a distance function as before. ι gives rise to a dragged distance function $d_\iota^\star(x, y) = d(\iota^{-1}x, \iota^{-1}y)$ for all $x, y \in S$. The distance structure of the space-time is symmetric concerning ι if and only if $d_\iota^\star(\cdot, \cdot) = d(\cdot, \cdot)$ everywhere. Space-time is *fully* symmetric under ι if this property holds for each function definitive of the space-time structure. As is well-known, the spatial part of full Newtonian space-time is invariant with respect to translations and rotations. By contrast, space-times in general relativistic theories need not display any symmetries.

---

[39]Some proponents of the Semantic View, e.g., BMS and [3] part company with me at this point because they include particular coordinate charts in their models. A potential reason is that, in practice, scientific work done with a theory often uses specific coordinates, and that a change in coordinates can have nontrivial consequences for this work. See [22], pp. 95–100 for illustrations.

To explain the significance of a symmetry, automorphisms of a manifold are often interpreted using active transformations, e.g., real translations. But this interpretation does not really make sense if applied to a whole space-time. It is not clear what it means to say that each event is, e.g., shifted in the same way. The point of applying an automorphism is rather to compare the space-time at different points. Symmetric space-times have the same characteristics at different points.

Potential and actual models.

Within the semantic framework, it is often useful to distinguish between potential and actual models of a theory. The potential models fix the ontology of the theory; figuratively speaking, they determine the sorts of properties and relations that can obtain in the systems. The actual models are further restricted and encompass those potential models in which the laws of the theory hold.[40]

We need laws to distinguish between merely potential and actual models. But what exactly are laws? BMS propose to say that laws characteristically connect several typified components s_i of the models of a theory (BMS, Sec. I.3). For instance, Newton's Second Law connects acceleration, mass and force. Granted this view, most assumptions about the space-time in our space-time theories are laws because they concern several typified sets in the models. E.g., Eq. (3) refers to the space-time manifold itself, the metric and the distance function. That Eq. (3) reflects a law is not implausible because it makes a substantial assumption about space. Concerning GPM, it is nontrivial that distances arise from a metric as described above.

The observational/theoretical distinction.

Proponents of the Semantic View distinguish between theoretical and observational components within the models. Sneed [33], Ch. II, particularly pp. 33–35, proposes that this distinction should be drawn in a theory-relative way, and in the following, I will focus on this relativized version of theoreticity. A typified component in the models of a theory T is *T-theoretical* if it can only be determined from other components within the class of *actual* models and thus using the laws of the theory. For instance, in CPM, the force can only be determined assuming the use of Newton's Second Law. Components that are not T-theoretical are either purely observational or imported from other theories via links that provide the meanings of the terms (see BMS, Sec. II.3.2).[41]

Which components in CPM_{ST} or GPM_{ST} are theoretical in this sense? The metric g is GPM_{ST}-theoretical as are the metrices of CPM_{ST}. To determine a metric, we have to relate it to distances and thus to other typified sets in the models. According to BMS, the pertinent relations between distances and the metric count as laws. Consequently, actual models are needed to fix the metric, and the metric is theoretical. The other components in the models, e.g., S and T or M, and the distance functions d_S, d_T from CPM_{ST}, by contrast, are not CPM_{ST}-theoretical.[42] Actual models are not needed to

---

[40]Nevertheless, some actual models of a theory may not match the actual world. The actual models are solutions to the equations of a theory, but these solutions may not describe our real world, e.g., due to unrealistic initial conditions.

[41]See BMS, Sec. II.3 for two formalizations of theoreticity with respect to a theory.

[42]Cf. BMS, pp. 51–52.

determine them. For instance, there are simply distances between the space points, and the metric would not help us to determine the distances. Rather, the distances have to be determined empirically, e.g., using rigid rods. Note though that the space points are theoretical in another sense because they are theoretical posits. We can pick some space points because they are occupied by point particles. Other space points each have their specific identities because they bear certain spatial relations to other points, e.g., because they are exactly in between two other points picked in some other way.

Constraints.

Physical systems are defined by humans. Several physical systems that are part of the same world can thus overlap and include the same objects. If the same theory holds true of these systems, then one theory has overlapping systems as its models. In this case, objects that belong to several systems should have the same properties in all these systems. Put differently, scientists focus on classes of systems that are consistent in that they assign the same properties to the same objects. BMS formalize this using so-called constraints (BMS, II.2.3). A constraint has subsets of models as its members. In each subset, the models are consistent in some respect. For instance, CPM has a mass constraint, which comprises subsets of models in which a specific particle has the same mass (BMS, p. 106). Constraints do not ensure that an object has the same characteristics in all models. They only group together models in which the object has the same characteristics. The intersection of all constraints comprises sets of models that are consistent in every relevant respect (BMS, p. 78).

Are there any constraints pertinent to the space-time? All models considered thus far include the whole space-time. If the models do not ascribe the same features to the space-time, e.g., if the metric properties of the space-times are different, then it is clear that two distinct worlds are pictured. There then is no need for constraints about one space-time. Put differently, whole space-times cannot overlap, which obliterates the use of constraints.

It is arguable, however, that the systems of CPM do not necessarily include the whole space-time, but only refer to portions of it.[43] If this is right, we can introduce constraints that express consistency constraints on overlapping models. The constraints group models that may describe parts of the same space-time.

The kernel.

The kernel of a theory comprises those components that are the same in every model.[44] So-called absolute objects are part of the kernel.[45] If the sets S and T in the models comprise all space points and instances of time, respectively, then these sets as well as sets of coordinate charts, distance functions and metrices are part of the kernel of CPM.[46] The reason is that CPM assumes a fixed metric structure (what

---

[43] BMS reconstruct CPM using local models; cf. our discussion below in the next section.

[44] [29], pp. 52, 66–68.

[45] Absolute objects are contrasted to dynamical objects. See [18], pp. 64–70 for the distinction.

[46] [29], pp. 52, 66–68. S and T cannot be part of the kernel if there are models that have only portions of the space-time as components.

should be the point of endowing the same space with different distance relationships between points?).

The kernel of general relativistic theories is less comprehensive. As the metric structure is not fixed, W as well as g can differ between the models and thus do not form part of the kernel. But of course, every model has to have a metric tensor field.

5   Space for Space-Time Theories: What May a Space-Time Theory Be?

So far, the notion of a space-time theory has not been introduced in a systematic way. We have assembled the assumptions that some theories make about space-time and tentatively called the results space-time theories. But which theories are *properly* called space-time theories? Intuitively, CPM is not, while GTR may well be. Is there a principled way to draw the distinction? Or does the distinction crumble under reflection? Our model-theoretic framework suggests a number of ways to draw the distinction. None of them is incorrect, but some may capture more faithfully than others what we take to be a space-time theory.

Suggestion 1: Cosmological import.

According to our first suggestion, the systems constitutive of space-time theories extend over the whole space-time. More technically, space-time theories have a full space-time in each of their models, while other theories do not because some of their models only include portions of space-time. This suggestion seems plausible because most theories are about small systems that only occupy a small part of the space-time, and such theories need not adopt assumptions about the whole space-time.

The suggestion is compatible with the idea that many solutions to equations from theories are defined for whole space and time, for instance because boundary conditions are assumed for infinite times, even though the theories are *not* space-time theories. The suggested criterion is that *every* model of a space-time theory contains the whole space-time, while other theories have at least some models that do not include the whole of it.

The suggestion may be formalized a bit more. A manifold equipped with a metric is extendable if it can be thought of as a proper sub-manifold of another manifold with a metric.[47] Space-time manifolds are sometimes required not to be extendable (the idea being that an extendable manifold can't be the whole space-time). We then can demand that the models of a space-time theory include an inextendable manifold of events.[48]

According to the suggestion, CPM would seem to count as a space-time theory. For we have here assumed that every model of CPM contains the whole space-time

---

[47]See e.g., [40], p. 215.

[48]Because extendability involves the manifold and the metric, a requirement of inextendability would have law-like status according to BMS.

(so does [29], Sec. II.1). But as already indicated, this assumption is not necessary. Many systems described by CPM are much confined in space and only studied for a small portion of time. They do not seem to assume that all of space is Euclidean. We could capture this by replacing the whole space-time manifold with suitable sub-manifolds that only contain a proper part of the space-time.

There are nevertheless problems with our first suggestion. First, the criterion seems too strict. GTR is a paradigm example of a space-time theory, but it seems artificial to say that each system constitutive of the theory comprises a whole space-time. Many applications of GTR concern subsystems of the universe such as certain binaries or black holes. Does one really need an inextendable space-time manifold to follow such a system using GTR?[49]

Second, the question of whether all models of a theory include the whole space-time as a component seems too technical to make a difference. It is often a matter of taste whether we reconstruct a particular application of the theory as a model about the whole space-time or not.

Finally, the suggestion does not capture the sense in which space-time theories are *about* space-time. A theory may accidentally, e.g., for some technical reason, only have models that include a whole space-time. Does it follow that it is a space-time theory? The best way to deal with this objection is to consider alternatives way of delineating space-time theories and to see whether they do better.

Suggestion 2: Exclusiveness.

According to a second suggestion, a space-time theory is about space-time iff it is about space-time *exclusively*. That is, the models contain only components that describe space-time and its properties, e.g., the metric. There are no particles, charged fields and so on. If we adopt this suggestion, then CPM_{ST} and GTR_{ST} are space-time theories, while GTR itself is not (see below).

The claim of such a theory would have to be that the metric returns physical distances between space-time points in a specific way, and, in the case of CPM_{ST}, that space is Euclidean. These claims do not have empirical significance, unless points are picked in some way, and this can only be done using physical objects. So the theory has to be linked to other theories. But if this is so, then exclusiveness can only be fulfilled in a formal sense: There are theories the models of which have only spatiotemporal notions as components, but the exclusiveness is deceptive because the theory is either not empirical or it has to be linked to other, more comprehensive

---

[49] One can rebut this objection though by saying that, even though every model of a space-time theory contains the whole space-time, the models need not be *used* to describe the whole space-time. For instance, formally, the Schwarzschild solution of GTR is a solution for the whole space-time, but it is often used as an idealization to understand subsystems of the Universe only. GTR would then still be a space-time theory.

theories. It then seems that exclusiveness is too formal a criterion to distinguish space-time theories form others.

Suggestion 3: Introduction of spatiotemporal notions as theoretical components.

A third suggestion has it that a space-time theory introduces spatiotemporal notions in the following sense: Spatiotemporal notions are T-theoretical with respect to a space-time theory T. Once this theoretical structure including the metric is available, other theories can draw on it. Technically speaking, they are linked to it.

The suggestion has to be precisified because there are several spatiotemporal notions. Which one is introduced by space-time theories? Consider first metrices. When we have reconstructed CPM_{ST} and GPM_{ST} using metrices, we have assumed that substantial assumptions relate the metric to measurable distances. Thus, the metric can only be determined using the theory itself, and thus is T-theoretical in both cases. Accordingly, CPM_{ST} and GPM_{ST} introduce the metric. Thus, if space-time theories introduce the metric, CPM_{ST} and GPM_{ST} are space-time theories, which is plausible.

By contrast, physical distances between points (i.e., between points in space, instances of time or events) can be determined empirically without recourse to CPM_{ST} and GPM_{ST}. This is at least so if the points are defined using physical objects. Thus, distances are not introduced by CPM_{ST} and GPM_{ST}, and the latter would not be space-time theories according to the third suggestion if space-time theories had to introduce distances. But CPM_{ST} and GPM_{ST} are space-time theories, at least intuitively speaking. It is thus not useful to require that a space-time theory introduces distance functions. We can thus conclude that the third suggestion should be understood as follows: A theory is a space-time theory if it introduces the metric in the sense explained above.

A problem with this suggestion is as follows. If reconstructed according to Def. 1, CPM_{ST} does not introduce a metric in the required sense because there is no metric in the models at all. Thus, understood in this way, CPM_{ST} would not count as a space-time theory according to the third suggestion, which sounds counterintuitive. But at least the spirit of the third suggestion seems right because most points themselves are theoretical posits, and theories from which CPM_{ST} imports notions may not yet include sets of all points.

Suggestion 4: Explanation of spatiotemporal structure.

To motivate another suggestion, we note that, often, GTR counts as a paradigmatic space-time theory while CPM does not. Why is this so? Plausibly because GTR allows for markedly different space-time structures while CPM does not. The idea here is that each theory is general and has several possible applications. If a space-time theory is about space-times, it should thus allow for different possible space-times. Since there is, presumably, only one space-time, it must at least allow that our space-time may be structured in alternative ways. That is, the metric structure of the space-time is not fixed in advance, but varies between the models.[50] In the terms of [29], p. 67, the metric is not part of the kernel of the theory.

Now this cannot be the whole story because, if it were, then even GPM_{ST} would have to count as a space-time theory, while CPM_{ST} would not. But this is not a sensible thing to say. Either CPM_{ST} and GPM_{ST} both are space-time theories, or both are not.

GTR has an additional benefit: It not only allows for alternative space-time structures, but also constrains the latter via Einstein's field equations and other quantities, notably the energy-momentum tensor. For this reason we can use GTR to explain why the space-time has those metric features it has. Let us thus strengthen our fourth suggestion and require that space-time theories must have explanatory power regarding our space-time in the sense that other components in the models constrain the space-time via axioms. GTR then is a space-time theory, while CPM_{ST} and GPM_{ST} are not.

Both the third and the fourth suggestion seem to make some sense, but they lead to incompatible results because CPM_{ST} and GPM_{ST} are space-time theories under the third suggestion while they are not under the fourth. This is not too much of a problem though. There may be no unambiguous notion of a space-time theory. It seems fair to say that, at some times when we talk about space-time theories, we have in mind something like the third suggestion, while, at other times, we require that a space-time theory be explanatory in accordance with the fourth suggestion. In this paper, the notion of space-time theory is normally reserved for theories that introduce a metric or other space-time notions in rough accordance with the third suggestion.

6   Space for Explanation: Explanatory Space-Time Theories

Let us now look at some theories that are explanatory in the sense of our fourth suggestion. According to such theories, the geometry of space-time is constrained by physical objects via laws. The aim of this section is to extend our model-theoretic analysis to such theories.

6.1   Einstein's GTR

Let us begin with Einstein's GTR. It is centered about Einstein's field equations (EFE) that relate the Einstein tensor G, which is built up from the metric and the connection,

---

[50] Here, the metric structure does not only mean the metric itself, but also physical relationships about events on world lines incorporated in the metric field tensor.

and the energy-momentum tensor to each other. The simplest way to account for the EFE is to expand the models of GPM. We add two typified components, viz.

- a symmetric, covariant tensor field T_{tot} of rank 2, the energy-momentum tensor and
- a symmetric covariant tensor G of rank 2, the Einstein tensor.

We then demand that the actual models obey the following requirements.

- G measures the curvature (i.e., it is a certain construction of the metric g and the connection Γ, see [40], pp. 40–41);
- Einstein's field equations are fulfilled:

$$G = 8\pi T_{tot} ,\tag{11}$$

As before, we neglect the freedom to choose one's units and assume units in which the gravitational constant is set to 1.

However, even though we have now used EFE to define a set-theoretic predicate that picks a set of models, our reconstruction does not yet capture the explanatory power of GTR. For we have only typified T_{tot}, and if T_{tot} is otherwise free, EFE do not restrict the space-time. But in fact, depending on the energy and matter components in the universe, T_{tot} can be obtained from the distribution of matter and energy. We could try to account for this using a link to another theory, viz., a theory that specifies the energy-momentum tensor for some matter model. For instance, if the universe is filled with a perfect fluid, then T_{tot} is a certain functional of the metric and of some characteristics of the fluid, as is well-known from general relativistic fluid dynamics, and we could link GTR to fluid dynamics. But such a link would be too inflexible because T_{tot} can also take different shapes for other types of matter and energy (e.g., electromagnetic fields). For this reason, Bartelborth [3] regards EFE and GTR as inter-theoretic in character. I will follow him even though his move is not necessary (as an alternative, one could link T_{tot} to several theories with a many-place relationship).

As a first step, we define several general relativistic dynamical theories, e.g., general relativistic perfect fluid dynamics (GPFD), general relativistic electrodynamics (GED), and so on. The former describes the motion of a perfect fluid on a given space-time, one law being that the energy-momentum tensor takes a certain form. GED describes the dynamics of the electromagnetic field and specifies the energy-momentum tensor for such fields. GPFD and GED are roughly on the same footing as GPM because they presume a space-time structure and define a dynamics on it. But contrary to GPM, GPFD and GED are field theories, i.e. the degrees of freedom are featured using fields.

As an example, consider briefly GPFD. We concentrate on those components that are most important for understanding the theory and drop, e.g., D_S.

Definition 4 x is a model of GPFD if and only if $x = \langle M, g, \Gamma, V, \mu, P, T_{pf} \rangle$, where

- M is a manifold of events;
- g is a symmetric covariant metric tensor field of signature $(-, +, +, +)$;
- Γ is an affine connection compatible with g;
- V is a vector field on M representing the velocity of the fluid;
- μ, P are scalar fields on M representing energy density and pressure, respectively;
- T_{pf} is a second-rank, contravariant tensor field on M.

Further, for all points of the manifold $q \in M$ we have

$$g(V(q), V(q)) = -1, \tag{12}$$

$$T_{pf}(q) = (P(q) + \mu(q)) \left(V(q) \otimes V(q) + P(q)g^\star(q) \right), \tag{13}$$

$$DT_{pf}(q) = 0. \tag{14}$$

where "$\otimes$" denotes the well-known tensor product.

This definition provides precise typifications, while the meanings of the components are indicated informally. For instance, g is the metric and returns distances. We could render the definition of GPFD more precise, e.g., by linking it to other theories.

We can proceed in an analogous way for GED and define models for it. One component in the models is the energy-momentum density due to the electrodynamic field, call this tensor T_{ed}. In the following, I restrict myself to GPFD and GED as examples of general relativistic dynamic theories.

In a second step, Einstein's field equations are used to link certain pairs of models of such theories. Call this link EFE. The intuitive idea is that certain models of GPFD and GED can be combined: They fit each other because they provide a consistent description of how things might go. The details are provided by the following definition. In this and the following, we label components from a model x with a sub-index x; for instance, M_x is the space-time manifold within model x.[51]

Definition 5 A model x of GPFD and another y of GED are linked to each other iff the following two conditions hold:

1. x and y agree on the space-time and on fields common to both GPFD and GED, most notably on the metric, i.e.,

$$M_x = M_y \quad \text{and} \quad \forall q \in M_x : g_x(q) = g_y(q). \tag{15}$$

2. The metric arises from the total energy momentum via Einstein's field equations: Einstein

$$G = 8\pi T_{tot}. \tag{16}$$

Here, T_{tot} contains all relevant contributions to energy and momentum, i.e.,

---

[51] According to the definition, the link is a two-place relation; it can be generalized to an n-place relation easily if there are n general relativistic dynamic theories. If there is no matter of a particular type, e.g., no perfect fluid component, the pertinent part of T_{tot} equals zero.

$$T_{tot} = T_{pfd} + T_{ed} \, . \tag{17}$$

G is the Einstein tensor as before.

The conditions in the definition can even be relaxed: We require identity up to certain transformations rather than strict identity. In mathematical terms, we demand only that there be a diffeomorphism ι from M_x to M_y that transforms the metric, etc., of x into those of y, respectively. In this case, Eqs. (16) and (17) have also to be adapted slightly.[52]

Formalized in this way, GTR is not a theory of its own, but rather a relationship between theories. It is not defined in terms of a set-theoretic predicate that fixes a class of models, but rather in terms of a relationship between models from two or more other theories. Nevertheless, we would like to assign a claim to GTR. To this effect, we think of GPFD, GED, etc., as forming a theory net. We associate the following literal claim with the theory net: When described in the right sort of way, every intended application is an actual model of each theory in the theory net and the pertinent actual models are mutually related to each other via the link EFE. An idealized empirical claim can be defined in a similar way.[53]

6.2 An Alternative Theory of Gravitation: The Brans–Dicke Theory

Let us now consider an alternative to GTR. The simplest, presumably, is the Brans–Dicke theory (BDT, for short; [6]). It is a special case of a scalar-tensor theory.[54] Many results of this section can easily be generalized to other scalar-tensor theories.

The Brans–Dicke theory introduces new dynamic degrees of freedom, viz., a scalar field ϕ. This field does not affect other matter directly, but it has a bearing on the metric, which in turn is crucial for the dynamics of all sorts of matter/energy. To reconstruct the Brans–Dicke theory we can thus again proceed in two steps.

First we define a theory for the dynamics of the scalar field on a given space-time, call this theory BDD. It is on the same footing as is GPFD. The models of BDD are of the following type: $x = \langle M, g, \Gamma, T_{tot}, \phi, V \rangle$, where M, g, Γ are as before; ϕ is a scalar field on the manifold (the Brans–Dicke field), and $V : \mathbb{R} \to \mathbb{R}$ is a function representing the potential for the Brans–Dicke field. The dynamical evolution of the Brans–Dicke field is also driven by the energy-momentum tensor of ordinary matter, T_{tot}, so we have to include it in the models as well and to link it to other theories.[55]

---

[52] See [40], pp. 437–439 for technical details about diffeomorphisms. We can relax the conditions in the definition because space-time manifolds that are identical up to certain structure-preserving maps are not supposed to picture distinct physical possibilities.

[53] This is effectively an amendment of Bartelborth's Def. 12.

[54] See [5] for a classic source about scalar-tensor theories and [41], Sec. 5.3 for an overview.

[55] Alternatively, we could regard BDD as inter-theoretical as we did with GTR before. However, this would complicate matters too much.

The following law characterizes the evolution of the scalar field ϕ and thus is crucial for the actual models of BDD: For all events on the manifold $q \in M$:

$$\Box\phi(q) = \frac{1}{2\omega + 3}\left(8\pi \operatorname{Tr}(T_{tot}(q)) + \phi(q)\frac{dV}{dy}(\phi(q)) - 2V(\phi(q))\right). \tag{18}$$

Here, $\Box$ is the covariant d'Alembert operator, ω is a constant parameter. The potential V is not specified by the theory.[56] According to this law, the dynamic evolution of the scalar field is fueled by all other fields via T_{tot} and via the potential.

The second step is to introduce BDT as an inter-theoretic link. We proceed as we did in the case of EFE, with the only difference that Einstein's field equations, Eq. (16), are replaced by the following equation ([8], p. 60, here cast in the abstract index notation, [40], pp. 23–26).

$$G_{\mu\nu} = \frac{8\pi}{\phi}(T_{tot})_{\mu\nu} + \frac{\omega}{\phi^2}\left(\nabla_\mu\phi\nabla_\nu\phi - \frac{1}{2}g_{\mu\nu}\nabla^\rho\phi\nabla_\rho\phi\right) + \frac{1}{\phi}\left(\nabla_\mu\nabla_\nu\phi - g_{\mu\nu}\Box\phi\right) - \frac{V}{2\phi}g_{\mu\nu}. \tag{19}$$

From a conceptual point of view, GTR and BDT do not differ much. Both can be conceptualized as inter-theoretic links. Further, regarding the description of a space-time, GTR and BDT do not differ. The space-time theories presupposed by both GTR and BDT in the sense of the third suggestion above are identical. Both times, space-time is characterized as a four-dimensional manifold with a metric tensor field with the same empirical significance.

Using a suitable conformal transformation,[57] g can be mapped to a variant such that the Brans–Dicke equations take the same form as Einstein's field equations (see [8], p. 361). This has motivated claims to the effect that GTR and Brans–Dicke theory are equivalent (see ibid., Sec. 3.6.4 for evidence). Such claims are problematic though. g is not an arbitrary tensor field on the manifold, but rather one with a specific meaning. If it is transformed using a conformal transformation, its meaning is changed. Thus, even though g, as considered in GTR, and its conformal transform, as considered in BDT, both solve equations of the same type, they are different things. This becomes also manifest when we consider GTR and BDT as parts of complete theories, respectively, because then g and its conformal transform play different roles in the models (e.g., the paths that freely falling test particles take are different [8], p. 88), which means that the theories are different.

Our results concerning BDT can be generalized to many other alternative theories of gravity. So-called metric $f(R)$-theories, which add expressions nonlinear in curvature to the action ([8], Sec. 3.2), do not even introduce new degrees of freedom. Other theories like the one by Rosen [28] do introduce new degrees of freedom, but the latter are only used to constrain the space-time structure, which is described using a metric as before ([7], p. 175). An exception may seem the tensor-vector-scalar the-

---

[56]Often, the Brans–Dicke theory is defined more narrowly with a zero potential.

[57]A conformal transformation maps the metric g to $\Omega^2 g$, where Ω is a strictly positive smooth function. $\Omega^2 g$ yields the same angles as does g, but other geometric features are not invariant under conformal transformations. See [40], p. 445.

ory by Bekenstein [4]. This theory contains a second metric tensor field over and above g, and it turns out that this tensor field determines lengths as measured by rods and clocks ([7], p. 175). But even in this theory, the space-time is described using a manifold and a metric, only that the metric plays otherwise a very different role than it does in Einstein's theory. In general, our results about BDT apply to all metric theories, which describe space-time structure using a metric.

7 In the Space of Space-Time Theories

We have now seen various theories that are space-time theories in one or the other sense. How are they related to each other? Previous work in the semantic framework has defined various inter-theoretic relationships, notably variants of reduction (see particularly [29, 30]). In the model-theoretic framework, two theories stand in an inter-theoretic relationship if their models are related in some way. For instance, the models of one theory may be built up using only components of models of another theory. The aim of this section is to explore inter-theoretic relationships between the space-time theories analyzed thus far. To simplify matters, I will largely bracket links and constraints. Properly speaking, reduction of one theory to another requires that their constraints and links be related, but I will neglect corresponding conditions.

Consider first the space-time theory incorporated in CPM, viz., CPM_{ST}, and that entailed in GPFD, call it $GPFD_{ST}$. In the first reconstruction presented here, CPM_{ST} is very different from $GPFD_{ST}$ because the models have mostly different components. However, CPM_{ST} was also reconstructed in an alternative way using terms from differential geometry. Both versions of CPM_{ST} are equivalent, where equivalence itself is an inter-theoretic relationship.

A differential geometric axiomatization of CPM_{ST} as expounded in Def. 2 brings it closer to GTR. But even then there are a lot of differences between the models of CPM_{ST} and GFD_{ST}. The models have components with different typifications, and they bear different relationships to experience, as stated in the axioms. Scheibe [30], Chs. VII–VIII implicitly concentrates on the former aspect and investigates the relationship between, e.g., a Newtonian space-time theory and the one behind the special theory of relativity. For this example, he constructs a limiting case reduction (his Sec. VII.2) and an asymptotic reduction (his Sec. VII.3).[58]

To establish the former, he expands the theory of Minkowski space. He introduces a parameter that is a function of the maximal velocity. Newtonian space-time is obtained, if this parameter approaches zero or, equivalently, if the maximal velocity approaches infinity. The parameter is not a variable in any theory considered, but the trick to establish a connection between a classical and a special relativistic space-time is to allow for models in which the parameter can take every value. It can then be shown that the models of the expanded theory of Minkowski space approach the only

---

[58]See [29], Sec. V.1, particularly pp. 175–177 for a general account of asymptotic reduction, and ibid., Sec. V.2 for limiting case reduction.

model of Newtonian space-time in the sense of point-wise convergence of the metric everywhere (ibid., p. 24). This is an important necessary condition for the envisaged limiting case reduction of Newtonian space-time theory to a special relativistic one. Since Minkowski space is but one space-time possible according to GTR, it should be possible to generalize Scheibe's result to general relativistic space-time theories, but the details are beyond the scope of this paper.[59]

Turn now to GTR and its explanatory rivals such as BDT. We simplify matters if we conceive of both as self-standing theories, and not as inter-theoretic links. To this effect, we include a component T_{tot} in the models of GPM$_{ST}$ as indicated earlier in Sec. 6.1. The models of GTR and those of its metric rivals then share the components that characterize the space-time, notably the four-dimensional manifold and the metric tensor field, and the meaning of the metric tensor is the same through the same geodesic connection (see the previous section).

A good candidate for characterizing the relationship between GTR and BDT then is the limiting case reduction in the terms of [29], Sec. V.1.[60] Intuitively, to "regain" GTR from BDT, we set the potential V to zero and let ω go to infinity. At least the latter limit takes us beyond BDT because ω is supposed to be constant according to BDT. We have thus to generalize BDT to BDT', which allows for several values of ω. BDT is a specialization of BDT'. The hope then is that every model of GTR is approached by a series of models from BDT'. There are two difficulties to establish this claim. First, to make good on claims to the effect that series of models converge to another model, we need to introduce a topology or, preferably, a metric on the models. We thus have to define distance measures between manifolds equipped with a metric. This is very difficult.[61] Second, we have to show that, according to the distance measure, each model of GTR is the limit of a sequence of models of BDT. There seem to be problems with such a claim (see [8], p. 61).

In practice, working scientists are interested in the different predictions of GTR and its rivals to test the empirical adequacy of GTR. To this end, they use the so-called Parameterized Post-Newtonian formalism (PPN formalism, for short; see [41], Chs. 4–6). The core of the formalism is a set of equations for the metric tensor field in specific coordinates (see ibid., Table 4.1 on pp. 103–104 for a succinct summary). The equations contain a set of free parameters. Different choices of the parameters yield solutions that are approximations to different rivals to GTR.

In the semantic framework, we can think of the PPN formalism as a theory net (see e.g., BMS, Def. DIV-2 on p. 172). In the special case of a theory net that is of interest here, several little theories are specializations of one and the same theory. Let us first consider this more general theory, call it PPN. We can define its models as follows:

---

[59]For the relationship between Newton's Theory of Gravitation and GTR see [30], Ch. VIII.

[60]GTR is not just a specialization of BDT because the latter has additional degrees of freedom, viz., the scalar field (see BMS, Def. DIV.1 on p. 170 for specialization).

[61]Scheibe's comparison between a classical space-time and Minkowski space-time is simpler because both manifolds are homogeneous.

Definition 6 x is a model of PPN iff $x = \langle M, \mathscr{P}, y, g, \rho, \mathbf{v}, \mathbf{w}, P, \Pi, \alpha, \beta, ..., y \rangle$, where

- M is a manifold of events;
- $\mathscr{P}$ is a set of particles;
- $y : M \to \mathbb{R}^4$ is a global coordinate chart on M with certain properties as indicated above.
- $g : \mathbb{R}^4 \to \mathbb{R}^{4 \times 4}$ is the metric in the coordinates;
- $\rho : \mathbb{R}^4 \to \mathbb{R}$ is the density of rest mass in the coordinates;
- $\mathbf{v} : \mathbb{R}^4 \to \mathbb{R}^3$ is the velocity field of matter in the coordinates;
- $\mathbf{w} \in \mathbb{R}^3$ is the velocity of the frame associated with the coordinates relative to the rest frame of the universe;
- $P : \mathbb{R}^4 \to \mathbb{R}$ is pressure in the coordinates;
- $\Pi : \mathbb{R}^4 \to \mathbb{R}$ is internal energy in the coordinates;
- $\alpha, \beta, ...$ are real numbers;
- r comprises the world lines of test particles.

Further, the functions are related as described in Table 4.1 on pp. 103–104 in [41]. For instance, the metric is expressed in terms of metric potentials that are calculated from ρ, $\mathbf{v}$ and $\mathbf{w}$.

As is evident, the models are coordinate-dependent, i.e., a particular coordinate system is assumed. This system allows us to distinguish between space and time. Assuming the coordinate system, the various fields simplify to functions from and to mathematical vector spaces. For instance, ρ is a function from $\mathbb{R}^4$ to $\mathbb{R}$. As a consequence, it is straightforward to define distances between different models of PPN.

We obtain different specializations of PPN if we set the parameters at certain values. The idea now is that certain parameter choices correspond to certain theories such as GTR or BDT because the models with these parameter choices are approximately models of GTR, BDT, etc. For instance, within the PPN formalism, we are lead to GTR when we set γ and β at 1 and the other parameters at zero ([41], p. 123).

But what exactly does it mean to say that a specialization of the PPN formalism, call it PPN(GTR), "corresponds" to GTR or "leads to it"? A first hope may be that PPN(GTR) approximately reduces to GTR. According to BMS, Def. DVII-22 on p. 373, an approximate reduction is a relation between the models of two theories.[62] The most important condition for a theory T_1 reducing to another, better one, T_2, approximately is that, for every model x of T_2, one can find a model of T_1 that comes arbitrarily close to x.[63]

---

[62] I here refer to BMS because their account of approximate reduction seems simpler than that by Scheibe.

[63] This does not require that a distance measure between models of both theories be defined, but it does require that there is a least a topology over models in one of the theories (cf. BMS, Ch. VII, particularly Sec. VII.3.1). At this point, BMS differ from Scheibe, who demands that a common space be defined in which the models of both theories are included.

PPN(GTR) does *not* approximately reduce to GTR in the sense just explained. First, the models of PPN(GTR) are very special because they are supposed to trace the solar system. They thus obey peculiar boundary conditions (according to which the metric approaches homogeneity and isotropy, see [41], pp. 91–92). Consequently, only some models of GTR have a chance of being approximated by models in PPN(GTR). In the terms of BMS, this is to say that the approximate reduction is at best indirect.[64]

But second, there is not even an indirect approximate reduction to GTR. The models of PPN(GTR) do not come arbitrarily close to models of GTR. The reason is that PPN is not an approximation scheme that can take into account arbitrary orders. Rather, it is restricted to fixed orders (see [41], Ch. 4). Thus, models of GTR are not approximated by models of PPN(GTR) to arbitrary precision, which is incompatible with there being an approximate reduction.

8 Conclusions

In this paper, we have discussed space-time theories from the semantic point of view and using reconstructions in terms of models. What are the payoffs? Let me mention four.

First, using a model-theoretic analysis, we can understand how concepts obtain their meanings and how they depend on each other. The notions of space-time and of spatiotemporal relationships are very interesting in this respect. On the one hand, they are very basic; most physical theories use these notions. On the other hand, space-time itself and its elements cannot be observed or measured. One way in which terms can obtain meaning in our framework is that they are introduced as theoretical terms by some theories. We have seen how the metric tensor field can be introduced as a theoretical term. The space-time manifold itself is more difficult to handle. It seems wrong that points are simply introduced as T-theoretical notions by some theory T because some points can be determined empirically without recourse to a space-time theory, e.g., as positions of events/particles. A precise account of how space-time points and their manifolds obtain meaning was beyond the scope of this paper.

Second, the framework considered in this paper suggests a way to define what a space-time theory is to begin with. I do not think that working scientists use the concept of a space-time theory always in the same way, but, often, a theory T that introduces spatial notions or the metric as T-theoretic component is called a space-time theory. Alternatively, we may call a theory space-time theory if it explains why the space-time has the structure it has.

Third, we can compare different conceptualizations of space-time and of spatiotemporal relationships. In this paper, we have seen two, viz., that known from Newtonian physics (Def. 1) and the general relativistic one (Def. 3). Both differ not

---

[64] See BMS, Defs. VI-5 and VI-6 on p. 277 for direct and indirect reduction. Scheibe's limiting case reduction is only indirect in the terms of BMS.

only in the mathematical apparatus used to represent space-time and spatiotemporal features such as distances. They also differ in the way the mathematical apparatus it related to experience. It is no surprise then that some philosophers take these notions to be subject to meaning-variance (in the sense defined by Feyerabend) or incommensurability.[65]

Fourth, in the framework adopted in this paper, we can study inter-theoretic relationships between space-time theories. Formal results were beyond the scope of this paper, but we have at least sketched some relationships between the theories. The relationship between classical space-time theories and relativistic ones is fairly complicated. It is possible to assimilate classical space-time theories to relativistic ones by recasting them in a covariant form using a four-dimensional manifold and a metric (see Def. 2). But even if this is done, it seems at most possible to obtain a limiting case reduction or an asymptotic reduction of a Newtonian space-time to Minkowski space-time, which is but one of many space-times allowed by GTR. The relationship between GTR and its alternatives, e.g., BDT, is more straightforward. Regarding the description of space-time, metric rivals do not differ at all from GTR: Space-time is featured using a four-dimensional manifold with a Lorentzian metric. The only difference is that the theories allow for different combinations of momentum-energy distribution and the metric.

Things become very different if we turn to theories that try to unify gravitation and quantum theory. Some of them suggest a new theoretical account of space-time, e.g., that the space-time emerges in some way. This should not come as a surprise. We have seen that the space-time manifold is largely a theoretical posit, so it seems possible to fancy alternative conceptualizations of spatiotemporal relationships. Proposals for an emerging space-time could certainly be considered in the model-theoretic framework; what is more, a related reconstruction could cast some light on the "emergence" of space-time.

There is space, and hopefully also time, for more then.

Acknowledgements I'm grateful for very helpful and constructive criticism by Dennis Lehmkuhl and Erhard Scholz. Thanks also to Raphael Bolinger for his comments.

References

1. Balzer, W., Moulines, C. U., & Sneed, J., *An Architectonic for Science. The Structuralist Program*, Reidel, Dordrecht, 1987.
2. Balzer, W., *Die Wissenschaft und ihre Methoden*, Karl Alber, Freiburg und München, 1997.
3. Bartelborth, T., *Hierarchy versus Holism: A Structuralist View on General Relativity*, Erkenntnis **39** (1993), no. 3, 383–412.
4. Bekenstein, J. D., *Relativistic Gravitation Theory for the Modified Newtonian Dynamics Paradigm*, Phys. Rev. D **70** (2004), no. 8, 083509.
5. Bergmann, P., *Comments on the Scalar-tensor Theory*, International Journal of Theoretical Physics **1** (1968), 25–36.

---

[65]See [25, 30], Sec. VII.3 for more discussion.

6. Brans, C. & Dicke, R. H., *Mach's Principle and a Relativistic Theory of Gravitation*, Phys. Rev. **124** (1961), 925–935.
7. Brown, H. R., *Physical Relativity. Space-time Structure from a Dynamical Perspective*, Oxford University Press, Oxford, 2005.
8. Capozziello, S. & Faraoni, V., *Beyond Einstein Gravity. A Survey of Gravitational Theories for Cosmology and Astrophysics*, Springer, Dordrecht, 2011.
9. Carrier, M., *Constructing or Completing Physical Geometry? On the Relation between Theory and Evidence in Accounts of Space-time Structure*, Philosophy of Science **57** (1990), no. 3, 369–394.
10. Carrier, M., *The Completeness of Scientific Theories. On the Derivation of Empirical Indicators within a Theoretical Framework: The Case of Physical Geometry*, Kluwer, Dordrecht, 1994.
11. Costa, N. C. A., Bueno, O., & French, S., *Suppes Predicates for Space-time*, Synthese **112** (1997), no. 2, 271–279.
12. da Costa, N. C. A. & French, S., *The Model Theoretic Approach in Philosophy of Science*, Philosophy of Science **57** (1990), 248–265.
13. Earman, J. & Norton, J. D., *What Price Spacetime Substantivalism? The Hole Story*, The British Journal for the Philosophy of Science **38** (1987), 515–525.
14. Earman, J., *A Primer on Determinism*, Kluwer, Dordrecht, 1986.
15. Earman, J., *World Enough and Space-time. Absolute versus Relational Theories of Space and Time*, MIT Press, Cambridge (MA), 1989.
16. Ehlers, J., Pirani, F. A. E., & Schild, A., *The Geometry of Free Fall and Light Propagation*, in: *General Relativity. Papers in Honour of J. L. Synge* (O'Raifeartaigh, L., ed.), Clarendon Press, Oxford, 1972, pp. 63–84.
17. French, S., *The Structure of Theories*, in: *The Routledge Companion to Philosophy of Science* (Psillos, S. & Curd, M., eds.), Routledge, 2008, pp. 269–280.
18. Friedman, M., *Foundations of Space-time Theories. Relativistic Physics and Philosophy of Science*, Princeton University Press, Princeton, 1983.
19. Hendry, R. F. & Psillos, S., *How to Do Things with Theories: An Interactive View of language and Models in Science*, in: *The Courage of Doing Philosophy. Essays Dedicated to Leszek Nowak* (Brzeziński, J., Klawiter, A., Kuipers, T. A. F., Łastowski, K., Paprzycka, K., & Przybysz, P., eds.), Rodopi, Amsterdam, 2007, pp. 123–157.
20. Hodges, W., *Model Theory*, in: *The Stanford Encyclopedia of Philosophy* (Zalta, E. N., ed.), fall 2009 ed., 2009, http://plato.stanford.edu/archives/fall2009/entries/model-theory/.
21. Hoefer, C., *The Metaphysics of Space-time Substantivalism*, Journal of Philosophy **93** (1996), no. 1, 5–27.
22. Humphreys, P., *Extending Ourselves: Computational Science, Empiricism, and Scientific Method*, Oxford University Press, New York, 2004.
23. Ludwig, G. & Thurler, G., *A New Foundation of Physical Theories*, Springer, Berlin etc., 2006.
24. Matveev, V. S., *Geodesically Equivalent Metrics in General Relativity*, Journal of Geometry and Physics **62** (2012), no. 3, 675–691, Recent Developments in Mathematical Relativity.
25. Mühlhölzer, F., *Science without Reference?*, Erkenntnis **42** (1995), no. 2, 203–222.
26. Quine, W. V. O., *On What There Is*, Review of Metaphysics **2** (1948), 21–38, nachgedruckt in Quine, From a Logical Point of View. Harvard 1961.
27. Rindler, W., *Essential Relatvity. Special, General, and Cosmological*, 2nd ed., J. Springer, New York, 1977.
28. Rosen, N., *A Bi-metric Theory of Gravitation*, General Relativity and Gravitation **4** (1973), 435–447.
29. Scheibe, E., *Die Reduktion physikalischer Theorien. Ein Beitrag zur Einheit der Physik. Teil I: Grundlagen und elementare Theorie*, Springer, Heidelberg, 1997.
30. Scheibe, E., *Die Reduktion physikalischer Theorien. Ein Beitrag zur Einheit der Physik. Teil I: Inkommensurabilität und Grenzfallreduktion*, Springer, Heidelberg, 1999.
31. Schlimm, D., *On the Creative Role of Axiomatics. The Discovery of Lattices by Schröder, Dedekind, Birkhoff, and Others*, Synthese **183** (2011), 47–68, DOI:10.1007/s11229-009-9667-9.

32. Scott, D. & Suppes, P., *Foundational Aspects of Theories of Measurement*, Journal of Symbolic Logic **23** (1958), 113–128.
33. Sneed, J., *The Logical Structure of Mathematical Physics*, Reidel, Dordrecht, 1979, revised edition; first edition 1971.
34. Suppe, F., *The Search for Philosophical Understanding of Scientific Theories*, in: *The Structure of Scientific Theories* (Suppe, F., ed.), University of Illinois Press, Urbana, IL, 1977, second edition.
35. Suppes, P., *Some Remarks on Problems and Methods in the Philosophy of Science*, Philosophy of Science **21** (1954), no. 3, pp. 242–248 (English).
36. Suppes, P., *Introduction to Logic*, van Nostrand Reinhold Company, New York etc., 1957.
37. Suppes, P., *Models of data*, in: *Logic, Methodology and Philosophy of Science: Proceedings of the 1960 International Congress* (Nagel, E., Suppes, P., & Tarski, A., eds.), Stanford University Press, Stanford, 1962, here quoted from reprint in: Patrick Suppes: Studies in the Methodology and Foundations of Science. Selected Papers from 1951 to 1969. Dordrecht: Reidel 1969, 24–35, pp. 252–261.
38. Synge, J. L., *Relativity: The General Theory*, North-Holland, Amsterdam, 1966.
39. van Fraassen, B., *The Scientific Image*, Clarendon Press, Oxford, 1980.
40. Wald, R. M., *General Relativity*, University of Chicago Press, Chicago, 1984.
41. Will, C. M., *Theory and Experiment in Gravitational Physics*, Cambridge University Press, Cambridge, 1993, revised edition.

The Relativity and Equivalence Principles for Self-gravitating Systems

David Wallace

Abstract I criticise the view that the relativity and equivalence principles are consequences of the small-scale structure of the metric in general relativity by arguing that these principles also apply to systems with non-trivial self-gravitation and hence non-trivial space-time curvature (such as black holes). I provide an alternative account, incorporating aspects of the criticised view, which allows both principles to apply to systems with self-gravity.

1 Introduction: Two Principles

The relativity principle—the observation that the laws of physics are the same in two reference frames in constant motion with respect to one another—is by now very well understood at least from a mathematical point of view. It can be understood as entailed by certain symmetries of the laws of physics: the Gallilei symmetry group in pre-relativistic physics, the Poincaré group in special relativity. Alternatively, and perhaps equivalently, it can be understood as a consequence of the structure of spacetime: in either neo-Newtonian[1] or Minkowski spacetime, boosts between reference frames are automorphisms of the background spacetime structure and so have no physically detectable consequences.

This raises a question: our world is apparently correctly described (or at any rate *better* described) by general relativity rather than by special relativity (let alone Newtonian physics). So the laws of physics do not have the Poincaré group as a symmetry (indeed, it is not even clear what that would mean), and the spacetime structure is described by a metric which does not, in general, have any automorphisms. So why is the relativity principle empirically correct, if its theoretical underpinnings are not?

The conventional wisdom appears to be that the relativity principle is recovered as a result of holding in regions which small enough that spacetime curvature can be

---

[1] See (e.g.) chapter 2 of [3] for an account of neo-Newtonian spacetime.

D. Wallace (✉)
Department of Philosophy, University of California, Berkeley, USA
e-mail: dmwallac@usc.edu

© Springer Science+Business Media, LLC 2017   257
D. Lehmkuhl et al. (eds.), *Towards a Theory of Spacetime Theories*,
Einstein Studies 13, DOI 10.1007/978-1-4939-3210-8_8

neglected. In terms of the metric: in a sufficiently small spacetime patch, curvature is negligible so the metric may be approximated as Minkowskian. In terms of the laws of physics: in a sufficiently small spacetime patch, the laws of special relativity hold to a high level of accuracy. (See the work of Harvey Brown, and in particular [1, pp. 169–172] for a particularly clear statement of this position.)

Just as the relativity principle was historically crucial in the development of special relativity, so the equivalence principle played a crucial part in developing general relativity.[2] A natural way to state it[3] is as the claim that a system falling freely in a uniform gravitational field will behave in exactly the same way as an isolated system. In Newtonian physics, this makes literal sense, and can be understood as the consequence of a symmetry of the theory: if we take the theory to be specified by a collection of point particles and by a gravitational potential $V(\mathbf{x}, t)$, where $V(\mathbf{x}, t)$ satisfies the Poisson equation and the particles obey Newton's laws (possibly with other distance-dependent or contact forces present), then the process of giving the masses uniform acceleration $\mathbf{a}(t)$ and adding a term $\mathbf{a}(t) \cdot \mathbf{x}$ to the potential is a process which takes solutions to solutions. While there is no geometric equivalent of this dynamical symmetry as long as we formulate Newtonian physics on neo-Newtonian spacetime, the very presence of this mismatch between dynamical and spacetime symmetries points to the existence of a more natural geometric arena for Newtonian gravity: Newton–Cartan spacetime, the automorphisms of which include arbitrary spatially constant accelerations.

It is somewhat harder to make sense of the equivalence principle in this way in general relativity, where there is no direct analogue to the gravitational potential. But a natural point of connection is the idea that physics should be the same in any freely falling reference frame (and therefore, in particular, the same in a freely falling reference frame as in a frame moving inertially in a region of spacetime where gravitational phenomena are negligible). Once that is understood, we are faced with a similar problem as for the relativity principle—how to define "reference frame" in a setting which has spacetime curvature—and a similar answer is available. Namely, the equivalence principle (says the conventional wisdom) is a corollary of the fact that sufficiently small regions of spacetime may be treated as flat (and hence isomorphic to regions of Minkowski spacetime). In particular, since in a sufficiently thin tube around an arbitrary geodesic the spacetime metric may be approximated as flat, physics within that tube ought to be indistinguishable from physics within a similarly shaped tube in Minkowski spacetime (or indeed in intergalactic space).

So we have a rather straightforward story: the relativity and equivalence principles both hold true, in general relativity, in sufficiently small spacetime patches, because of the small-scale behaviour of the metric, and in particular because in any metric,

---

[2]Equally crucial was Einstein's desire to extend the relativity principle to cover non-uniform motion; cf [6]. (I am grateful to an anonymous referee for stressing this point.)

[3]There are many ways to state it, and much controversy about just what was meant by it historically; see, e.g., [9] for more details. The formulation I use here is approximately that of [8, p. 386] and [1, p. 169] (his 'SEP') and [7, p. 874] (her 'NSEP'). I hold no brief that this is 'the' or 'the right' equivalence principle, simply that it is a principle worthy of study.

a sufficiently small spacetime patch (or a sufficiently thin spacetime tube around some geodesic) may be idealised as metrically flat. Neither principle, therefore, is concerned with the *large*-scale behaviour of the metric in solutions to the Einstein field equations; neither can apply in situations where curvature cannot be neglected.

The point of this paper is to challenge this story: it is not right, or at any rate it is not at all the whole truth, and there is an alternative and preferable account. In section 2 I will give examples to show just why there must be more to say; in the rest of the paper, I will try to say it.

2 Galileo's Black Hole

The folklore of physics (correctly, so far as I know) attributes the first statement of the relativity principle to Galileo, in his famous thought-experiment:

> Shut yourself up with some friend in the main cabin below decks on some large ship, and have with you there some flies, butterflies, and other small flying animals. Have a large bowl of water with some fish in it; hang up a bottle that empties drop by drop into a wide vessel beneath it.
>
> With the ship standing still, observe carefully how the little animals fly with equal speed to all sides of the cabin. The fish swim indifferently in all directions; the drops fall into the vessel beneath; and, in throwing something to your friend, you need to throw it no more strongly in one direction than another, the distances being equal; jumping with your feet together, you pass equal spaces in every direction.
>
> When you have observed all of these things carefully (though there is no doubt that when the ship is standing still everything must happen this way), have the ship proceed with any speed you like, so long as the motion is uniform and not fluctuating this way and that. You will discover not the least change in all the effects named, nor could you tell from any of them whether the ship was moving or standing still. [4]

Tragically (if unsurprisingly), Galileo did not live to see the development of general relativity. If he had, who can doubt that he would have quickly penned a sequel, which no doubt would have become known as *Galileo's black hole*:

> Put yourself together with some friend in orbit around some large black hole, and have with you there some planetoids, interstellar dust, and other solid matter. Have a long rope with a clock at one end of it; place a light source into an orbit below yours. With the black hole standing still, observe carefully how the redshift from the light source is of equal magnitude at all points in its orbit. The clock slows equally whatever side of the hole it is lowered towards; and, in allowing the solid matter to fall inward and form an accretion disk, the disk forms no more strongly on one side than another; observing the high-energy jets above and below its plane, they are as like to emerge in one direction as in the other.
>
> When you have observed all these things carefully (though there is no doubt that when the black hole is standing still everything must happen in this way), have the hole proceed with any speed you like (staying in orbit around it all the while), so long as the motion is uniform and not fluctuating this way and that. You will discover not the least change in all the effects named, nor could you tell from any of them whether the hole was moving or standing still.

The relativity principle, in other words, had better apply to all manner of systems, including those which are strongly self-gravitating. More to the point (for what does

Nature care what "had better" be true?), there is abundant evidence that it *does* apply to such systems. Observations of black holes and neutron stars are commonplace in high-energy astrophysics these days, and astrophysicists do not even consider the "absolute velocity" of the system they study (whatever that would be) in their analysis, except insofar as they apply standard length contraction and time dilation formulae to translate phenomena in the system's rest frame into phenomena in our rest frame. And at a more mundane level, even planets have non-trivial self-gravity, and the relativity principle manifestly applies to that self-gravity itself. Objects fall under Earth's gravity in just the same way in summer as in winter, despite the ~ 60 km/s velocity difference.

But the analysis I gave of the relativity principle in section 1 cannot possibly apply to self-gravitating systems. For that analysis required the principle to apply only in regions where the metric could be idealised as flat, and what is "a region where the metric can be idealised as flat", if not a region in which self-gravitational phenomena can be idealised away? Insofar as we idealise the metric of, say, the Earth-Moon system as flat, we idealise away the gravitational binding between Earth and Moon. And more starkly, regions of spacetime in which the metric is approximately flat do not as a rule contain black holes.

And the same is true of the equivalence principle. The Earth-Moon system is in orbit around the Sun, and thus moving freely in the Sun's gravitation field, yet the Moon orbits the Earth, and objects fall under Earth's gravity, just as they would were the Sun not there.[4] Drop a star—or even a neutron star or black hole—deep into the gravitational field of a supermassive black hole. As long as tidal forces remain insignificant on the lengthscale of the star, the star's own physics—including the star's gravitational physics—will show no sign that it is not in open space far from the supermassive hole.

So: physically speaking both the equivalence principle and the relativity principle make sense not only in flat regions of spacetime but for systems with significant self-gravity. In the next two sections I will consider how to make sense of this fact mathematically speaking, and thus show what makes it correct. The framework I adopt is a special case of that developed more generally in [5]. I should stress that the level of mathematical rigour I adopt is about that in the mainstream theoretical physics literature (at the level, say, of [2, 8, 10]); those who prefer their general relativity to be completely rigorous should regard my claims as heuristic rather than as theorems, though they should also recognise that "general relativity" in this sense falls very short of encompassing all that is actually done with the theory in contemporary physics.[5]

---

[4]To a good approximation, at any rate; the only observable effect of which I am aware (other than astronomical observations) is the monthly variation in the strength of the tides as Sun and Moon move in and out of alignment.

[5]On this note, see in particular Zee's comments on mathematical rigor in the introduction to [10].

3 The Relativity and Equivalence Principles for Non-gravitating Systems

To get a version of the relativity and equivalence principles applicable to systems with appreciable self-gravity, I need a rather operationalised version of the principles, relatively close in content to the thought-experiments of the previous section. I interpret the operational content of the relativity principle as follows:

> Given a collection of isolated systems each moving in Minkowski spacetime (or in some region of a more general spacetime which can be approximated as Minkowskian), the physics internal to each system as described in inertial coordinates comoving with that system is independent of the velocity of that system.

While the operational content of the equivalence principle is as follows:

> Given a general spacetime, and a geodesic within that spacetime, there is a local inertial frame along that geodesic such that to an arbitrarily good approximation, the physics of a sufficiently small system freely falling along that geodesic (and thus isolated from other systems) is indistinguishable from that of the same system in Minkowski spacetime.

These operational results follow, in non-gravitating systems, respectively from the Poincaré-covariance of the non-gravitational interactions and from the fact that a sufficiently small region of a (pseudo-)Riemannian manifold can be approximated as flat. Let me review briefly *how* they follow. First, a little terminology: I assume that we are considering (general- or special-) relativistic spacetimes with some given matter fields and dynamical equations. If one such spacetime $\mathcal{M}$ is foliated by spacelike hypersurfaces, a *temporal segment* of $\mathcal{M}$ is the part of the spacetime between any two such hypersurfaces. A *tube* in a temporal segment $\mathcal{T}$ of $\mathcal{M}$ is a smooth map $\varphi : B^3 \times [0, 1] \to \mathcal{M}$ (where B^3 is the closed unit ball in R^3) such that

1. φ is a diffeomorphism onto a closed subset of $\mathcal{T}$;
2. Each of the curves $\varphi(\{x\} \times [0, 1])$ is a segment of a timelike curve;
3. Each of the surfaces $\varphi(B^3 \times \{x\})$ is spacelike;
4. Each of the surfaces in (3) is orthogonal to each of the curves in (2);
5. $\varphi(B^3 \times \{0\}$ and $\varphi(B^3 \times \{1\}$ lie, respectively, in the initial and final bounding surfaces of $\mathcal{T}$.

In Minkowski spacetime coordinatised in the usual way, the canonical example of a tube is given by $\varphi(x, y, z, t) = (\lambda x, \lambda y, \lambda z, \mu t)$ for positive real λ and μ. A *tube around* a timelike curve segment $\gamma : [0, 1] \to \mathcal{M}$ is a tube φ such that $\varphi(0, x) = \gamma(x)$. The *surface* of a tube is $\varphi(S^3 \times [0, 1])$, where S^3 is the boundary of the unit ball. And two tubes are *surface-compatible* if there is a diffeomorphism of neighbourhoods of their respective surfaces that maps the metric and matter fields of one to the other. (So in vacuum, this reduces to the requirement that the tubes' surfaces have isometric neighbourhoods.)

Now let $\mathcal{T}$, $\mathcal{T}'$ be temporal segments of some spacetime, and let φ, φ' be tubes in $\mathcal{T}$, $\mathcal{T}'$ that are surface-compatible. We can construct another temporal segment $\mathcal{T}''$

by replacing the matter and metric fields inside φ by those inside φ', via the diffeo-morphism $\varphi \cdot \varphi'^{-1}$. Since the tubes are surface-compatible, there is a neighbourhood of the tube whose matter and metric fields are diffeomorphically related to those in a neighbourhood of $\mathscr{T}'$, and a neighbourhood of the tube whose matter and metric fields are diffeomorphically related to those in a neighbourhood of $\mathscr{T}$. So there is an open covering of $\mathscr{T}''$ by sets on each of which the dynamical equations are satisfied. And so $\mathscr{T}''$ is a temporal segment of a dynamically possible spacetime.

This ability to patch one tube into another allows us to derive both the relativity and equivalence principles in the form I give above. To see this for the relativity principle, consider an isolated system moving in Minkowski spacetime, with its centre of mass following some given inertial trajectory. If that system is indeed *completely* isolated, that is to say that its evolution would be unchanged if all other bodies were absent—which is to say that there is another Minkowski spacetime in which nothing is present except (a copy of) the system, but where the system itself is unchanged. Which is in turn to say that there is some tube along (any given segment of) the body's trajectory, and some tube within (a temporal segment of) another Minkowski spacetime such that nothing is present outside the tube, where the fields inside the two tubes are the same (i.e., diffeomorphically related).

In idealisation, a sufficient condition for all this is that each of the isolated bodies has no fields associated with it outside some tube containing any finite period of its evolution, so that the spacetime is empty in a neighbourhood of the surfaces of each tube. From this it follows that each tube is surface-compatible with a tube in otherwise empty Minkowski space. But then given another inertial trajectory having velocity v compared to the first, the Poincaré covariance of the Minkowski metric means that any given segment of that trajectory has a tube along it that is surface-compatible with the first tube. And so the contents of the first tube can be pasted into the second tube, creating a new spacetime (or temporal segment thereof) in which the system is boosted relative to its original state while remaining intrinsically unchanged.

Of course, in realistic systems isolation is never complete: there will be fields arbi-trarily far from the system's centre of mass trajectory still associated with the system (think of long-range Coulomb fields, for instance). So the result of this patching is only an exact solution of the dynamical equations in the idealised limit of perfect isolation; otherwise, there will be a slight discontinuity at the boundary. However, we have good (albeit heuristic) grounds to think that extremely small perturbations of the solutions within and without the tube can be made that will remove this dis-continuity. (Insofar as this is *not* the case, the claim that the systems are isolated becomes questionable.)

For the equivalence principle, let γ be a geodesic in some spacetime. If there were some tube surrounding the geodesic in which the metric was *exactly* flat, we could take a tube in Minkowski spacetime, surrounding some isolated system, and patch it onto the first tube: this would show that local physics along that geodesic was intrinsically identical to the same local physics playing out in Minkowski spacetime.

Exactly flat tubes are going to be hard to come by, though: in an exactly flat tube, the Riemann tensor vanishes, and it's typically non-zero pretty much everywhere on nontrivial spacetimes. However, this tensor is dimensionful—it has the dimensions of $1/\text{length}^2$—which suggests that in sufficiently small tubes it can be treated as negligible. We can flesh out this suggestion by noting that the physical significance of the Riemann tensor comes through its contribution to the holonomy—the effect of parallel transporting a vector around a closed loop and that, if the maximum value of a given component of the Riemann tensor with respect to some tetrad is R_{max}, the maximum value of the component of holonomy associated with that component is the area of the loop around which it is evaluated times R_{max}. So (given that the length of the tube is fixed) for any degree of approximation desired there will be a lengthscale such that a tube narrower than that lengthscale can be treated as flat to that degree of approximation. Given a dynamical system in Minkowski space that can be enclosed within a tube of that lengthscale, to that degree of approximation the tube can be pasted onto the geodesic.

One way to quantify this is to note that for a sufficiently narrow tube the linear approximation to general relativity will be valid to any given degree of accuracy, so that we can understand the physics of the tube interior as playing out on a Minkowski spacetime with a symmetric tensor field representing gravity. If there is a regime in which that gravity field is dynamically negligible compared to the internal interactions relevant to the system, in that regime the system can be treated as interacting on flat spacetime. Of course, for any given physical system there will be levels of curvature so great that this cannot be done—for instance, there are curvatures sufficient to tear apart atoms, and still greater curvatures sufficient to tear apart nuclei. It cannot be expected that any operationalised version of the equivalence principle holds for *arbitrarily* high curvatures and for any given sort of non-gravitational interaction.

4 Extending the Principles to Self-gravitating Systems

Note that in both the relativity principle and the equivalence principle case, the "isolated systems" being studied are modelled as systems alone in Minkowski spacetime. What is required of them is (i) that their dynamical equations are Poincaré-covariant (so that they can indeed be modelled on that spacetime); (ii) that they can indeed be treated as isolated, i.e. contained within some finite width tube to any given degree of accuracy.

Condition (i) does not hold in any straightforward sense for self-gravitating systems, which are not modelled on a background Minkowski spacetime. However, we can consider a temporal segment of a more general spacetime which to any given degree of approximation is Minkowskian outside, and on the boundary of, some tube of given size and shape. For instance, the spacetime around a black hole has this character: asymptotically at spatial infinity, the metric tends to the Minkowski metric. So for any given level of tolerable deviation from the Minkowski metric, we can find a tube around the black hole such that the spacetime outside that tube approximates the Minkowski metric to that level. The same will very plausibly be true for more complex self-gravitating systems such as the Solar system. Such systems also satisfy condition (ii). So these systems, too, can be pasted into tubes in existing spacetimes provided those tubes are themselves flat to a sufficient degree of approximation (as discussed above).

A self-gravitating system of mass m, if its angular momentum, charge, and radiative emissions can be neglected, has a metric which at large distances from the system tends towards the Schwarzschild metric

$$ds^2 = -(1 - 2m/r)dt^2 + (1 - 2m/r)^{-1}dr^2 + r^2(d\theta^2 + \sin^2\theta d\phi^2)$$

(indeed, this can be taken as *definitional* of the system's mass). (Defining the Schwarzschild radius $r_s = 2m$, we have the result that the metric of such a system is approximately Minkowskian when $r \gg r_s$.) Such a system, asymptotically, is a Minkowski spacetime with a preferred standard of rest: that in which the lines $(r, \theta, \phi) = $ constant are at rest.

Given one such system, geodesics sufficiently far from its centre will approximate straight lines in Minkowski space (for given finite period of time) to any desired degree of accuracy. Another such system can then be pasted onto a tube around any such geodesic, and the intrinsic physics of that second system will be independent of the velocity of the second system with respect to the preferred rest frame of the first. Thus, this combination of two systems demonstrates a form of the relativity principle.

Now consider the case where one spherical black hole is falling into a much larger one (as happens in astrophysics when galactic centre black holes consume stellar-mass black holes). The effective size of the smaller hole—the radius, in Schwarzschild coordinates, of the tube we wish to place around it—is $\sim \lambda r_s$, for some dimensionless $\lambda \gg 1$: at distances greater than λr_s the spacetime around the black hole can be treated as approximately Minkowskian (with the exact value of λ being dependent on the level of approximation required and the scale of the black hole dynamical processes of interest—for accretion λr would need to be the width of the accretion disc, for instance, or possibly the full size of the binary system in which the accretion is occurring). Then we can paste the smaller hole onto a geodesic of the larger hole if its curvature times $(\lambda r_s)^2$ is sufficiently small.

The tetrad components of the curvature of a Schwarzschild black hole of mass M, at radial distance R, have magnitude $\sim M/R^3$ [8, p. 822]. So we require

$$\frac{M}{R^3}(\lambda r_s)^2 \ll 1.$$

Rearranging, and defining R_s as the Schwarzschild radius of the second black hole, gives

$$\frac{R}{R_s} \gg \lambda \left(\frac{m}{M}\right)^{2/3}$$

as a condition for the validity of the equivalence principle in this case. So if the larger black hole has a mass many times that of the former, we would expect local physics as observed by those closely orbiting the smaller hole to continue undisturbed until well after that smaller hole has crossed the event horizon of the larger one.

5 Conclusions

We can coherently talk about isolated systems in general relativity because, as a matter of dynamics, there exist a large number of solutions to the equations—including ones which represent stars, planets, black holes, etc., as well as interacting sets of these—where the curvature and matter are concentrated in some finite region and far outside that region the spacetime is approximately empty and flat. This allows us to paste such solutions together, to form regions of spacetime consisting of a number of isolated subsystems embedded in approximately flat spacetime. Because of the Poincaré symmetry of flat spacetime, we can perform a Poincaré transformation on one of the subsystems without violating the boundary conditions between subsystems; hence, the relativity principle applies for collections of such subsystems.

In turn, regions of effectively flat spacetime can always be found in a given spacetime, provided we are prepared to make those regions sufficiently small. If "sufficiently small" is nonetheless large compared to the effective size of the subsystems we are interested in, then (a) we can apply the above argument for the relativity principle to isolated systems in a curved spacetime; (b) we can embed such systems in any such effectively flat region without affecting their internal dynamics, since their Minkowski boundary conditions are compatible with any region flat on sufficiently large lengthscales.

In effect, then, the equivalence principle applies in general relativity because the metric of isolated systems at sufficiently *large* distances is the same as the metric of any system at sufficiently *small* distances. The relativity principle applies because, in addition to this, that metric has the Poincaré group as a symmetry group.

To some extent, of course, this is definitional of isolated systems, which in this paper I have *defined* as systems which are asymptotically Minkowskian. But it is interesting to note that *this* definition of 'isolated' does not obviously coincide with the common definition of "isolated" as "completely alone in the Universe". Systems isolated in *that* sense might be expected to have spacetimes that were asymptotically anti-de-Sitter, or asymptotically Friedmannian, or asymptotically whatever spacetime fits our current best cosmology. "Isolated" here—and, I'd suggest, in physics more generally—actually means something more mundane: "such that the details of their internal dynamics don't depend on the details of the dynamics of other systems".

I began this paper by criticising the 'conventional wisdom' that the relativity and equivalence principles apply to general relativity because of the flatness of the metric at sufficiently short lengthscales. I hope that I have ended by showing that this is at most half the story: the flatness of the metric of certain systems at sufficiently large lengthscales has an equally important role. There is therefore an important dynamical aspect to both principles which does not seem to have been widely recognised: whether or not the small-scale behaviour of the metric should be understood as 'mere kinematics', it is a nontrivial dynamical fact that there exist asymptotically flat solutions of the Einstein field equations appropriate for the description of the isolated systems in our Universe.

Acknowledgements This paper has benefitted greatly from conversations with Harvey Brown, Eleanor Knox, Dennis Lehmkuhl, Oliver Pooley and Jim Weatherall, and from extensive feedback from two anonymous referees.

References

1. Brown, H. R. (2005). *Physical Relativity*. Oxford: Oxford University Press.
2. Carroll, S. (2003). *Spacetime and Geometry: an Introduction to General Relativity*. San Francisco, CA: Addison Wesley.
3. Earman, J. (1989). *World Enough and Space-Time*. Cambridge, Massachusetts: MIT Press.
4. Galileo (1632/1967). *Dialogue Concerning Two Chief World Systems* (2nd revised ed.). Berkeley, California: University of California Press. Translated by S. Drake.
5. Greaves, H., and D. Wallace (2014). Empirical consequences of symmetries: a new framework. *British Journal for the Philosophy of Science 65*, 59–89.
6. Janssen, M. (2014). 'no success like failure...': Einstein's quest for general relativity, 1907-1920. In *The Cambridge Companion to Einstein*, pp. 167–227. Cambridge: Cambridge University Press.
7. Knox, E. (2014). Newtonian spacetime structure in light of the equivalence principle. *British Journal for the Philosophy of Science 65*, 863–880.
8. Misner, C. W., K. S. Thorne, and J. A. Wheeler (1973). *Gravitation*. New York: W.H. Freeman and Company.
9. Norton, J. (1985). What was Einstein's principle of equivalence? *Studies in the History and Philosophy of Science 16*, 203–246.
10. Zee, A. (2013). *Einstein Gravity in a Nutshell*. Princeton, NJ: Princeton University Press.

The Physical Significance of Symmetries from the Perspective of Conservation Laws

Adán Sus

1 Introduction

The empirical significance of symmetries in physical theories has been the subject of considerable discussion in recent times. Although there seems to be no problem with the interpretation of global spacetime symmetries, there is no consensus in relation to the empirical import of gauge symmetries and local spacetime symmetries. Nonetheless, it is usually assumed that global, but not local, symmetries have some special empirical significance due to the fact that global, but not local, transformations have an active interpretation. The physical intuition linked to this is that some (gauge and local spacetime) symmetries connect different mathematical representations of the same physical situation while in general global symmetries can connect genuinely different physical states.

Furthermore, it is well known that there is a relationship between symmetries and conservation laws which, for Lagrangian theories, is encoded by Noether's theorems. Here conventional wisdom holds the following: it is global symmetries, through Noether's first theorem (NFT), that are related to conservation laws. Less well known is the fact that for theories with local symmetries, because they necessarily have global subgroups as symmetry groups, NFT is also applicable, but this time it produces conservation laws with a less clear physical interpretation. Noether herself introduced the terminology of improper and proper conserved currents to distinguish between those found in theories with and without local symmetries. There is a sense in which the presence of local symmetries trivialises the conserved quantities obtainable, but recent work shows that things are not so simple; even in theories with local symmetries that have certain boundary conditions some conserved quantities can be defined that resemble those obtained in theories for which the global symmetry group is not extended by a local one.

This paper aims to bring together these two discussions and shows that such a conjunction produces interesting results. On the one hand, introducing the relationship with conservation laws into the discussion of the empirical significance of symme-

A. Sus (✉)
Department of Philosophy, University of Valladolid, Valladolid, Spain
e-mail: adansus@fyl.uva.es; adansus@gmail.com

© Springer Science+Business Media, LLC 2017
D. Lehmkuhl et al. (eds.), *Towards a Theory of Spacetime Theories*,
Einstein Studies 13, DOI 10.1007/978-1-4939-3210-8_9

tries can help to discriminate between symmetries that in principle will or will not have direct empirical significance. The rationale behind this is that proper conservation laws are a good indication of empirical significance; to show this I use and modify a well-known analysis of what symmetries with direct empirical significance are. The original idea is simple: Noether's theorems provide formal relations between symmetries and conservation laws; differences in the physical status of symmetries should be reflected in differences in the physical interpretation of conserved quantities. On the other hand, introducing the reference to the empirical significance of symmetries might provide a good and novel perspective for the ongoing discussion about the physical meaning of the connection between symmetries and conservation, together with introducing a much needed philosophical discussion about the status of the different types of conserved currents obtained in physical theories with local symmetries.

The paper is laid out as follows. First, I briefly review the discussion of the connection between symmetries and conserved currents established by Noether's theorems (Section 2). I concentrate on the significance of the classification of the conserved currents associated with different types of symmetries. This is done, partly, through the discussion of some standard examples and leads naturally to the question regarding the physical significance of the symmetries themselves (Sections 3 and 4). Then, I assess whether the notion of direct empirical significance of symmetries can be clarified through the previously discussed connection to conserved currents (Sections 5 and 6). As I mentioned above, the intuition that motivates the introduction of such a notion is the attempt to capture those symmetry transformations that are transformations of subsystems of the universe and which produce empirically distinct situations. I defend the idea that both conserved currents and symmetries that are physically relevant can also be connected to another intuition; the existence of a physically relevant background. This is taken, in Section 7, as a guide to identify the formal features of symmetries that can have direct empirical significance and which, at the same time, mean that the conserved currents associated with them are not physically trivial. I end the paper by considering some potential problems for the proposal (Section 8).

2 Conserved Currents for Lagrangian Theories. Noether's Theorems

NFT provided a systematic general account of the relationship between symmetries and conservation laws for Lagrangian theories, although it had been recognised long before that such a connection should exist in the context of Newtonian mechanics. Nevertheless, when talking about the relationship between symmetries and conservation laws, one must be careful about what may be meant by the extremes of that relation. Strictly speaking, what NFT tells us is that, for field theories derivable from an action invariant under a group of global symmetries, one can derive a number of relations which, under certain conditions (satisfying the field equations for the fields that are arguments of the Lagrangian), express the conservation of a current. Usually, but not always, these conserved currents can be expressed in the form of an integral conservation law; as a quantity that remains constant in time. So the first complication for the simple connection between symmetries and conservation laws that is referred to so often, comes from understanding under what conditions the differential continuity equations can be turned into the usual integral conservation laws; a complication, by the way, that is essential if we are to tackle the intricate debate about the status of energy conservation in GR.

There is what I will call a second complication that is concerned with the question of when the conserved currents can be derived and under what conditions they can be said to be conserved; an issue that is intimately related to the physical interpretation of the conserved quantities. It was shown long ago [3] that any Lagrangian theory that has a local group as a symmetry group, also has an infinite number of one-parameter global subgroups as symmetry groups and, therefore, under the appropriate conditions, an infinite number of conserved currents should be derivable. Moreover, for such theories, it is also true that there exists an infinite number of conserved currents that do not need any conditions for them to be conserved—that is, their continuity equations are mathematical identities and such currents are said to be conserved *off-shell*.

Before reflecting on the physical interpretation of such currents, it is necessary to revisit the application of Noether's theorems to a Lagrangian field theory with a local symmetry group. This presentation of Noehter's theorems follows [5]. Let us consider a field theory whose field equations are derivable through Hamilton's principle from an action $S = \int d^4 x L(\varphi_i, \partial_\nu \varphi_i, x^\mu)$, where L is a Lagrangian density that depends on the independent variables x^μ, together with the fields φ_i and their derivatives. I will use the term *variational symmetries* to denote those transformations that leave the Lagrangian density form invariant up to a divergence term (note that this is a sufficient condition for the Euler–Lagrange equations to be invariant under such transformations). This is the type of symmetries for which Noether's theorems apply.

Noether's First Theorem (NFT): If the variational symmetry transformations form a continuous group depending on p constant parameters ($k = 1, \ldots, p$), such that $\delta x^\mu = \epsilon^k \xi_k^\mu(x)$, $\delta_0 \varphi_i(x) = \varphi_i'(x) - \varphi_i(x)$, then the following p relations hold:

$$\sum_i L_{\varphi_i} \frac{\partial(\delta_0 \varphi_i)}{\partial \epsilon_k} = \partial_\mu j_k^\mu \qquad (1)$$

where L_{φ_i} are the Euler–Lagrange expressions given by the variational derivatives of the Lagrangian with respect to the corresponding fields $(\frac{\delta L}{\delta \varphi_i})$ and j_k^μ is the current associated with the parameter ϵ_k. From this we obtain a conserved current when the Euler–Lagrange equations for all the fields are satisfied: $\partial_\mu j^\mu = 0$.

Noether's Second Theorem (NST): If the symmetry transformations form a continuous group depending on η arbitrary spacetime functions $\varepsilon^k(x)(k = 1, \ldots, \eta)$

and their first derivatives, $\delta x^\mu = \varepsilon^k(x)\xi_k^\mu(x)$, $\delta_0\varphi_i = \sum_k(a_{ki}(\varphi_i, \partial_\mu\varphi_i, x)\varepsilon_k(x) + b_{ki}^\nu(\varphi_i, \partial_\mu\varphi_i, x)\partial_\nu\varepsilon_k(x))$, then the following η relations hold:

$$\sum_i L_{\varphi_i} a_{ki} = \sum_i \partial_\nu(b_{ki}^\nu L_{\varphi_i}) \tag{2}$$

We are now going to assume that we have a global symmetry group that is a subgroup of a local one. Then we can write $\varepsilon^k(x) = \epsilon^m\zeta_m^k(x)$ and $\delta_0\varphi_i(x) = (a_{ki}\zeta_m^k(x) + b_{ki}^\nu\partial_\nu\zeta_m^k(x))\epsilon^m$ for the variation of the fields. Now applying NFT and NST (combining (1) and (2)) we obtain

$$\sum_i \partial_\nu(b_{ki}^\nu L_{\varphi_i}\zeta_m^k) = \partial_\nu j_m^\nu \tag{3}$$

From equation (3) we derive the existence of conservation laws that hold identically (independently of whether any field equation is satisfied)

$$\partial_\nu(j_m^\nu - \sum_i b_{ki}^\nu L_{\varphi_i}\zeta_m^k) = 0 \tag{4}$$

The functions in brackets have identically vanishing divergences, from which one can infer the existence of antisymmetrical functions, the so-called superpotentials $U_m^{[\nu\rho]}$ (whose divergences vanish identically, $\partial_\nu\partial_\rho U_m^{[\nu\rho]} = 0$) such that

$$j_m^\nu = \sum_i b_{ki}^\nu L_{\varphi_i}\zeta_m^k + \partial_\rho U_m^{[\nu\rho]}. \tag{5}$$

The Noether current obtained from the NFT is now expressed as a term that vanishes on-shell plus one whose divergence vanishes identically. In the next section, I will start to reflect on the physical meaning of such currents.

3 Proper and Improper Conservation Laws

Noether (following Hilbert) introduces the following classification of conservation laws. She uses the term *improper conservation laws* to denote those where the conserved current can be written in the above form (5): as a combination of Euler–Lagrange expressions plus an identically conserved quantity. This implies that, when the Euler–Lagrange equations are satisfied (on-shell), the conserved current is given by the divergence of an arbitrary superpotential. Such conservation laws are always obtained in a theory with a local symmetry group. In contrast, she reserves the name *proper conservation law* for those in which the conserved current cannot be decomposed in this way. Trautman [10, p. 179] argues that proper conservation laws are obtained in theories where the global symmetry group cannot be enlarged so that the theory possesses a local symmetry group without introducing auxiliary, non-dynamical fields.

There is another common term that is used in discussions about conservation laws. Bergmann names *weak conservation laws* those that are only conserved if all the field equations are satisfied. In contrast, *strong conservation laws* are divergences that vanish whether or not the Euler–Lagrange equations are satisfied. Although the term *strong* might have positive connotations, the fact that in many cases such expressions are mathematical identities does not confer any robust physical significance on them.

Now, the focus of our interest is on the physical significance, if any, of such classifications. I will start by briefly reviewing some of the positions expressed with respect to this issue in the physics literature. Here we find what has been named the Noether-charge puzzle for gauge symmetries [1] which surfaces when one tries to define a charge related to a gauge symmetry using NFT, as one does for global symmetries. As we have seen, the presence of the local symmetry makes the Noether current vanish on-shell up to the divergence of a superpotential (a term whose divergence vanishes identically, that is, which is completely arbitrary). So, if we use the current to define the charge, as one does following the usual procedure this results in an undefined Noether charge (the surface integral of an arbitrary function):

$$Q[\varphi(x)] = \int_{\Sigma} j \mid_{\varphi(x)} = \int_{\delta\Sigma} U \mid_{\varphi(x)} \tag{6}$$

This problem was introduced by Bergmann and collaborators. They showed that a theory with local symmetries produces an infinite number of strong conservation laws (generated by the existence of the one-parameter global subgroups of the local ones (see eq. 4)) which induce the definition of the superpotentials that are at the origin of the above-mentioned puzzle. So, superpotentials are going to come together with local (gauge) symmetries and this, in some cases, engenders improper conservation laws. One of the first tasks of this paper will be to clarify why improper conservation laws do not appear with every local symmetry and the consequences of this for the discussion of the significance of these quantities. A second question that we will need to address is whether, and in what precise sense, the mere fact of having improper conservation laws has any devastating effect on our usual interpretation of symmetries. It seems to be the case that the existence of improper conservation laws trivialises the definition of charges, and we will need to see why the definition of charges might be important for the interpretation of symmetries. Before that, however, it has been pointed out that the connection between improper conservation laws and the impossibility of defining respectable charges is not correct; or at least, it must be tempered. In section 4, I will look at the solutions that have been proposed to escape the puzzle.

3.1 Some Examples

In order to start thinking about the meaning of the different types of conserved currents, it will be useful to consider some concrete examples.[1] First, I will illustrate how weak proper conservation laws appear in theories with global symmetries that are not embedded in a local symmetry group. Consider a Lagrangian density that produces, through Hamilton's principle, the Klein–Gordon equation for a complex scalar field

$$L_{kg} = \partial_\mu \varphi \partial^\mu \varphi^* - m^2 \varphi \varphi^* \tag{7}$$

This Lagrangian is invariant under the following global transformations:

$$\begin{aligned} \varphi &\to \varphi' = \varphi e^{i\theta} \\ \varphi^* &\to \varphi^{*'} = \varphi^* e^{-i\theta} \end{aligned} \quad \text{(with } \theta \text{ constant)} \tag{8}$$

Taking that the phase transformation is infinitesimal, to first order, we have

$$\begin{aligned} \delta\varphi &= i\theta\varphi \\ \delta\varphi^* &= -i\theta\varphi^* \end{aligned} \tag{9}$$

Applying NFT (1), we obtain:

$$\partial_\mu j_{kg}^\mu = i\varphi^* E_{\varphi^*} - i\varphi E_\varphi \tag{10}$$

with

$$j_{kg}^\mu = i(\varphi^* \partial^\mu \varphi - \varphi \partial^\mu \varphi^*). \tag{11}$$

When the Euler–Lagrange equations are satisfied ($E_\varphi = E_{\varphi^*} = 0$), then j_{kg}^μ is a weak proper conserved current. The continuity equation $\partial_\mu j_{kg}^\mu = 0$ can be transformed, through imposing suitable boundary conditions involving the rapid fall-off of fields at spatial infinity, into the conservation of a quantity (charge)

$$Q_{kg} = i \int_V (\varphi^* \partial^\mu \varphi - \varphi \partial^\mu \varphi^*) d^3 x \tag{12}$$

It is easy to transform the Lagrangian L_{kg} into one that is invariant under local phase transformations by introducing a four-vector field $A^\mu(x)$ and a covariant derivative associated with it, $D_\mu = (\partial_\mu + A_\mu)$

$$L_{KG} = D_\mu \varphi D^\mu \varphi^* - m^2 \varphi \varphi^* \tag{13}$$

L_{KG} is invariant under the following transformations:

---

[1]See Brading [4] for a similar treatment of these examples.

$$\varphi \to \varphi' = \varphi e^{i\theta}$$
$$\varphi^* \to \varphi^{*\prime} = \varphi^* e^{-i\theta}$$
$$A_\mu \to A'_\mu = A_\mu + \partial_\mu \theta \qquad \text{(with } \theta = \theta(\mathbf{x}) \text{ arbitrary functions)} \qquad (14)$$

Now we can apply both of Noether's theorems and obtain a strong conservation law (substituting in (4)). Nonetheless, when we write the conserved current, we realise that it is not truly an improper one, due to the fact that $A^\mu(x)$ is not a dynamical field and its Euler–Lagrange expression does not vanish; current (5) takes the following form on-shell:

$$j^\nu = L_A + \partial_\rho U^{[\nu\rho]} \qquad (15)$$

Notice that this would not be a conserved current were it not the case (as in fact it is) that the non-dynamical field is invariant under the global subgroup of constant phase transformations. This can be seen directly on the application of NFT; divergence (1) in this case becomes on-shell:

$$\partial_\mu j^\mu = L_{A_\nu} \frac{\partial(\delta_0 A_\nu)}{\partial\theta} \qquad (16)$$

which vanishes despite the Euler–Lagrange expression being non-zero, because: $\frac{\partial(\delta_0 A_\nu)}{\partial\theta} = 0$. So in this case, the currents associated with the transformations that are symmetries of the non-dynamical field cannot be suspected of generating trivial charges (we will see that this is also the case for generally covariant spacetime theories with a non-dynamical symmetrical metric; in those cases, the properly conserved currents are associated with the Killing vectors of the background metric).

Finally, let us consider an example of local symmetries where there are no non-dynamical fields. For this, I am going to complete the Lagrangian L_{KG} with a term that provides field equations for A_μ

$$L_{EM} = D_\mu \varphi D^\mu \varphi^* - m^2 \varphi \varphi^* - \frac{1}{4} F^{\mu\nu} F_{\mu\nu} \qquad (17)$$

with $F_{\mu\nu} = \partial_\mu A_\nu - \partial_\nu A_\mu$. This Lagrangian is invariant under the local transformations (14). As above, we can apply both theorems. However, as opposed to what happened previously, we now obtain an improper conservation law because on-shell all the Euler–Lagrange expressions become zero. We seem to be in a potentially problematic situation: through NFT we obtain a weakly conserved current; but by applying both theorems, we discover that this current is improper. This leaves the charge, in principle, ill-defined: $j^\nu = \partial_\rho U^{[\nu\rho]}$. Now, we know that the Euler–Lagrange equations for A_μ are

$$\partial_\nu F^{\mu\nu} = j^\mu \qquad (18)$$

with

$$j^\mu = i(\varphi^* D^\mu \varphi - \varphi D^\mu \varphi^*). \qquad (19)$$

From which, simply by definition of $F^{\mu\nu}$:

$$\partial_\mu \partial_\nu F^{\mu\nu} = \partial_\mu j^\mu = 0 \tag{20}$$

So even if the application of both theorems indicates to us the possible triviality in the definition of charge, the dynamics of the theory offers us a plausible physical interpretation. If we take $F^{\mu\nu}$ to be the superpotential, then the definition of charge through the superpotential that we considered before as problematic, just expresses this quantity as the flux of a field quantity through a tree-dimensional surface

$$Q[\varphi(x)] = \int_{\delta\Sigma} F^{\mu\nu} \mid_{\varphi(x)} \tag{21}$$

This makes the situation here not so desperate. To see why, let us have a look at a possible definition of superpotential and the associated charge. By looking at the result of applying NFT (or simply the Euler–Lagrange equations for A) we realise that one possibility for the superpotential is the electromagnetic field F; this undermines the claim that the definition of a charge that is based on this quantity should have a problematic physical status. We must not forget, though, an important feature of this theory, one that not every theory with local symmetries is going to posses: the global transformations that are associated with the conserved current through NFT are still symmetries of the field A_μ (as they were for the theory with Lagrangian L_{KG}). This makes the expression for the current the same as before. Therefore, the physical status of the charge defined through that current should be as respectable as before. Nonetheless, we still have the result from the application of both theorems that obscures the physical meaningfulness of the conserved current. From this perspective, we might say that, even if the current is given by the divergence of an arbitrary superpotential on-shell, a natural choice for the superpotential, the current and the subsequent charge is available; thanks to the relationship between the electromagnetic field and the sources provided by the field equations. Moreover, this choice produces quantities that are invariant under the transformations that leave the A field invariant. As we will see in the next section, this procedure is somehow generalisable; at least to some solutions of other theories with local symmetries.

To sum up, the fact that the solutions of the theory are globally symmetrical seems to be tied up with the current obtained through NFT not being trivial; even if it has the form of a superpotential on-shell, the same field equations suggest the definition of the superpotential and suggest the interpretation of charge as field flux. Moreover, the superpotential is symmetrical under the global transformations that leave the solutions invariant and this means that the surface integral that enters into the charge definition is well defined. All this will generalise as desirable features for superpotentials involved in the definition of physically meaningful charges for theories with local symmetries.

4 Conservation Laws in Gauge Theories: Asymptotic Symmetries

A potential problem for theories with local symmetries—what has been called the Noether-charge puzzle for gauge symmetries—is that the usual definition of charges associated with global symmetries yields quantities that can be said to be physically trivial. We need to consider the relevance of this result for the interpretation of symmetries; but before that we must make this claim more precise, following the hints extracted from the discussion of the examples above. The following is what one can extract from the discussion of application of Noether's theorems to theories with local symmetries.

First, the presence of local symmetries is not a sufficient condition for the theory to have improper conservation laws. This is a direct consequence of (5). Note that local symmetries always imply strong conservation laws and therefore superpotentials (see (4)); but by itself this does not mean that there is a conserved current associated with a global symmetry that is a superpotential on-shell. In other words, we might have strong conservation laws that are trivial in a sense (this is not a surprise if one considers that no dynamical conditions were imposed to arrive at such laws) but we do not necessarily have weak improper conservation laws; when there are absolute variables in the theory, if there are weak conservation laws, they are going to be related to the symmetries of the absolute objects and are going to be proper laws. So we can conclude that in such cases, local symmetries do not trivialise conserved quantities that have their origin in global symmetries.

The situation is different when all the variable fields in the Lagrangian are dynamical. Nonetheless, one must distinguish between theories with symmetrical fields and theories where none of the fields have global symmetries. In both cases (5) is going to be an improper, weakly conserved current, but in the first case it is possible to think of a theory where the symmetrical field that produces the same conserved current is non-dynamical but is, as above, proper. This may be interpreted as indicating that the improperness of the current in the original theory is accidental and that the definition of non-trivial charges is possible.

Finally, we are left with the most general case: theories for which all fields are dynamical and whose solutions are not generally symmetrical (this group includes General Relativity and Yang–Mills theories). These theories produce improper currents that are weakly conserved and the charges defined through them fall prey to the accusation of a lack of meaningfulness. Nonetheless, as Barnich et al. [1] point out, the problem can be, at least partially, solved. The idea is that although there are no non-trivial conserved currents in these theories, one can define non-trivial asymptotically conserved currents, and well-defined charges, when solutions are asymptotically symmetrical. According to Barnich et al. [1] it is enough that the "theory becomes asymptotically linear near the boundary when expanded around a suitable background." When this is the case, they claim to prove that the asymptotically non-trivial conserved currents are generated by transformations that are symmetries of such a background.

The details are not important for our discussion but it might be useful to reflect on the origins of the problem and this solution to it. Recall that a consequence of the presence of local symmetries is to make the definition of charge dependent on a surface integral of an arbitrary identically conserved function (the superpotential). This makes charge, in general, undefined; but it also indicates that, if we have a criterion that allows us to choose (construct) the right superpotential, the definition of charge will depend only on the properties of the superpotential around the boundary. If, as happens for Maxwell's theory, we can choose a physically meaningful superpotential, then charge can be interpreted as the flux of a field through a closed surface. According to the proposed solution, non-trivial conserved superpotentials exist when the theory can be decomposed around a symmetrical background at the boundary. Meaningful conserved charges are then linked to symmetries of the background; this is the same as we found for electromagnetism but, while in that case the background was an exact solution, in general it is enough if such a background exists at the boundary and the solutions tend asymptotically to it. What this seems to suggest is that there is a connection between meaningful conserved charges and global symmetries of a background.

5 The Physical Significance of Local Symmetries

It is widely recognised that symmetries and symmetry arguments have played a very important role in theorising in physics. Some of the most influential physical theories of the twentieth century have brought the concept of symmetry to the fore, which makes a correct understanding of the physical importance of such a feature of theories essential. In fact, of the different ways in which one can classify the symmetries present in physical theories, there is one that is intended precisely to distinguish between symmetries that are a sub-product of the formalism that the theory uses to describe phenomena, and others that capture a feature belonging to the phenomena themselves. Such a distinction is what is going to occupy the rest of this discussion. In different contexts it has been labelled using various pairs of names (such as analytical/physical, passive/active, with/without empirical significance, gauge/non-gauge) and equated, many times erroneously, to other ends of different criteria of classification (internal/external, local/global). So, before moving on to the remainder of the debate, it is necessary to revisit the terms in which this discussion will take place.

An analysis of the empirical significance of symmetries has been set out (see Kosso [9], Brading and Brown [6]). In it, there is a distinction between the direct and indirect empirical significance of symmetries. Symmetries would have indirect empirical significance if they had consequences that are observable; as an example of this authors often refer to conservation laws. On the other hand, symmetries with direct empirical significance are characterised by the two following conditions:

(1) Transformation Condition: the transformation of a subsystem of the universe with respect to a reference system must yield an empirically distinguishable scenario; and

(2) Symmetry Condition: the internal evolution of the untransformed and transformed subsystems must be empirically indistinguishable.

For Brading and Brown, symmetries that have direct empirical significance (those that therefore meet the conditions just quoted) are those that correspond to transformations that can, in principle, actively transform effectively isolated with respect to the rest of the universe; this involves symmetry transformations that connect two, in principle, empirically distinct scenarios. Although Kosso, and Brading and Brown coincide in their account of what the empirical significance of symmetries is, they differ on the verdict as to the empirical significance of some specific symmetries. According to Brading and Brown, it is only global spacetime symmetries that can have direct empirical significance; some instances of them have an active interpretation when they are applied to effectively isolated subsystems of the universe. Note that it is not enough to say that global spacetime symmetries have direct empirical significance because they include transformations of the whole universe, which cannot receive the same kind of active interpretation (this is almost by definition as the whole cannot be a subsystem). So, according to this approach, it seems that there must be something in global spacetime transformations that is always lacking in global internal transformations and this is the fact that under the former, but not the latter, it is possible to effectively isolate subsystems to which the transformation is applied. Brading and Brown hint at the dynamical reason behind this difference in the following passage:

> In so far as internal global symmetries and local symmetries are perfect symmetries (i.e., there are no other interactions that fail to respect the symmetry in question), they have no direct empirical significance, only indirect empirical significance.
> [6]

This idea is stressed in the discussion about the possible empirical significance of general covariance

> Active arbitrary coordinate transformations in General Relativity involve transformations of both the matter fields and the metric, and they are symmetry transformations having no observable consequences. Coordinate transformations applied to the matter fields alone are no more symmetry transformations in General Relativity than they are in Newtonian physics (whether written in generally covariant form or not). Such transformations do have observational consequences. Analogously, local gauge transformations in locally gauge invariant relativistic field theory are transformations of both the particle fields and the gauge fields, and they are symmetry transformations having no observable consequences. Local phase transformations alone (i.e. local gauge transformations of the matter fields alone) are no more symmetries of this theory than they are of the globally phase invariant theory of free particles.
> [6]

I believe that the central idea behind this analysis, i.e. that for transformations to have empirical significance it must be possible that they change only some of the fields in the theory and not others, is correct; but this idea has not been implemented with enough care. Although there might be more direct roads to this conclusion, I will argue that our previous discussion of the status of conservation laws helps to clarify the physical significance of some symmetries with doubtful status that do not have empirical significance according to Brading and Brown. Furthermore, this will also make clear why some global spacetime symmetries cannot have empirical significance.[2]

6 Physical Significance of Conservation Laws

I have argued, following Kosso, and Brading and Brown, that the empirical significance of symmetries is linked to the possibility of effectively isolating a subsystem from its environment and being able to perform a transformation that renders an empirically distinguishable scenario but one for which the same physical laws apply. In other words, physical experiments inside the subsystem cannot tell whether a symmetry transformation has been performed, but there must be some physical differences between the two situations connected by the transformation which are observable in principle. Those differences must come from the relation that exists between the subsystem and the rest of the universe, in other words, the transformation must change the relationship between a given subsystem and other subsystems. This involves change between subsystems, but at the same time invariance in the relationship that holds between the transformed subsystem and the physically relevant environment, and which is expressed through the dynamical laws. This can be conceptualised by the existence of a background (a common framework) that stands in the same relationship to both the untransformed and transformed subsystems while allowing a distinct relationship between those subsystems. It must be the case that with respect to (some of) the transformations that are symmetries of the dynamical laws that encode the evolution of the subsystem, the background is symmetrical; but at the same time that such transformations are not symmetries of possible interactions between the subsystem and other subsystems (or the subsystem and the background). Therefore, the background must be symmetrical with respect to such a transformations.

Let us go back to our previous discussion of the different types of conservation laws and their connection to the symmetries of theories. We have seen that there is a correlation, shown by Noether's theorems, between a subset of the dynamical symmetries (the variational symmetries) and continuity equations that can sometimes be transformed into conservation laws. Moreover, depending on the type of symmetry present in the theory, we will obtain different types of conserved quantities. I said before that some of these conservation laws are arguably trivial; we must see exactly which ones and in what sense that is the case.

---

[2]In a recent paper, Greaves and Wallace [7], a new theoretical framework is offered to try to capture this same idea and solve some of the problems of the original framework. Part of this manoeuvre involves being able to express isolation for a subsystem and using that to distinguishing between different types of local symmetries. I believe that the verdict with respect to the empirical significance of local symmetries arrived at following such a path is in agreement with the one defended here.

Two different types of trivial conservation laws appeared in the foregoing discussion. The first type consists of having a conserved current that vanishes, up to an identically conserved term, when all the field equations are satisfied. These are what I have called improper weakly conserved currents and they appear in theories with local symmetries where all the dependent (field) variables are dynamical. Such currents are associated with symmetry transformations that transform all the fields. The second type of triviality consists of the divergences that vanish identically (without the need to impose any field equation). I have called these strong conservation laws and they are present in any Lagrangian theory with a local symmetry, irrespective of whether it has variables to which Hamilton's principle applies or not.

Of the two types of triviality, the second is linked to the formal character of some symmetry transformations. The kind of continuity equations that I have called strong conservation laws have nothing to do with the dynamics of the theory and appear in theories where the symmetry is merely a formal feature of the field equations and the action. One can say that no physical content is involved in that kind of expression. The first kind of triviality, however, is of quite a different nature; it involves having conserved currents that reduce to identically conserved currents on-shell. So, the difference seems to be that, in the first case, the trivial conservation laws appear after taking into account the dynamics of the theory. One would expect symmetries with direct empirical significance to have consequences that depend on some specific features of the space of solutions, rather than just holding in any kinematically possible model. This seems to suggest the following paradoxical claim: although for theories with local symmetries and no absolute objects the number of symmetries associated with conserved currents on-shell is higher, in reality they are all trivial (in the first sense). However, as we have seen, for a class of solutions with certain properties at the boundary, the problem of the triviality of such currents can be solved, which results in a higher number of potentially physically meaningful currents.

Although arguably[3] these two different types of triviality of conservation laws can have some consequences for the interpretation of the symmetries to which they are related, it is the case that non-trivial conservation laws share a common feature: they are somehow related to the presence of some kind of symmetrical background. In the case of having non-dynamical variables in the theory, the weakly conserved currents are associated with the symmetries of such non-dynamical variables. For theories with only dynamical variables, we have seen that non-trivial charges are definable through superpotentials for solutions that, at the boundary, have a globally symmetrical background.

---

[3] More about this below.

7   Physical Symmetries and Conservation Laws

Let us go back to the question regarding the empirical significance of symmetries. It seems clear that in the talk about symmetries one must distinguish between when one is referring to an empirical situation (an empirical system) and when one is attributing the symmetry to theoretical formalism, even if this is done through the consideration of a model that is a solution of the equations of the theory (we must not forget that a model always involves idealisation of some kind). Any discussion of the empirical consequences of symmetries must, therefore, make clear when one is talking about symmetries as a feature of the world, which are independent of which theory might be used to describe such empirical situation; and when one is talking about symmetries as a feature of a given theory. It seems to me that the attempt to characterise symmetries with direct empirical significance introduced above is ambiguous in this respect; it is not clear whether one is talking about models of theories or empirical systems.

I propose to modify the proposal in the following way. First, one must explicitly declare that symmetries with direct empirical significance must be a type of theoretical symmetry (symmetries of some theoretical models). In consideration of this not being the intention of the original proposal, I call such symmetries *symmetries with, in principle, direct empirical significance* or, for short, *symmetries with physical significance*. The conditions for a symmetry to be one of these must be modified as follows:

(1) Transformation Condition: the transformation of a subsystem of the universe with respect to a reference system must yield an, in principle, empirically distinguishable scenario; and

(2) Symmetry Condition: the transformations are, at least, dynamical symmetries (they produce physically indistinguishable evolutions).

The second condition is the one that ensures that the transformations in question are in fact symmetries in the relevant way: with respect to any experiments performed inside a subsystem and that test the evolution of such subsystem according to certain laws, the two situations at the end of the transformation are indistinguishable. Meanwhile, the first condition is the one that provides the empirical significance; but now we are viewing the system as a theoretical model and the question is whether the theory behind such a model has enough resources to distinguish between the two situations. Naturally, this question cannot be answered by looking at the mathematical formalism alone (from the mathematical point of view, the distinction is certainly possible) and will be dependent on the interpretation of the theory (which obviously does not mean that any interpretation will do for a given formalism).

The presence, in the models, of *geometrical objects* that are invariant under the symmetry transformations can be taken as the indication of, in principle, direct empirical significance; that such objects are symmetrical under the transformations in question permits us to use them as a marker of the change suffered by the other objects in the model. Naturally, this is originally a mathematical difference that will only be physically significant if the interpretation of the theory allows us to see change relative to the symmetrical object as physically relevant. In any case, what can be said so far is that symmetrical objects are a sign of, in principle, direct empirical significance of symmetries.

In the previous section, we saw that non-trivial conservation laws are also related, in one way or another, to the existence of symmetrical backgrounds. My claim is that such a coincidence is not accidental, but expresses a deeply rooted fact about what is involved in performing an, in principle, directly empirically significant transformation. From the point of view of both the symmetry transformation and the conserved current, the presence of a symmetrical background indicates that one can effectively isolate a subsystem from the rest of the universe.

8 Do Non-trivial Conservation Laws Always Relate to Empirically Significant Symmetries?

The discussion presented here can, at the most, make the connection between empirical symmetries and non-trivial conservation laws plausible through the existence of symmetrical backgrounds. A more systematic study would be needed in order to establish the connection on firmer grounds. Nonetheless, even without more rigorous consideration, which is beyond the scope of this paper, we must make sure that the proposal does not have any obvious drawbacks.

Here is a positive way of presenting the task ahead: we may assume that a certain connection between physical symmetries and non-trivial conservation laws has been established. So, we look at the symmetries that, according to the grounds for that connection, should have this kind of empirical significance, and discuss whether they are, mainly, the ones we expect and what this approach says about the cases which are usually considered as controversial in discussions about the empirical significance of symmetries. To this end, I will concentrate on applying the proposal to global symmetries that are a subgroup of local ones, and global symmetries applied to the universe as a whole.

8.1 The Empirical Significance of Global Symmetries

The first inconvenience for my claim is the existence of an obvious potential counterexample. Global symmetries are related to physically respectable conservation laws and these can be applied to the universe as a whole, but it seems obvious that this type of symmetry transformation cannot have any direct empirical significance. In this section, I argue that the global symmetry transformations that NFT relate to proper conservation laws should not, in reality, be interpreted as transformations of the universe as a whole.

To see this, let us look at a simple example; a Lagrangian formulation of a special relativistic Klein–Gordon scalar field

$$L_{kg}(\phi, x) = \partial_\mu \phi \partial^\mu \phi - m^2 \phi^2 \tag{22}$$

The variational symmetries of this theory are given by the transformations of the Poincaré group: global symmetries that are linked, through NFT, to ten weakly conserved currents. Each of these transformations applies to the only field present in the theory, the scalar field, and in this sense they are universe transformations with no possible empirical significance. My assertion, then, that the existence of non-trivial conserved currents is linked to empirical symmetries, seems to go completely astray in this example.

Nonetheless, this interpretation might be misleading; it obscures the existence of the Minkowski metric that is invariant under the symmetry transformations. Therefore, the symmetry transformations leave a metrical structure invariant; a structure that can play the role of background that I mentioned above. Moreover, the existence of such a background can make it physically possible to distinguish between the two states separated by the transformation. Therefore, the symmetries under consideration can, in principle, have empirical significance. Of course, whether it does or not will depend on whether one is ready to endow physical significance on the spacetime points represented by the Minkowski metric.

However, if the foregoing argument is correct, there is a question that we should answer: what happens if we have a relational theory, understanding by this one for which the dynamics is not formulated as dependent on spacetime points, but in terms of relational quantities? Arguably, a Lagrangian for such a theory is going to have, let us say, translations as variational symmetries; but the field variables are themselves also going to be invariant under such transformations. The consequence of this is that the currents conserved according to NFT are going to be strongly conserved ones, and therefore they will be affected by one of the two types of triviality explained above. The verdict is that for a relational theory, one for which no background structure can exist, the conservation laws associated with transformations of the universe are trivial. In contrast, for a theory in which there is a structure that can be interpreted as a background that remains unaltered while the other fields are transformed, conserved currents associated with what one might have thought were universe transformations are non-trivial and have, in principle, direct empirical significance. If this result seems anti-intuitive, one must remember that such empirical significance is a property of symmetries relative to theoretical models, which indicates that the theory has the resources to describe the symmetry transformation as connecting two physically distinct situations.

8.2 The Empirical Significance of Local Symmetries

Conservation laws linked to symmetry transformations that are arguably not observable are affected by some type of triviality and in this paper, I have tried to argue that we can extrapolate physical meaning from this fact. Furthermore, there appear to be different kinds of trivial conservation laws and, I argue here, they seem to be linked to two different types of unobservable symmetry transformations.

For Lagrangian theories with local symmetries that have non-dynamical variables, proper weakly conserved quantities, non-trivial ones then, can be derived that are linked to the symmetries of the non-dynamical object. Moreover, we have strong conservation laws associated with the one-parameter global subgroups of the local symmetry group. Such conservation laws are trivial in a very strong sense; they are identically conserved off-shell. In other words, the conserved current is identical to a superpotential (it differs from zero only by this arbitrary, identically conserved quantity: it belongs to the null equivalence class).

Let us consider in which cases such currents appear. Theories with absolute variables that are nonetheless locally symmetrical can be said to have a merely formal or passive variant of local symmetry; think for instance of coordinate independent theories with an absolute spacetime structure. Obviously, not all theories with absolute variables are going to have this high degree of formal symmetry, but as a consequence the number of symmetry transformations related to trivial conservation laws will also vary.

Trivial conservation laws are also present in theories with no absolute variables (for which all the variables varied in the Lagrangian are subject to Hamilton's principle). But in this case, we have conserved currents that are arbitrarily identically conserved only on-shell; what, following Noether, I have called improper conservation laws. We have seen how these currents, for solutions with symmetrical background around the boundary, allow the definition of non-trivial charges. One can argue that this feature is connected to these quantities having a milder form of triviality. The symmetries associated with this type of conservation laws are local symmetries that, although they are often clustered under the term gauge, one would not want to see as merely formal. Examples of these are general covariance in GR or the gauge symmetry of Yang–Mills theory. These symmetries, non-observable in general, become observable when they are asymptotical symmetry transformations of the background at the boundary.

What is the best way to interpret these facts? The following is what can be derived from the strategy adopted in this paper. We can think of symmetries in physics originally as a formal feature of theories: they are formal features of field equations or of their models. In some cases, such transformations can correspond to some particular physical transformations, and the ones that can be so must be the symmetries that we want to call with in principle direct empirical significance. In such cases, it must be the case that the physical systems to which one can apply the transformations have some specific physical requirements and that the mathematical apparatus of the theory has the power to capture some of them (in the form of conservation laws, for instance). Conservation laws (non-trivial ones) would then be a feature indicating that a physical transformation corresponding to a given symmetry can be performed and that the interpretation of the mathematical apparatus of the theory can capture the transformation as physical. In this sense, we would be able to distinguish between merely formal and substantive versions of symmetries depending on the theory in which they are implemented.

With this to hand, we can make sense of cases in which global symmetries, understood as a subgroup of local ones, can have direct empirical significance. According to Brading and Brown, only global spacetime symmetries can have direct empirical significance. As I said above, this poses the problem of how to understand the fact that global symmetries applied to subsystems of the universe can have this status in theories in which such symmetries are part of a local symmetry group: as an example of this, just think of how from GR one would explain a situation of the type of Galileo's ship. From our perspective, this seems an easy task. GR has some theoretical models in which the spacetime metric has certain symmetries at the boundary. In those models, the transformations that at the boundary correspond to those symmetries are the ones that would have, in principle, empirical significance; and the ones for which non-trivial conserved currents can be defined.

9 Concluding Remarks

No doubt that the discussion regarding the physical significance of symmetries is one of the most venerable in the foundations of physics. I have argued that by introducing conservation laws into the discussion, one sheds light on the much-debated issue of the physical/empirical significance of symmetries in physical theories. The main purpose of this paper is to show just how this can be so.

We can summarise the virtues of using the connection between symmetries and conserved currents (encoded by Noether's theorems for Lagrangian theories) for the discussion in the following manner. First, it shows that there is an issue that has been greatly ignored regarding the physical status of different conserved currents that runs parallel to the question about the physical/empirical status of the associated symmetries. On the side of currents, the issue is elucidated by noting that physically significant conserved currents are such in virtue of reference to some physical background. Second, we can use this same idea to modify the characterisation of symmetries with direct empirical significance. The result indicates a property of formal symmetries that identifies those that can have direct empirical significance; the formal imprint of physical symmetries is that the models have absolute objects that, while being symmetrical, can be used to distinguish situations linked by the symmetry transformation. Third, this strategy provides a certain natural understanding of how global symmetries with direct empirical significance can be a subgroup of local symmetries (which are always under suspicion of being merely formal): what is relevant, as in the Noether-charge puzzle, is whether some models of the theory contain objects with certain symmetries (perhaps only introduced as boundary conditions) that permit us to define empirical transformations. This is compatible with the idea that local symmetries are *prima facie* just formal features of the theory.

Acknowledgements Research for this paper has been supported by Spanish Ministry of Science and Innovation (MICINN) Grants FFI2011-29834-C03-03 and FFI2012-37354.

References

1. Barnich G and Brandt F 2002 Covariant theory of asymptotic symmetries, conservation laws and central charges. *Nucl. Phys.* B 633 3Ð82 (Preprint hep-th/0111246)
2. P. G. Bergmann, Non-Linear Field Theories, *Phys. Rev.* 75 (1949), 680–685.
3. Bergmann, P. (1958) Conservation Laws in General Relativity as the Generators of Coordinate Transformations. *Phys. Rev.* 112, 287.
4. Brading, K. (2002). Which symmetry? Noether, Weyl and the conservation of electric charge. *Studies in History and Philosophy of Modern Physics* 33, 3–22.
5. Brading, K., & Brown, H. R., (2003). Symmetries and Noether's theorems. In K. A. Brading & E. Castellani (Eds.), *Symmetries in Physics: Philosophical Reflections* (pp. 89–109). Cambridge: Cambridge University Press
6. Brading, K. and H. Brown (2004). Are gauge symmetry transformations observable? *British Journal for the Philosophy of Science* 55, 645–665.
7. Greaves, H. and Wallace, D (2014). Empirical consequences of symmetries. *British Journal for the Philosophy of Science.* 65(1), 59–89.
8. Noether, E. (1918). Invariante Variationsprobleme. *Nachrichten von-der Königl. Gesellschaft der Wissenschaften zu Göttingen, Mathematisch-Physikalische Klasse.* (pp. 235–257). (English translation: Tavel, M. A. (1971). *Transport Theory and Statistical Mechanics*, 1(3), 183–207.)
9. Kosso, P. (2000). The empirical status of symmetries in physics. *British Journal for the Philosophy of Science* 51, 81–98.
10. Trautman, A. (1962). Conservation laws in general relativity. In L. Witten (Ed.), *Gravitation: An Introduction to Current Research.* New York: Wiley.

Does Time Exist in Quantum Gravity?

Claus Kiefer

Abstract Quantum theory and general relativity contain different concepts of time. This is considered as one of the major obstacles to constructing a quantum theory of gravity. In my essay, I investigate those consequences for the concept of time in quantum gravity that may be drawn without a detailed knowledge of the final theory. The only assumptions are the experimentally supported universality of the linear structure of quantum theory and the recovery of general relativity in the classical limit. Among the consequences are the fundamental timelessness of quantum gravity, the approximate nature of a semiclassical time and the correlation of entropy with the size of the Universe.

1 Time in Physics

On 14 December, 1922, Albert Einstein delivered a speech to students and faculty members of Kyoto University in which he summarized how he created his theories of relativity [1]. As for the key idea in finding special relativity in 1905, he emphasized: "An analysis of the concept of time was my solution". He was then able to complete his theory within five weeks.

An analysis of the concept of time may also be the key for the construction of a quantum theory of gravity. Such a hope is supported by the fact that a change of the fundamental equations in physics is often accompanied by a change in the notion of time. Let me briefly review the history of time in physics.

Before the advent of modern science, time was largely associated with periodic motion, notably the motion of the 'Heavens'; time was considered 'the measure of change'. It was Newton's great achievement to invent the notion of an absolute and continuous time, described by an external parameter t. Such a concept was needed for the formulation of his laws of mechanics and universal gravitation. Although Newton's concepts of absolute space and absolute time were heavily criticized by

C. Kiefer (✉)
Institute for Theoretical Physics, University of Cologne,
Zuelpicher Strasse 77, 50937 Koeln, Germany
e-mail: kiefer@thp.Uni-Koeln.DE

© Springer Science+Business Media, LLC 2017
D. Lehmkuhl et al. (eds.), *Towards a Theory of Spacetime Theories*,
Einstein Studies 13, DOI 10.1007/978-1-4939-3210-8_10

some contemporaries as being unobservable, alternative relational formulations were only constructed after the advent of general relativity in the 20th century [2].

In Einstein's theory of special relativity, time was unified with space to form a four-dimensional spacetime. But this 'Minkowski spacetime' still constitutes an absolute background in the sense that there is no *reactio* of fields and matter—Minkowski spacetime provides only the rigid stage for their dynamics. Einstein considered this lack of back reaction as very unnatural.

Minkowski spacetime provides the background for relativistic quantum field theory and the Standard Model of particle physics. In the non-relativistic limit, it yields quantum mechanics with its absolute Newtonian time t. This is clearly seen in the Schrödinger equation,

$$i\hbar \frac{\partial \psi}{\partial t} = \hat{H}\psi. \tag{1}$$

It must also be noted that the presence of t occurs on the left-hand side of this equation together with the imaginary unit, i; this fact will become important below. In relativistic quantum field theory, (1) is replaced by its functional version.

The Schrödinger equation (1) is, with respect to t, deterministic and time-reversal invariant. As was already emphasized by Wolfgang Pauli, the presence of both t and i are crucial for the probability interpretation of quantum mechanics, in particular for the conservation of probability *in* time.

But the story is not yet complete. It was Einstein's great insight to see that gravity is a manifestation of the geometry of spacetime; in fact, gravity is geometry. This led him to his general theory of relativity, which he completed in 1915. Because of this identification, spacetime is no longer absolute, but dynamical. There *is* now a *reactio* of all matter and fields onto spacetime and even an interaction of spacetime with itself (as is, e.g. the case in the dynamics of gravitational waves).

So, time is absolute in quantum theory, but dynamical in general relativity. What, then, happens if one seeks a unification of gravity with quantum theory or, more precisely, seeks an accommodation of gravity into the quantum framework? To paraphrase Einstein's words from above: an analysis of the concept of time is needed. In this context, one often speaks about the 'problem of time' [3–6].

But does one really have to unify gravity with quantum theory into a theory of quantum gravity? In the next section, I shall give a concise summary of the main reasons for doing so. I shall then argue that one can draw important conclusions about the nature of time in quantum gravity *without* detailed knowledge of the full theory; in fact, all that is needed is the semiclassical limit—general relativity. I shall then describe the approximate nature of any time parameter and clarify the relevance of these limitations for the interpretation of quantum theory itself. I shall finally show how the direction of time can be understood in a theory which is fundamentally timeless.

2 The Disappearance of Time

The main arguments in favour of quantizing gravity have to do with the *universality* of both quantum theory and gravity. The universality of quantum theory is encoded in the apparent universality of the superposition principle, which has passed all experimental tests so far [7, 8]. There is, of course, no guarantee that this principle will not eventually break down. However, I shall make the conservative assumption, in accordance with all existing experiments, that the superposition principle does hold universally. Arbitrary linear combinations of physical quantum state do again lead to a physical quantum state; in general, the resulting quantum states describe highly entangled quantum systems. If the superposition principle holds universally, it holds in particular for the gravitational field.

The universality of the gravitational field is a consequence of its geometric nature: it couples equally to all forms of energy. It thus interacts with all quantum states of matter, suggesting that it is itself described by a quantum state. This is not a logical argument, though, but an argument of naturalness [9].

A further argument for the quantization of gravity is the incompleteness of general relativity. Under very general assumptions, one can prove singularity theorems that force us to conclude that time must come to an end in regions such as the big bang and the interior of black holes. This is, of course, only possible because time in general relativity is dynamical. The hope, then, is that quantum gravity will be able to deal with these situations.

It is generally argued that quantum gravity effects can only be seen at a remote scale—the Planck scale, which originates from the combination of the three fundamental constants c (speed of light), G (gravitational constant) and $\hbar$ (quantum of action). The Planck length, for example, is given by

$$l_{\mathrm{P}} = \sqrt{\frac{\hbar G}{c^3}} \approx 1.62 \times 10^{-35}\,\mathrm{m}, \tag{2}$$

and is thus much smaller than any length scale that can be probed by the Large Hadron Collider (LHC).

This argument is, however, misleading. One may certainly expect that quantum effects of gravity are always important at the Planck scale. But they are not restricted to this scale a priori. The superposition principle allows the formation of nontrivial gravitational quantum states at any scale. Why, then, is such a state not observed? The situation is analogous to quantum mechanics and the nonobservability of states such as a Schrödinger–cat state. And the reason why such states are not found is the same: decoherence [7, 8]. The interaction of a quantum system with its ubiquitous environment (that is, with unaccessible degrees of freedom) will usually lead to its classical appearance, except for micro- or mesoscopic situations. The process of decoherence is founded on the standard quantum formalism, and it has been tested in many experiments [8].

The emergence of classical behaviour through decoherence also holds for most states of the gravitational field. But there may be situations where the quantum nature of gravity is visible—even far away from the Planck scale. We shall encounter such a situation in quantum cosmology. It is directly related to the concept of time in quantum gravity.

Due to the absence of a background structure, the construction of a quantum theory of gravity is difficult and has not yet been accomplished. Approaches are usually divided into two classes. The more conservative class is the direct quantization of general relativity; path-integral quantization and canonical quantum gravity belong to it. The second class starts from the assumption that a consistent theory of quantum gravity can only be achieved within a unified quantum theory of all interactions; superstring theory is the prominent (and probably unique) example for this class.

In this essay, I want to put forward the view that the concept of time in quantum gravity can be discussed without having the final theory at one's disposal; the experimentally tested part of physics together with the above universality assumptions suffice.

The arguments are similar in spirit to the ones that led Erwin Schrödinger in 1926 to his famous equation (1). Motivated by Louis de Broglie's suggestion of the wave nature of matter, Schrödinger tried to find a wave equation which yields the equations of classical mechanics in an appropriate limit, in analogy to the recovery of geometric optics as a limit to the fundamental wave optics. To achieve this, Schrödinger put classical mechanics into the Hamilton–Jacobi form, from which the desired wave equation could be easily guessed [10].

The same steps can be followed for gravity. One starts by casting Einstein's field equations into Hamilton–Jacobi form. This was already done by Asher Peres in 1962 [11]. The wave equation behind the gravitational Hamilton–Jacobi equation is then nothing but the Wheeler–DeWitt equation, which was derived by John Wheeler [12] and Bryce DeWitt [13] in 1967 from the canonical formalism. It is of the form

$$\hat{H}_{\text{tot}} \Psi = 0, \tag{3}$$

where here $\hat{H}_{\text{tot}}$ denotes the full Hamilton operator for gravity plus matter. The wave functional Ψ depends on the *three-dimensional* metric plus all non-gravitational fields.[1]

The Wheeler–DeWitt equation (3) may or may not hold at the fundamental Planck scale (2). But as long as quantum theory is universally valid, it will hold at least as an approximate equation for scales much bigger than l_{P}. In this sense, it is the most reliable equation of quantum gravity, even if it is not the most fundamental one.

The wave function Ψ in the Wheeler–DeWitt equation (3) does not contain any time parameter t. Although at first glance surprising, this is a straightforward consequence of the quantum formalism. In classical mechanics, the trajectory of a particle consists of positions q in time, $q(t)$. In quantum mechanics, only probability ampli-

---

[1]There also exist the so-called diffeomorphism constraints, which state that Ψ is independent of the choice of spatial coordinates, see e.g. [4] for details.

tudes for those positions remain. Because time t is external, the wave function in (1) depends on both q and t, but not on any $q(t)$. In gravity, three-dimensional space is analogous to q, and the classical spacetime corresponds to $q(t)$. Therefore, upon quantization, spacetime vanishes in the same manner as the trajectory $q(t)$ vanishes. But as there is no absolute time in general relativity, only space remains, and one is left with (3).

We can thus draw the conclusion that quantum gravity is timeless solely from the validity of the Einstein equations at large scales and the assumed universality of quantum theory. Our conclusion is independent of additional modifications at the Planck scale, such as the discrete features that are predicted from loop quantum gravity and string theory. The latter two approaches do, however, lead to additional modifications in the concept of space [4]. The AdS/CFT correspondence in string theory, for example, suggests that laws including gravity in three spatial dimensions are equivalent to laws excluding gravity in two spatial dimensions. It has even been claimed that gravity thus is an illusion [14].

3 Time Regained

In August 1931, Neville Mott submitted a remarkable paper to the Cambridge Philosophical Society [15]. He discussed the collision of an alpha-particle with an atom. The remarkable thing is that he considered the time-independent Schrödinger equation of the total system and used the state of the alpha-particle to *define* time and to derive a time-dependent Schrödinger equation for the atom alone. The total quantum state is of the form

$$\Psi(\mathbf{r}, \mathbf{R}) = \psi(\mathbf{r}, \mathbf{R})e^{i\mathbf{k}\mathbf{R}}, \tag{4}$$

where $\mathbf{r}$ and $\mathbf{R}$ refer to the atom and the alpha-particle, respectively. The time t is then defined from the exponential in (4) through a directional derivative,

$$i\hbar\frac{\partial}{\partial t} \propto i\mathbf{k} \cdot \nabla_{\mathbf{R}}. \tag{5}$$

This leads to the time-dependent Schrödinger equation for the atom. Such a viewpoint of time as a concept derived from a fundamental timeless equation is rarely adopted in quantum mechanics. It is, however, the key step to understanding the emergence of time from the timeless Wheeler–DeWitt equation (3). While the alpha-particle in Mott's example corresponds to the gravitational part, the atom corresponds to the non-gravitational degrees of freedom. The time t of the Schrödinger equation (1) is then *defined* by a directional derivative similar to (5). Various derivations of such a 'semiclassical time' have been given in the literature (reviewed, e.g. in [4]), but the general idea is always the same. Time emerges from the separation into two different subsystems: one subsystem (here: the gravitational part) defines the time with respect to which the other subsystem (here: the non-gravitational part) evolves.[2] Time is thus only an approximate concept. A closer investigation of this approximation scheme then reveals the presence of quantum-gravitational correction terms [16]. Such terms can in principle be observed in the anisotropy spectrum of the Cosmic Microwave Background radiation [17].

I have remarked above that the Hilbert-space structure of quantum theory is related to the probability interpretation, and that the latter seems to be tied to the presence of t. In the light of the fundamental absence of t, one may speculate that the Hilbert-space structure, too, is an approximate structure and that different mathematical structures are needed for full quantum gravity.

I have also remarked above that the time t in the Schrödinger equation (1) occurs together with the imaginary unit i. The quantum-mechanical wave functions are thus complex, which is an essential feature for the probability interpretation. Since the Wheeler–DeWitt equation is real, the complex numbers emerge together with the time t [18, 19]. Has this not been put in by hand through the i in the ansatz (4)? Not really. One can start with superpositions of complex wave functions of the form (4), which together give a real quantum state. But now, again, decoherence comes into play. Irrelevant degrees of freedom distinguish the complex components from each other, making them dynamically independent [7]. In a sense, time is 'measured' by such irrelevant degrees of freedom (gravitational waves, tiny density fluctuations). Some time ago, I estimated the magnitude of this effect for a simple cosmological model [20] and found that the interference terms between the complex components can be as small as

$$\exp\left(-\frac{\pi m c^2}{128 \hbar H_0}\right) \sim \exp\left(-10^{43}\right) , \qquad (6)$$

where H_0 is the Hubble constant and m the mass of a scalar field, and some standard values for the parameters have been chosen. This gives further support for the recovery of time as a viable semiclassical concept.

There are, of course, situations where the recovery of semiclassical time breaks down. They can be found through a study of the full Wheeler–DeWitt equation (3). One can, for example, study the behaviour of wave packets. Semiclassical time is only a viable approximation if the packets follow the classical trajectory without significant spreading. One may certainly expect that a breakdown of the semiclassical limit occurs at the Planck scale (2). But there are other situations, too. One occurs for a classically recollapsing universe and is described in the next section. Other cases follow from models with fancy singularities at large scales. The 'big brake', for example, corresponds to a universe which classically comes to an abrupt halt with infinite deceleration, leading to a singularity at large scale factor. The corresponding quantum model is discussed in [21]. If the wave packet approaches the classical singularity, the wave function will necessarily go to zero there. The time t then loses

---

[2]More precisely, some of the gravitational degrees of freedom can also remain quantum, while some of the non-gravitational variables can be macroscopic and enter the definition of time.

its meaning, and all classical evolution comes to an end before the singularity is reached. One might even speculate that not only time, but also space disappears [22].

The ideas presented here are also relevant to the interpretation of quantum theory itself. They strongly suggest, for example, that the Copenhagen interpretation is not applicable in this domain. The reason is the absence of a classical spacetime at the most fundamental level, which in the Copenhagen interpretation is assumed to exist from the outset. In quantum gravity, the world is fundamentally timeless and does not contain classical parts. Classical appearance only emerges for subsystems through the process of decoherence—with limitations dictated by the solution of the full quantum equations.

4 The Direction of Time

A fundamental open problem in physics is the origin of irreversibility in the Universe, the recovery of the arrow of time [23]. It is sometimes speculated that this can only be achieved from a theory of quantum gravity. But can statements about the direction of time be made if the theory is fundamentally timeless?

The answer is yes. The clue is, again, the semiclassical nature of the time parameter t. As we have seen in the last section, t is defined via fundamental gravitational degrees of freedom. The important point is that the Wheeler–DeWitt equation (3) is *asymmetric* with respect to the scale factor that describes the size of the Universe in a given state. It assumes a simple form for a small universe, but a complicated form for a large universe. For small scale factor, there is only a minor interaction between most of the degrees of freedom. The equation then allows the formulation of a simple initial condition [23]: the absence of quantum entanglement between global degrees of freedom (such as the scale factor) and local ones (such as gravitational waves or density perturbations). The local variables serve as an irrelevant environment in the sense of decoherence.

Absence of entanglement means that the full quantum state is a product state. Tracing out the environment then has no effect; the state of the global variables remains pure. There is then a vanishing entropy (as defined by the reduced density matrix) connected with them; all information is contained in the system itself. The situation changes with increasing scale factor; the entanglement grows and the entropy for the global variables increases, too. As soon as the semiclassical approximation is valid, this growth also holds with respect to t; it is inherited from the full equation. The direction of time is thus defined by the direction of increasing entanglement. In this sense, the expansion of the Universe is a tautology.

There are interesting consequences for a classically recollapsing universe [24]. In order to produce the correct classical limit, the wave function of the quantum universe must go to zero for large scale factors. Since the quantum theory cannot distinguish between the different ends of a classical trajectory (such ends would be the big bang and the big crunch), the wave function must consist of many quasi-classical components with entropies that increase in the direction of a larger universe;

one could then never observe a recollapsing universe. In the region of the classical turning point, all components have to interfere destructively in order to fulfil the final boundary condition of the wave function going to zero. This is a drastic example of the relevance of the superposition principle far away from the Planck scale—with possible dramatic consequences for the fate of our Universe: the classical evolution would come to an end in the future. Such a quantum end could occur also in dark-energy models that do not recollapse [21].

Let me finally emphasize again that all the consequences presented in this essay result from a very conservative starting point: the assumed universality of quantum theory and its superposition principle. Unless this assumption breaks down, these consequences should hold in every consistent quantum theory of gravity. We are able to understand from the fundamental picture of a timeless world both the emergence and the limit of our usual concept of time, at least in principle.

Acknowledgements I am grateful to Dennis Lehmkuhl for inviting me to an exciting conference. This contribution is a slightly revised version of my essay with the same title, which in 2009 received a second prize in the *The Nature of Time* essay context organized by the Foundational Questions Institute (www.fqxi.org). I thank Marcel Reginatto and H.-Dieter Zeh for their comments on an earlier version of this essay.

References

1. A. Einstein, How I created the theory of relativity. Translated by Y. A. Ono. *Physics Today*, August 1982, 45–47.
2. J. B. Barbour, Leibnizian time, Machian dynamics, and quantum gravity. In: *Quantum Concepts in Space and Time*, edited by R. Penrose and C. J. Isham (Oxford University Press, Oxford, 1986), pp. 236–246.
3. C. J. Isham, Canonical quantum gravity and the problem of time. In: *Integrable Systems, Quantum Groups, and Quantum Field Theory*, edited by L. A. Ibort and M. A. Rodríguez) (Kluwer, Dordrecht, 1993), pp. 157–287.
4. C. Kiefer, *Quantum Gravity*, third edition (Oxford University Press, Oxford, 2012).
5. K. V. Kuchař, Time and interpretations of quantum gravity. In: *Proceedings of the 4th Canadian Conference on General Relativity and Relativistic Astrophysics*, edited by G. Kunstatter, D. Vincent, and J. Williams (World Scientific, Singapore, 1993), pp. 211–314.
6. E. Anderson, Problem of Time in Quantum Gravity. *Annalen der Physik*, **524**, 757–786 (2012).
7. E. Joos, H. D. Zeh, C. Kiefer, D. Giulini, J. Kupsch, and I.-O. Stamatescu, *Decoherence and the Appearance of a Classical World in Quantum Theory*, second edition (Springer, Berlin, 2003).
8. M. Schlosshauer, *Decoherence and the Quantum-to-Classical Transition* (Springer, Berlin, 2007).
9. M. Albers, C. Kiefer, and M. Reginatto, Measurement analysis and quantum gravity. *Physical Review D*, **78**, 064051 (2008).
10. E. Schrödinger, Quantisierung als Eigenwertproblem II. *Annalen der Physik*, **384**, 489–527 (1926).
11. A. Peres, On Cauchy's problem in general relativity-II. *Nuovo Cimento*, **XXVI**, 53–62 (1962).
12. J. A. Wheeler, Superspace and the nature of quantum geometrodynamics. In: *Battelle rencontres*, edited by C. M. DeWitt and J. A. Wheeler (Benjamin, New York, 1968), pp. 242–307.
13. B. S. DeWitt, Quantum theory of gravity. I. The canonical theory. *Physical Review*, **160**, 1113–1148 (1967).

14. J. Maldacena, The illusion of gravity. *Scientific American*, November 2005, 56–63.
15. N. F. Mott, On the theory of excitation by collision with heavy particles. *Proceedings of the Cambridge Philosophical Society*, **27**, 553–560 (1931).
16. C. Kiefer and T. P. Singh, Quantum gravitational correction terms to the functional Schrödinger equation. *Physical Review D*, **44**, 1067–1076 (1991).
17. C. Kiefer and M. Krämer, Can effects of quantum gravity be observed in the cosmic microwave background? *International Journal of Modern Physics D*, **21**, 1241001 (2012).
18. J. B. Barbour, Time and complex numbers in canonical quantum gravity. *Physical Review D*, **47**, 5422–5429 (1993).
19. C. Kiefer, Topology, decoherence, and semiclassical gravity. *Physical Review D*, **47**, 5414–5421 (1993).
20. C. Kiefer, Decoherence in quantum electrodynamics and quantum gravity. *Physical Review D*, **46**, 1658–70 (1992).
21. A. Y. Kamenshchik, C. Kiefer, and B. Sandhöfer, Quantum cosmology with a big-brake singularity. *Physical Review D*, **76**, 064032 (2007).
22. T. Damour and H. Nicolai, Symmetries, singularities and the de-emergence of space. *International Journal of Modern Physics D*, **17**, 525–531 (2008).
23. H. D. Zeh, *The Physical Basis of the Direction of Time*, fifth edition (Springer, Berlin, 2007).
24. C. Kiefer and H. D. Zeh, Arrow of time in a recollapsing quantum universe. *Physical Review D*, **51**, 4145–4153 (1995).

Raiders of the Lost Spacetime

Christian Wüthrich

Abstract Spacetime as we know and love it is lost in most approaches to quantum gravity. For many of these approaches, as inchoate and incomplete as they may be, one of the main challenges is to relate what they take to be the fundamental non-spatiotemporal structure of the world back to the classical spacetime of general relativity (GR). The present essay investigates how spacetime is lost and how it may be regained in one major approach to quantum gravity, loop quantum gravity.

Many approaches to quantum gravity (QG) suggest or imply that space and time do not exist at the most fundamental ontological level, at least not in anything like their usual form. Thus deprived of their former status as part of the fundamental furniture of the world, together, perhaps, with quarks and leptons, they merely 'emerge' from the deeper physics that does not rely on, or even permit, their (fundamental) existence, rather like tables and chairs. The extent to which the fundamental structures described by competing approaches to QG diverge from relativistic spacetimes varies, along different dimensions [22]. That modern physics puts time under pressure is widely accepted. One can read the history of modern physics from the advent of relativity theory to the present day as a continuing peeling away of the structure that time was initially believed to exemplify ([23] §2.1). But at least in some approaches, spacetime as a whole comes under siege. This may occur in the relatively mild sense that the

I thank audiences at the University of London's Institute of Philosophy and at the spacetime workshop at the Bergisch University of Wuppertal. I am also grateful to my fellow Young Guns of General Relativity, Erik Curiel, John Manchak, Chris Smeenk and Jim Weatherall for valuable discussions and comments, and to Francesca Vidotto for correspondence. Finally, I owe thanks to two perceptive referees for their comments. Work on this project has been supported in part by a Collaborative Research Fellowship by the American Council of Learned Societies, by a UC President's Fellowship in the Humanities, and by the Division for Arts and Humanities at the University of California, San Diego. Some of the material in this essay has its origin in [47].

C. Wüthrich (✉)
Département de philosophie, Université de Genève, Rue de Candolle 2,
CH-1211 Geneva, Switzerland
e-mail: Christian.Wuthrich@unige.ch

© Springer Science+Business Media, LLC 2017
D. Lehmkuhl et al. (eds.), *Towards a Theory of Spacetime Theories*,
Einstein Studies 13, DOI 10.1007/978-1-4939-3210-8_11

fundamental structure turns out to be discrete; or it may be discrete and non-local, as it happens in loop quantum gravity (LQG); or the reality of some dimensions of space is questionable altogether, as it is in theories with certain dualities; or it may exhibit non-commutativity among different dimensions, obliterating the usual geometric understanding that we routinely have of spacetime.

Just how radical the departure from the spacetime we know and love is remains to be seen, but it is likely to have profound implications. For instance, it may render some of our cherished philosophical theories not just of space and time, but also of persistence, causation, laws of nature and modality obsolete, or at least in need of revision [48]. But this paper will be concerned with the consequences for the physics, rather than the metaphysics. Two urgent, and related, issues arise. First, one might worry that if it is a necessary condition for an empirical science that we can at least in principle measure or observe something *at some location at some time*. The italicized locution, in turn, seems to presuppose the existence of space and time. If that existence is now denied in quantum theories of gravity, one might then fear that these theories bid adieu to empirical science altogether. It thus becomes paramount for advocates of these theories to show that the latter only threaten the *fundamentality*, but not the *existence* of space and time. To discharge this task means to show how relativistic spacetimes re-emerge and how measurable quantities arise from the fundamental structure as postulated by the theory at stake.

This first issue is closely related to a second problem: a novel theory can supplant an incumbent theory only if it recreates at least most of the empirical success of the old theory. The way in which this requirement is typically met in physics is by showing how the newer theory offers a more general framework than the older one, and that therefore the older is a special case of the newer, which can be regained, or at least mocked in formally suggestive ways, in some limit or to some approximation. For instance, it was important to Albert Einstein to be able to show that one obtains from general relativity (GR), in a weak-field limit, a theory which returns essentially all the same empirical results in the appropriate regime as Newtonian gravitational theory. This recovery mattered because the Newtonian theory garnered impressive empirical successes over the more than two centuries preceding Einstein's formulation of GR. For the very same reason, present-day quantum theories of gravity must eventually prove that they relate, in physically salient ways, to the classical GR that the last century of observations has found to be so accurate.[1] In fact, given the complete absence of direct empirical access to the quantum-gravitational regime, establishing this link with 'old' physics arguably constitutes the single most important constraint on theorizing in the quantum-gravitational realm.

Consequently, in theories of lost spacetime, relativistic spacetimes must be regained from the fundamental structure in order to discharge the tasks of securing both the theory's empirical coherence and its account of why the theory it seeks to supplant was as successful as it was. It is the goal of this essay to show just

---

[1]Given this formidable success of the classical theory, one might wonder why we need a quantum theory of gravity at all. There are good reasons to think that we do, but they do not fully align with the standard lore one finds in the physics literature ([49] §1).

how spacetime vanishes and how it might be seen to re-emerge in one important approach to quantum gravity, LQG. Since the emergence of spacetime from a non-spatiotemporal structure is often thought to be impossible, establishing the mere *possibility* of such emergence assumes vital importance.[2]

The next section, Section 1, explicates how time, rather than spacetime, disappears in a class of approaches to QG, the so-called 'canonical' theories. Canonical QG casts GR in a particular way, and the section will show how time and change vanish already at the level of GR so cast. Section 2 then investigates the fundamental structures as they are described by LQG and discusses the two main ways in which they differ from relativistic spacetimes, viz. in their discreteness and their non-locality. The following section, Section 3, starts to clear the path for the re-emergence of relativistic spacetime by arguing how the emergence relation should *not* be construed in the present case. Specifically, it argues against a non-reductive understanding of emergence and an attempt to cash out the relation between the structures in terms of unitary equivalence as both inadequate to the task at hand. Next, Section 4 sketches a way in which the relationship between fundamental spin networks and relativistic spacetimes might be worked out and tries to understand what it would generally take to relate them. Section 5 offers brief conclusions.

1 The Problem of Time in Canonical General Relativity

Casting GR as a Hamiltonian system with constraints has many advantages, as John Earman [16] affirmed: it gives the vague talk about 'local' and 'global' transformations a more tangible meaning, it explains how the fibre bundle formalism arises in the cases it does, it has a sufficiently broad scope to relate GR to Yang–Mills gauge theories, it offers a formalization of the gauge concept, and it connects to foundational issues, such as the nature of observables and the status of determinism in GR and in gauge theories. Moreover, the Hamiltonian formulation affords a natural affinity to the initial-value problem in GR.[3] The real gain of a Hamiltonian formulation, however, arises when one tries to quantize the classical theory. Typically, prescriptions to find a quantum theory from a classical theory require either a Lagrangian (e.g. for the path integral method) or a Hamiltonian (e.g. for canonical quantization) formulation of the theory. LQG relies on a canonical quantization procedure and thus uses a Hamiltonian formulation of GR as a starting point.[4,5]

---

[2]For a very recent critical view, see, e.g. [25].

[3]Cf. ([46], Appendix E.2). A *locus classicus* for the Cauchy problem in GR is [12]; a more recent survey article is [19].

[4]A useful introduction to the Lagrangian and the Hamiltonian formulation of GR is given in ([46], Appendix E). Wald's textbook of 1984 only deals with the ADM version of Hamiltonian GR and, as time travel was not yet invented in 1984, does not treat Ashtekar's version, pioneered in 1986.

[5]Of course, for most cases we care about, Hamiltonian theories afford a corresponding equivalent Lagrangian theory, and vice versa. Currently, a debate rages in philosophy of physics over which

However, forcing GR, to use the words of Tim Maudlin ([29], 9) "into the Procrustean bed of the Hamiltonian formalism" also comes, as conveyed by the quote, at a cost. The cost arises from the fact that the Hamiltonian formalism tends to construe the physical systems it describes as spatially extended three-dimensional objects evolving over an external time, and this is no different for the Hamiltonian formulation of GR.[6] Recasting GR in a Hamiltonian formalism thus reinterprets the four-dimensional spacetimes of standard GR as three-dimensional 'spaces' which evolve in a fiducial 'time' according to the dynamics governed by Hamilton's equation. Pulling space and time asunder in this way, of course, contravenes the received view of what many take to be the deepest insight of relativity, viz. that no separation of the fundamental *spacetime* into space and time can in any physically relevant way be privileged. This blatant violation of four-dimensionalism, of course, gets mathematically mended in the formalism through the imposition of constraints. But we are getting ahead of ourselves. What this brief paragraph should suggest is that having a philosophically closer look at the dynamics of this reformulation of classical GR is worth our while.[7]

A spacetime is an ordered pair $\langle \mathcal{M}, g_{ab} \rangle$ consisting of a four-dimensional pseudo-Riemannian manifold $\mathcal{M}$ and a metric tensor field g_{ab} defined on $\mathcal{M}$. Starting out from the Einstein–Hilbert action $S[g_{ab}]$ for gravity without matter,

$$S[g_{ab}] = \frac{1}{16\pi G} \int_{\mathcal{M}} d^4 x \sqrt{-g} R, \tag{1}$$

where G is Newton's gravitational constant, g the determinant of the metric tensor g_{ab}, and R the Ricci scalar, one can gain a Lagrangian formulation of GR with the dynamical Euler–Lagrange equations in terms of a Lagrangian function $L(q, \dot{q})$ of generalized coordinates q and the generalized velocities $\dot{q}$. The Lagrange function is essentially the integrand in the action integral (1) integrated over the three spatial dimensions. This action leads to the (vacuum) field equations of GR if one varies (1) with respect to the metric g_{ab}. Thus, Einstein's vacuum field equations can be recognized as the equations of motion of the Lagrangian formulation of GR, i.e. as the Euler–Lagrange equations. They are second-order differential equations. The solutions to the Euler–Lagrange equations will be uniquely determined by $q, \dot{q}$ just in case the so-called 'Hessian' matrix $\partial^2 L(q, \dot{q}) / \partial \dot{q}^{n'} \partial \dot{q}^n$ of $L(q, \dot{q})$, where n labels the degrees of freedom, is invertible. This is the case if and only if its determinant,

---

(Footnote 5 continued)
of the two, if any, is more fundamental or more perspicuous. Nothing I say here should be taken to entail a stance in that debate.

[6]There are, of course, purely internal degrees of freedom of particles, such as classical spin, which admit of a Hamiltonian treatment without the system necessarily being extended in space. Now, even a point particle with internal degrees of freedom is at least a physical system *in* space, and it certainly also evolves over external time.

[7]In connection with what follows, Chapter 1 of [21] is recommended reading. For a less formal and hence more accessible treatment of the problem of time, cf. ([23], §2) and references therein. Cf. also Kiefer's contribution to this collection.

confusingly sometimes also called 'Hessian', does not vanish. In case the determinant of the Hessian vanishes, which means the Hessian is 'singular', the accelerations $\ddot{q}$ will not be uniquely determined by the positions and the velocities and the solutions to the Euler–Lagrange equations are not only not unique in q and $\dot{q}$, but also contain arbitrary functions of time. Thus, the impossibility of inverting $\partial^2 L(q, \dot{q})/\partial \dot{q}^{n'} \partial \dot{q}^n$ is an indication of gauge freedom. How such gauge freedom arises in constrained Hamiltonian systems is the topic of the next subsection, §1.1, followed by an analysis in §1.2 of how this lesson carries over into the context of Hamiltonian GR and leads to the problem of time.

1.1 Hamiltonian Systems with Constraints

Finding a Hamiltonian formulation amounts to putting the Euler–Lagrange equations in the form of Hamiltonian equations of motion, $\dot{q} = \partial H/\partial p$ and $\dot{p} = \partial H/\partial q$, which are of first order. This can be achieved by the introduction of canonical momenta via

$$p_n = \frac{\partial L}{\partial \dot{q}^n}, \tag{2}$$

where $n = 1, ..., N$, N being the number of degrees of freedom of the system at stake. These momenta are not all independent when we are faced with a system exhibiting gauge freedom—i.e. just in case the Hessian is singular. These dependencies get articulated in constraint equations

$$\phi_m(q, p) = 0, \quad m = 1, ..., M, \tag{3}$$

where M is the number of dependencies. The relations (3) between q and p are called *primary constraints* and define a submanifold smoothly embedded in phase space called the *primary constraint surface*. The phase space Γ is defined as the space of solutions of the equations of motion. Assuming that all equations (3) are linearly independent, which may not be the case, this submanifold will be of dimension $2N - M$. Equations (3) imply that the transformation map between the Lagrangian phase space $\Gamma(q, \dot{q})$ and the Hamiltonian phase space $\Gamma(q, p)$ is onto but not one-to-one. Equations (2) define a mapping from a $2N$-dimensional manifold of the q's an $\dot{q}$'s to the $(2N - M)$-dimensional manifold defined by (3). In order to render the transformation bijective and thus invertible, the introduction of extra parameters— 'gauge fluff'—is required.[8]

Next, one introduces a Hamiltonian H as a function of position and momentum variables as

$$H(q, p) = \dot{q}^n p_n - L(q, \dot{q}). \tag{4}$$

---

[8]For more details on how the constraints arise in some Hamiltonian systems, see ([21], Ch. 1). My exposition largely follows this reference.

This canonical Hamiltonian is uniquely defined only on the primary constraint surface but can arbitrarily be extended to the rest of phase space. The 'Legendre transformation' defined by (2) turns out to be invertible just in case $\det(\partial^2 L / \partial \dot{q}^{n'} \partial \dot{q}^n) \neq 0$. Should the determinant of the Hessian vanish, as above, one can add extra variables u^m and thus render the Legendre transformation invertible. In this case, the Hamiltonian equations corresponding to the Euler–Lagrange equations become

$$\dot{q}^n = \frac{\partial H}{\partial p_n} + u^m \frac{\partial \phi_m}{\partial p_n},$$

$$\dot{p}_n = -\frac{\partial H}{\partial q^n} - u^m \frac{\partial \phi_m}{\partial q^n},$$

$$\phi_m(q, p) = 0.$$

These Hamilton equations lead via arbitrary variations δq^n, δp_n, δu^m (except for the boundary conditions $\delta q^n(t_1) = \delta q^n(t_2) = 0$ and that they must conserve H) to the Hamiltonian equations of motion for arbitrary functions $F(q, p)$ of the canonical variables

$$\dot{F} = \{F, H\} + u^m \{F, \phi_m\}, \tag{5}$$

where $\{,\}$ is the usual Poisson bracket

$$\{F, G\} := \frac{\partial F}{\partial q^i} \frac{\partial G}{\partial p_i} - \frac{\partial F}{\partial p_i} \frac{\partial G}{\partial q^i}.$$

Consistency requires that the primary constraints ϕ_m be preserved over time, i.e. that $\dot{\phi}_m = 0$. As primary constraints are phase space functions, equation (5) then implies

$$\{\phi_m, H\} + u^{m'} \{\phi_m, \phi_{m'}\} = 0. \tag{6}$$

This equation has one of two possible forms: either it embodies a relation only between the q's and p's, without any u^m, or it results in a relation including u^m. In the latter case, we just end up with a restriction on u^m. In the former case, however, (6) leads to additional constraints, called *secondary constraints*, on the canonical variables and thus on the physically relevant region of the phase space. These secondary constraints must also fulfill the consistency requirement of being preserved over time, which leads to new equations of the type (6), which again are either restrictions on the u^m or constraints on the canonical variables, etc. Once the process is finished, and we have all secondary constraints,[9] denoted by $\phi_k = 0$ with $k = M + 1, ..., M + K$, all constraints can be rewritten as $\phi_j = 0$ with $j = 1, ..., M + K =: J$. The full set of constraints $\phi_j = 0$ defines a 'subsubmanifold' in the phase space Γ, i.e. a submanifold of the primary constraint surface $\phi_m = 0$, called the *constraint surface* $\mathscr{C}$. The relevant difference between primary and secondary constraints is that primary con-

---

[9]They are *not* referred to as tertiary, quaternary etc. constraints, but only collectively as 'secondary' constraints.

straints are direct consequences of equation (2), whereas the secondary constraints only arise once the equations of motion (5) are given.

Any two functions F and G in phase space that coincide on the constraint surface are said to be *weakly equal*, symbolically $F \approx G$. In case they agree throughout the entire phase space, their equality is considered *strong*, expressed as usual as $F = G$. Above, I have introduced the qualification of constraints as primary. However, there is a more important classification of constraints into first-class and second-class constraints, defined as follows:

Definition 1 (*First-class constraints*) A function $F(q, p)$ is termed *first class* if and only if its Poisson bracket with every constraint vanishes weakly,

$$\{F, \phi_j\} \approx 0, \quad j = 1, ..., J. \tag{7}$$

If that first-class function is a constraint itself, then we call it a *first-class constraint*. A function in phase space is called *second class* just in case it is not first class.

The property of being first class is preserved under the Poisson bracket, i.e. the Poisson bracket of two first-class functions is first class again.

The fact that arbitrary functions u^m enter the Hamilton equations (or, equivalently, the Hamiltonian equations of motion) implies that a physical state is uniquely determined by a pair (q, p), i.e. by a point in (Hamiltonian) phase space $\Gamma(q, p)$, but not vice versa. In other words, these arbitrary functions encode the gauge freedom which arises for systems with a singular Hessian. It can be shown that a dynamical variable F, i.e. a function on Γ, differs in value from time t_1 to time $t_2 = t_1 + \delta t$ by

$$\delta F = \delta v^a \{F, \phi_a\} \tag{8}$$

where the ϕ_a range over the complete set of first-class primary constraints and the v^a are the totally arbitrary part of the u^m, with $\delta v^a = (v^a - \tilde{v}^a)\delta t$ where v^a and $\tilde{v}^a$ are two different choices of v^a at t_1.[10] In a deterministic theory, the transformation (8) does not modify the physical state and is thus considered a gauge transformation. In this sense, the first-class primary constraints generate gauge transformations. The famous 'Dirac conjecture' attempts to extend this result to include all first-class constraints as generating gauge. In general, however, the conjecture is false as the existence of some admittedly contrived counter examples illustrates.[11] There is no harm for present purposes, however, if we assume that all first-class constraints generate gauge transformations. The restriction of a phase space function F to $\mathscr{C}$ is gauge-invariant just in case $\{F, \phi_a\} \approx 0$, in which case (8) implies $\delta F \approx 0$. The first-class constraints are thus seen to generate motions within $\mathscr{C}$. In contrast, second-class constraints generate motions leading outside of $\mathscr{C}$.[12] This distinction permits the explication of another important concept: the gauge orbit. A *gauge orbit* is a

---

[10]Cf. ([21], §1.2.1).
[11]Cf. ([21] §1.2.2).
[12]Cf. ([8] §10.2.2).

submanifold of $\mathscr{C}$ which contains all those points in $\mathscr{C}$ which form an equivalence class under a gauge transformation. The sets of these points are path-connected in $\mathscr{C}$ since gauge transformations that connect these points are continuous and do not leave $\mathscr{C}$. They form a curve in $\mathscr{C}$. The gauge motion produced by the first-class constraints can thus be seen to be the tangents to these curves. The points of the gauge orbits in $\mathscr{C}$, equipped with a projection $\mathscr{C} \rightarrow \Gamma_{phys}$, constitute the so-called *reduced* or *physical phase space* Γ_{phys}. The physical phase space Γ_{phys} is defined as the set of points representing gauge equivalence classes of points in Γ. In other words, the physical phase space is obtained by identifying all points on the same gauge orbits. This means that the bundle of admissible dynamical trajectories passing through a particular point $x \in \mathscr{C}$ is mapped to the physical phase space such that the bundle is projected onto a single dynamical trajectory through the point in Γ_{phys} representing the gauge equivalence class in which x falls.

Assume a Hamiltonian system with constraints is given. Assume further that all constraints are first-class.[13] Constraint equations are equations which the canonical variables must satisfy in addition to the dynamical equations of the system. If a set of variables were to determine one and only one physical state, then, given the existence and uniqueness of the solutions of the dynamical equations, one could plug the set of variables uniquely specifying the state into the dynamical equations and could thus obtain the full deterministic dynamical evolution of the physical degrees of freedom. If constraints are present, however, a set of variables does not uniquely describe a physical state. Solving the constraints thus means to use these additional equations to explicitly solve for a variable. This permits the elimination of this variable (and the now solved constraint equation). Solving the constraints of the constrained Hamiltonian system thus amounts to the reduction of the number of variables used to specify the physical state of the system. Once all constraint equations are solved and thus eliminated, the remaining canonical variables are ineliminable for the purpose of uniquely specifying a physical state. In this case, we are back to an unconstrained Hamiltonian system in the sense that its phase space is its *physical* phase space. In the absence of any second-class constraints, the total number of canonical variables $(=2N)$ minus twice the number of first-class constraints equals the number of *independent* canonical variables. Equally, the number of *physical* degrees of freedom is the same as half the number of independent canonical variables, or the same as half the number of canonical variables minus the number of first-class constraints.[14]

---

[13] Second-class constraints can be regarded as resulting from fixing the gauge of a 'larger' system with an additional gauge invariance. They can be replaced by a corresponding set of first-class constraints which capture the additional gauge invariance. Second-class constraints are thus eliminable. In fact, in some cases, it may prove advantageous to thus 'enlarge' a system as this permits the circumvention of some technical obstacles ([21] §1.4.3), albeit at the price of introducing new 'unphysical' degrees of freedom. Without loss of generality, we can thus consider a Hamiltonian system whose constraints are all first-class.

[14] This manner of counting the physical degrees of freedom is well defined for any finite number of degrees of freedom, and perhaps for countably many too. For uncountably many degrees of freedom, new subtleties arise. Cf. ([21] §1.4.2).

1.2 Gauge Freedom in Hamiltonian General Relativity

Hamilton's equations, at least in the narrower standard sense, explicitly solve for the time derivatives. This can only be achieved within GR if its original four-dimensional quantities are broken up into (3+1)-dimensional quantities, with time accruing in the one single dimension. Similar coercion must be exercised upon the four-dimensional structure of spacetime, nota bene, when we wish to consider an initial-value formulation of GR. In order to find a Hamiltonian or an initial-value formulation, GR must be regarded as describing the dynamical evolution *of* something. Breaking up spacetime into 'space' that evolves in 'time' in order to determine whether a well-posed initial-value formulation exists, i.e. whether the physical degrees of freedom enjoy an at least minimally stable deterministic evolution, becomes manageable once we impose a gauge condition to weed out any unphysical degrees of freedom. The traditional formulation of GR as a constrained Hamiltonian system entertains 12 dynamical variables, the six independent components of the three-metric q_{ab} and the six independent components of the corresponding conjugate momentum π^{ab}. Half this number is six, and there are four first-class constraint equations, which leaves the gravitational field with two physical degrees of freedom per point in space. Fortunately, this is the same number of degrees of freedom as one gets for a linear spin-2 field propagating on a flat spacetime background, which can be considered as a weak-field limit of GR.[15] With a gauge condition enforced, Einstein's field equations can be massaged into a form of hyperbolic second-order differential equations defined on manifolds which admit existence and uniqueness theorems. Even in an appropriate gauge fix, however, GR allows for ways in which the field equations may fail to uniquely determine their solutions.[16]

The conceptually most momentous consequence of casting GR as a constrained Hamiltonian system is that the Hamiltonian H is itself a constraint bound to vanish on the constraint surface of the phase space. This is what ultimately leads to the 'problem of time', a conceptual tangle in the foundations of Hamiltonian GR and of quantizations relying thereon, consisting of essentially two strands, the disappearance of time as a fundamental magnitude and the 'freezing' of the dynamics. The first aspect, the vanishing of time as a fundamental physical magnitude, is suggested at the classical level by the increasing elimination of time in classical physics, leading up to Hamiltonian GR, as it is retraced in ([23] §2.1 and §2.2). However, there is a sense in which it only comes to full fruition in quantum theories, as will be elaborated below.

---

[15]See ([46] §4.4b); cf. also ([46] 266) for a slightly different way of calculating the degrees of freedom of the gravitational field.

[16]For an explanation of the failures of determinism in this setting, cf. ([47] §4.1), on which the past few pages have been based. Also, and at the peril of burying an absolutely central point in a footnote, this severance of space and time threatens the general covariance so central to GR. How general covariance gets implemented in Hamiltonian GR and the subtleties that arise in doing so are discussed in ([47] §4.4). What follows explicates the gist of this implementation.

The freezing of the dynamics—more aptly called the 'problem of change'—, however, fully appears at the classical level. A crucial premise of the argument leading to the problem of change is that only gauge-invariant quantities can capture the genuinely physical content of a theory. This premise is justified by pointing to the fact that two distinct mathematical models of a theory describe the same physical situation just in case they are related by maps which are interpreted as 'gauge' transformations. Of course, it may be controversial for any given theory just which maps ought to be considered 'gauge', but I take the justificatory fact invoked in the previous sentence to be analytic of what it means to be 'gauge', viz. to capture a representational redundancy not reflective of the true physical situation. In other words, the premise stipulates that the physical content of a theory is exhausted by the gauge-invariant quantities as codified by the theory. The concept of 'Dirac observables' tries to capture this idea in the context of constrained Hamiltonian theories:

Definition 2 (*Dirac observables*) A(n equivalence class of) *Dirac observable(s)* is defined as the (set of those) function(s) in phase space that has (have) weakly vanishing Poisson brackets with all first-class constraints (and coincide on the constraint surface). Equivalently, Dirac observables are functions in phase space which are constant along gauge orbits on the constraint surface.

Thus, if the premise is true, and if the gauge-invariant quantities of a constrained Hamiltonian theory are precisely its Dirac observables as defined in Definition 2, then the physical content of a constrained Hamiltonian theory is exhausted by its Dirac observables.

In order to determine the physical content of Hamiltonian GR, thus, it becomes paramount to identify its first-class constraints. I will not execute this task here with the mathematical precision it deserves but rest content with a conceptual motivation.[17] The vantage point is the principle of general covariance so central to GR. This principle demands that the Einstein equations' dynamical symmetry group $Diff(\mathcal{M})$ of active spacetime diffeomorphisms is the gauge group of GR.[18,19] In other words, active spacetime diffeomorphisms, which map a solution of the dynamical equation to another solution, ought to be considered relating two mathematically distinct solutions describing one and the same physical situation.[20] Thus, general covariance is spelled out as gauge invariance under active spacetime diffeomorphisms.

---

[17]For a somewhat rigorous execution in the case of the so-called ADM and Ashtekar-Barbero versions of Hamiltonian GR, cf. [47], §4.2.1 and §4.2.2, respectively.

[18]A *spacetime diffeomorphism* is a one-to-one and onto C^∞-map from $\mathcal{M}$ onto itself which has a C^∞-inverse. Diffeomorphisms induce transformations in the fields defined on the manifolds. Intuitively, a map between manifolds is active if it 'moves around' the points without recourse to any coordinate system. Thus, an active transformation is not a change in coordinate systems, but a transformation pushing around the physical fields on the manifold. But this metaphorical picture should be enjoyed with the adequate mathematical caution.

[19]This is the received view, but it should be noted that there has been recent dissent, e.g. in ([14] §3).

[20]For a detailed analysis and justification, cf. ([47] §3, particularly §3.2).

In the Hamiltonian formalism, the dynamical symmetry of GR gets encoded as constraints which generate the spacetime diffeomorphisms in the sense explained in §1.1. In the standard formulation of GR, the elements of the symmetry group $Diff(\mathcal{M})$ are defined as maps between four-dimensional manifolds. The Hamiltonian formalism breaks this four-dimensionality down to a three-plus-one-dimensional rendering; accordingly, $Diff(\mathcal{M})$ breaks down into a group of three-dimensional 'spatial' diffeomorphisms and a group of one-dimensional 'temporal' diffeomorphisms. This move is not without subtleties, as expounded in ([47] §4.2): the symmetry group in Hamiltonian versions of GR differs from that in the usual articulation of the theory, thus distinguishing Hamiltonian GR from its standard cousin in yet another way from those given at the end of the section. In the exemplary ADM version of Hamiltonian GR, the spacetime diffeomorphisms are generated by normal and tangential components of the Hamiltonian flow. Since the constraints generating the diffeomorphism must vanish (weakly), these components of the Hamiltonian vanish (weakly). Furthermore, in a Hamiltonian theory, it is the Hamiltonian which generates the dynamical evolution via the Hamilton equations. Since the Hamiltonian is constrained to vanish, the dynamics gets 'frozen'.

More specifically, (the normal component of the) Hamiltonian is a first-class constraint. Thus, the Dirac observables must have weakly vanishing Poisson brackets with the Hamiltonian and thus turn out to be constants along the gauge orbits generated by the Hamiltonian. This accords with the stipulation above that the physical-content-capturing Dirac observables must be invariant under gauge transformations, here constituted by active spacetime diffeomorphisms. Since the Dirac observables are constant along orbits generated by the Hamiltonian, all genuinely physical magnitudes must be constants of the motion, i.e., they must remain constant over time. In other words, any supposed change is purely a representational redundancy, and not a physical fact. Thus, the argument concludes, there is no change! Since GR, or any quantum theory of gravity replacing it, is a fundamental theory, we are saddled with the uncomfortable task of explicating how time and change can arise phenomenologically—which they undoubtedly do—in a fundamentally changeless world. *O quam cito transit gloria temporis.*[21]

Avoiding this unpalatable conclusion might be all too easy by simply brushing aside Hamiltonian GR as a failed articulation of the theory. But this move is not readily available, at least not without some considerable cost. A prima facie justification for brushing it aside points out that Hamiltonian GR is not theoretically equivalent to the standard formulation of GR. It is true: Hamiltonian GR presupposes that spacetimes can always be sliced up to conform to its $(3 + 1)$-dimensional framework, but this is demonstrably false in GR. Thus, Hamiltonian GR at best captures the sector of GR containing sliceable, globally hyperbolic spacetimes. Furthermore, known articulations of Hamiltonian versions of GR exclude any matter content from the spacetimes and thus only codify vacuum spacetimes. It is not clear, however, that this inequivalence suffices to evade the strictures of the above argument. And most importantly, Hamiltonian formulations of GR serve as the basis for one of the most

---

[21] For a discussion of philosophical reactions to this situation, cf. ([23] §2.3).

important family of approaches to formulating a quantum theory of gravity. By virtue of this fact alone, they deserve to be taken seriously, not just mathematically, but also philosophically.

2 How Spacetime Dissolves in LQG

Once the classical theory is cast in a Hamiltonian fashion, then it can be subjected to the powerful canonical quantization technique. This procedure, pioneered by Paul Dirac, converts the canonical variables of the classical theory into quantum operators defined on an appropriately chosen Hilbert space. The Poisson bracket structure of the classical level is thereby transposed to give rise to the canonical commutation relations obtaining between the basic operators in the quantum theory. From these basic operators, more complex operators can be built up. The classical constraint functions get translated into such complex operators acting on elements in the Hilbert space, thus turning the constraint equations into wave equations. Since they are constraint equations, the constraint operators annihilate the states on which they are acting. Only those states which are so annihilated by the constraints operators are considered *physical* states. As usual in quantum mechanics, the Hamiltonian operator $\hat{H}$ generates the dynamics via a Schrödinger-type equation.

As we have seen in §1.2, in Hamiltonian formulations of GR, the Hamiltonian itself becomes a constraint. In the quantum theory, we get

$$\hat{H}|\psi\rangle = 0 \qquad (9)$$

which is demanded to hold for all physical states $|\psi\rangle$. The 'physical' Hilbert space $\mathscr{H}$ consists just of those states, which satisfy all constraints, i.e., are annihilated by all constraint operators in the theory. Equation (9), also called the 'Wheeler–DeWitt equation', gives a very direct intuition of both the problem of time and that of change. Concerning the problem of time strictly so-called, comparing (9) to the ordinary Schrödinger equation,

$$\hat{H}|\psi\rangle = i\hbar \frac{\partial}{\partial t}|\psi\rangle, \qquad (10)$$

we notice the absence of the time parameter t in (9). This is indicative of the problem of time: the absence of time from the fundamental picture. Quite literally, time drops out of the equation in Hamiltonian quantum gravity.

Given that (9) plays the role of the dynamical equation in quantum Hamiltonian GR just as (10) does for ordinary quantum mechanics, we also glean the first traces of the quantum version of the problem of change by recognizing that the time derivative vanishes. Analogous to the classical case, constraint operators generate the gauge symmetries of the theory. Accordingly, the criterion for the gauge-invariant observables, the Dirac observables defined in Definition 2 of the quantum theory, gets translated as requiring that functions $\hat{F}$ of operators represent Dirac observables just in case they commute with all the constraint operators $\hat{C}_i$

$$[\hat{F}, \hat{C}_i]|\psi\rangle = 0,$$

for all $i = 1, ..., m$, where m is the number of constraints, and for all $|\psi\rangle$ in $\mathcal{H}$. This entails that every Dirac observable must commute with the Hamiltonian. Since the Hamiltonian is what generates the dynamical evolution of the states, all Dirac observables must thus be constants of the motion, i.e., not changing over time. However, the Dirac observables also exhaustively capture the physical content of the theory, at least according to the premise stated in §1.2. Thus, no genuine physical magnitude changes over time. Hence, the dynamics of the world described in canonical quantum gravity is 'frozen' in time. There simply is no change at the most fundamental level described by these Hamiltonian quantum theories of gravity! Change, as it turns out, only arises as a representational artefact—'gauge'—with no physical counterpart in the fundamental theory.

Unlike at the classical level, where arguably the strictures of the argument can be evaded, at least to some extent, by avoiding Hamiltonian formulations of GR, this is evidently not possible for quantizations based on them as the problem is built right into the framework. Perhaps we ought to have expected such an outcome—after all, GR teaches us that time is not external to the physical systems of interest but itself partakes as part of spacetime in dynamical interactions with the material content of the universe, which constitute the usual physical systems physics describes. In other words, time is part of the physical system we are trying to quantize.

In fact, indications persist that quite generically in quantum gravity space and time, at least as standardly understood in GR, no longer form part of the fundamental ontology. Instead, space and time, or at least one or the other, are 'emergent' phenomena that arise from the basic physics. As it is used in the present essay, 'emergent' should not be taken as the terminus technicus in philosophy that designates properties which are not even weakly reducible. Rather, it should be considered as an umbrella term for a relationship that may well turn out to be reductive, as will be argued in §3.1. In fact, to characterize the exact nature of this relationship is the ultimate goal of the research addressing the issue at stake. In the language of physicists, spacetime theories such as GR are 'effective' theories trading in 'emergent' phenomena, much like thermodynamics is an effective theory dealing with the emergent phenomenon of temperature, as it is built up from the collective behaviour of gas molecules. However, quite unlike the fact that temperature is emergent, the idea that the universe and its material content is not *in* space and time shocks our very idea of physical existence as profoundly as any previous scientific revolution did.

So there is at least a sense in which time vanishes in canonical approaches to quantum gravity. It has been argued that because string theory contains GR "in some limit... [t]he disappearance of external time should... also hold in string theory" ([24] 10). As a consequence of the holographic principle, space as well can be considered emergent in string theory [23]. Furthermore, the fundamental structures postulated by various quantum theories of gravity diverge significantly from the familiar spacetimes of GR. For instance, so-called non-commutative geometry replaces the basic geometric picture we have of spacetime by algebraic relations between temporal and spatial coordinates or directions and generalizes multiplicative relations among them so that they no longer commute. This generalization has weird consequences and renders the basic structure conceptually quite different from spacetime [23]. As another example, the fundamental structure generically turns out to be discrete rather than continuous. For a vast class of quantum theories of gravity, Lee Smolin, takes discreteness to be "well established." ([42] 549).

Of course, one might react to these developments as John Earman did, at least concerning LQG, and insist that

> although classical general relativistic spacetime has been demoted from a fundamental to an emergent entity, spacetime per se has not been banished as a fundamental entity. After all, what LQG offers is a quantization of classical general relativistic spacetime, and it seems not unfair to say that what it describes is quantum spacetime. This entity retains a fundamental status in LQG since there is no attempt to reduce it to something more fundamental. ([17] 21)

If this is just a quarrel over words, I have no appetite to engage in it. We are free to call LQG's fundamental structure, to be described in the remainder of this section, 'quantum spacetime' all right, but given the profound departures from relativistic spacetimes, the use of a different term is not only warranted, but also preferable, as I have argued elsewhere [48]. Let us leave this debate to one side and delve into the physics in order to get a sense of what it is LQG theorizes about.

2.1 Introducing LQG

Canonical quantum gravity generally, and LQG in particular, attempt to transpose the central lesson of GR into a quantum theory. The pertinent key innovation of GR is the recognition that spacetime does not passively offer a fixed 'background' which determines the inertial 'forces' acting on the physical content of the universe, but instead a dynamical structure which interacts with matter. To repeat, LQG is based on a reformulation of GR as a 'Hamiltonian system', which reinterprets spacetimes as $(3 + 1)$-dimensional rather than 4-dimensional, with constraints. Thus, recasting GR as a Hamiltonian theory forces a 'foliation' of its spacetimes by an equivalence relation into three-dimensional 'spatial' hypersurfaces, parametrized by a one-dimensional 'time'. The natural interpretation of the Hamiltonian system would be that of a three-dimensional 'space' considered as a dynamical physical system which evolves over 'time', where the three-dimensional hypersurfaces would represent the instantaneous state of the dynamical theory.

LQG is thus a canonical quantization of Hamiltonian GR.[22] Before we proceed, let it be noted that the particular formulation required entails a substantive limitation

---

[22]For a thorough introduction to LQG, cf. [34]; for the mathematical foundations, cf. [44]. [35] is a recent review article.

of the approach: only 'globally hyperbolic' spacetimes of the classical theory are considered. If a spacetime is globally hyperbolic, then it is topologically '3 + 1', i.e., the topology of $\mathcal{M}$ is $\Sigma \times \mathbb{R}$, where Σ is a three-dimensional submanifold of $\mathcal{M}$.[23] Just how severe this limitation is is debatable; many physicists do not consider it troubling, some philosophers have dissented. To impose global hyperbolicity as a necessary condition for physically reasonable spacetimes amounts to asserting a strong form of the merely conjectured, but not proven, cosmic censorship hypothesis. Dissenting voices cautioning against stipulating global hyperbolicity as necessary include Earman [18], Erik Curiel [13], Chris Smeenk and Wüthrich [41], and John Manchak [27]. Manchak ([27] 414) proves that as long as a spacetime is not "causally bizarre", it is observationally indistinguishable from another spacetime, not isometric to the first and not globally hyperbolic, yet with exactly the same local properties. From this, Manchak concludes that "[i]t seems that, although our universe may be... globally hyperbolic..., we can never know that it is." (ibid.) In the light of this result, it appears brash to enthrone global hyperbolicity as a sine qua non of physical reasonability. Having said that, however, if LQG were to be a huge empirical success, its premises would be vindicated. Note the future subjunctive tense in the previous sentence.

There currently still persists another, uncontroversially problematic, limitation of the approach: only vacuum spacetimes are considered, i.e., the classical vantage point of the approach is the vacuum sector of GR with everywhere vanishing energy-momentum. This technical simplification comes at the price of rendering it unclear whether the resulting quantum theory can deal with a non-zero energy and matter content of the universe, presumably a necessary condition for giving an empirically adequate account of the actual world. The situation may not be quite as bleak for LQG as this may suggest, for three reasons. First, vacua are physically important states and their theoretical understanding may shed decisive light on the necessary steps leading to a more general theory encompassing matter. Second, the assessment as to whether or not models of a theory or vacuum states of the universe contain matter may come apart for classical and quantum theories. In other words, the quantum theory which started out from classical vacuum states may be interpreted to contain matter. This possibility does not come without further complications, though: the emerging matter may well be highly non-local and may violate most or all energy conditions. Third, and most speculatively, matter, just as space and time, may emerge from the— perhaps topological or combinatorial—properties of the fundamental structure and hence not be present at the fundamental level.

The goal of the quantization is to find the Hilbert space corresponding to the physical state space of the theory and to define operators on the Hilbert space representing the relevant physical magnitudes. The hope would naturally be that some of the latter make contact to the empirically testable. In order to get the quantization started, one chooses a pair of canonically conjugate variables which coordinatizes the relevant sector of the classical phase space. Different choices lead to differ-

---

[23]For a more systematic explication of global hyperbolicity and neighbouring concepts, see ([41] 593).

ent quantum theories: geometrodynamics's choice is the induced three-metric on the three-hypersurface and its conjugate momentum constructed from the external curvature of the three-hypersurface, LQG starts out from Abhay Ashtekar's 'new variables' of a connection A_a^i and its conjugate, a densitized triad 'electric field' E_i^a and constructs a 'holonomy' and its conjugate 'flux' variables from them. The geometrical structure of the classical phase space is encapsulated in the canonical algebra given by the Poisson brackets among the basic variables. This structure gets transposed into a quantum theory by first defining an initial functional Hilbert space of quantum states $|\psi\rangle$. The basic canonical variables are turned into operators whose algebra is determined by their commutation relations arising from the classical Poisson brackets. The classical constraints, which are functions of the canonical variables, now become operators constructed 'isomorphically' as functions of the basic operators. Classically, the constraint functions are set to zero; in the quantum theory, they annihilate the states. Thus, by imposing the constraints, the theory effectively demands that only states which are annihilated by *all* constraint operators are considered physical. Dynamical equations, as was already clear at the outset of this section, play a somewhat different role. In a sense, given that the 'Schrödinger-like' equation of the quantum theory is the constraint equation (9), there is no additional dynamical equation governing any 'dynamics' of the theory.

In LQG, three families of constraints arise. First, the so-called 'Gauss constraints' indicate a rotational gauge freedom of the triads and generate an infinitesimal $SU(2)$ transformation in the internal, as opposed to spacetime, indices (indicated by letters from the middle of the alphabet). These are comparatively straightforward to solve. Next, we find three '(spatial) diffeomorphism' constraints, which generate the spatial diffeomorphisms on the three-hypersurfaces. These constraints are hard to solve, but it has been done. The resulting Hilbert space, i.e., the Hilbert space we obtain from the states which get annihilated by the Gauss and diffeomorphism constraints, is called the 'kinematical Hilbert space' and will here be denoted by $\mathcal{H}_K$. Finally, there is the Hamiltonian constraint which has so far defied solution. In fact, it is not even clear what the concrete form of the formal equation (9) is. In this sense, LQG is not yet a complete theory. As will hopefully become clear later in the essay, there remain plenty of reasons not to walk away from LQG, at least not just yet.

Given the technical and conceptual difficulties with the 'dynamics' (9), various authors have sought ways to circumvent the standard conceptualization of dynamics in a Hamiltonian theory. One main approach conceives of the dynamics in ways similar to perturbative approaches to quantum field theory, taking elements of $\mathcal{H}_K$ as three-dimensional 'initial' and 'final' 'spaces' and compute transition amplitudes between them ([35] §3). Or alternatively, as Carlo Rovelli has suggested, the states in the physical Hilbert space may not be 'states at some time'; instead, they are 'boundary states', i.e., states describing quantum space surrounding a four-dimensional region of spacetime.[24]

Because (9) is not solved yet, all results must remain preliminary. One way to see this immediately is to remind the reader that all Dirac observables must commute

---

[24] A more detailed analysis of dynamics in LQG can be found in ([47] §5.3).

Fig. 1 A spin network state
is characterized by an
abstract graph with
'spin'-representations on the
nodes and the links between
them

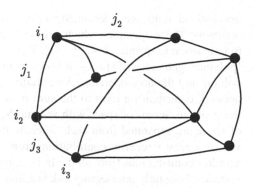

with all constraints. If we accept that the set of Dirac observables is identical to the
set of genuine physical magnitudes, as arguably we should on pain of introducing
gauge-dependent quantities, then we cannot determine the physical magnitudes yet,
as we do not know the explicit form of $\hat{H}$ and so cannot determine which operators
commute with it. It thus remains open whether any of the geometric operators to be
introduced shortly really corresponds to a genuine physical magnitude.

Let us study the structure of $\mathcal{H}_K$ then. It turns out that so-called 'spin network
states' provide a useful basis in $\mathcal{H}_K$.[25] These spin network states are interpreted to be
the quantum states of the gravitational field. Since physical 'space' will be in a state
in $\mathcal{H}_K$, as §4.2 will suggest, it will generally be in a quantum superposition of spin
network states. Spin network states can be represented by *abstract labelled graphs* as
in Figure 1,[26] as they are completely characterized and uniquely identified by three
types of 'quantum numbers'. The first label characterizes the abstract graph Γ, the
second the irreducible $SU(2)$-(hence 'spin') representations j_l on the links and a third
the $SU(2)$-representations on the nodes, denoted by i_n. It should be emphasized that
the abstractness of the graph is central to the correct interpretation of the emerging
picture here: the spin network states are not quantum states of a physical system *in*
space; rather they are the quantum states *of* physical space.

The spin network states $|\Gamma, j_l, i_n\rangle$ are eigenstates of the so-called area and volume
operators defined on $\mathcal{H}_K$. The spectra of these operators yield important information
concerning the geometrical interpretation of the spin network states, although it must
be emphasized that the interpretation of the states relies, in turn, on an *interpretation*
of these operators as geometric. Since we study the properties of the gravitational
field via the geometry of the physical space, the properties of (three-dimensional)

---

[25]For the technical background of this basis and its interpretation, cf. ([35] §2.3).

[26]More precisely, they are represented by labelled graphs embedded in some background space.
Thus, they are not invariant under spatial diffeomorphisms, i.e., when they are 'pushed around'
on the embedding manifold. In order to fully solve the diffeomorphism constraints, then, we need
equivalence classes of spin network states under three-dimensional diffeomorphisms on the background manifold. Sometimes, these equivalence classes, represented by *abstract labelled graphs*,
are called 's-knot states' in the literature. So I am being slightly sloppy by using the locution 'spin
network states' ambiguously.

gravitational fields are determined by the spectral properties of the area and volume operators. These operators, which will be discussed in greater detail in §4.2, turn out to have discrete spectra [2–4, 36, 37]. The granularity of the spatial geometry—the 'polymer' geometry of space—follows from the discreteness of the spectra of the volume and the area operators. Essentially, each node (and only the nodes) in the network contributes a term to the sum of the volume of a region. On each node, there sits an 'atom' of space with volume V_n, as it were. These elementary grains of space are separated from each other by their surfaces of contiguity. Just as the volume operator receives contributions from the nodes of a region, the area operator acquires contributions from all the links that intersect the surface. For instance, the surface whose only intersecting link is a link with quantum number j_l has a surface area of $A_l \propto \sqrt{j_l(j_l + 1)}$ ([34] §6.7). Thus, the 'size' of the surface connecting adjacent 'chunks' of 'space' is constructed from the spin representations sitting on the relevant links. Thus, the smooth space of the classical theory is supplanted by a *discrete quantum structure* displaying the granular nature of space at the Planck scale. Continuous space as we find it in classical theories such as GR and as it figures in our conceptions of the world is a merely *emergent phenomenon*.[27]

Physical three-space, in Rovelli's interpretation, is a quantum superposition of spin network states, analogously to the physical electromagnetic field consisting of a superposition of n-photon states. LQG *predicts* the existence of indivisible quanta of volume, area, and length, as well as their spectra (up to a constant). Importantly, this discreteness was a *result* of the loop quantization, rather than an *assumption*. According to LQG, measurements of the Planck geometry of space must therefore yield one of the values in the spectrum of the concerned operator.

As mentioned above, the 'dynamics' of canonical LQG are only known in formal outline. As in any Hamiltonian theory, the dynamics of the theory is generated by the Hamiltonian operators $\hat{H}$, which is defined on $\mathcal{H}_K$, via the Wheeler–DeWitt equation (9). The space of the solutions of (9) will constitute the *physical Hilbert space $\mathcal{H}$*. But since there exist several inequivalent versions of $\hat{H}$—all of which may be false—the Hilbert space $\mathcal{H}$ has not yet been constructed and the theory remains incomplete.

Before we start to consider how spacetime emerges from the fundamental structures of LQG—spin network states—, let us make sure that it has indeed vanished from the fundamental ontology. Of course, as Earman suggested in the quote above, we might simply call the spin-network structure 'quantum spacetime' and move on with it. To use homonyms, or near-homonyms, for two rather different structures, however, promises to create more confusion than comprehension. The spin networks diverge from classical relativistic spacetimes in at least two crucial points. First, unlike the continua of classical physics, they are discrete. As was observed above, many expect the fundamental structure in quantum gravity to be discrete and this expectation is certainly borne out in many of the extant approaches. This is

---

[27]It should be kept in mind, however, that these operators are not Dirac observables and should therefore be taken with a grain of salt. They are partial observables in the sense of [33].

a significant departure, but may not sway everyone to discontinue considering the fundamental structure a 'spacetime'.

Arguably, however, the deeper divergence from classical spacetimes arises from the 'non-localities' that we find in spin networks (and in many other quantum structures).[28] How these fundamental structures can be 'non-local' needs a bit of explaining, given that (non-)locality is a spatiotemporal, or anyway a spatial, concept. To appreciate the sense in which the spin networks do contain 'non-localities', consider a fundamental relational structure consisting of a set of basal atoms, which exemplify, in pairs, a basal 'adjacency' relation. Together with an intrinsic 'valence' attributed to each of the atoms and each of the exemplified relations, this yields a connected structural complex of the kind we find in LQG. Contrast this with the spatiotemporal structure we find in GR, where the spatiotemporal, indeed metric, relations obtaining between the spacetime events give rise to a locality and neighbourhood edifice. Now, these two structures are supposed to be related by an emergence relation. More specifically, the idea is that the exemplified fundamental structure is related, in some limit or in some approximation or at some scale, to a relativistic spacetime. Given two particular structures related in this way, one can map the atoms of the fundamental structure onto events in the spacetime. What it then means to say that there are 'non-localities' present in the fundamental structure is that some pairs of adjacent basal atoms, i.e., pairs of atoms exemplifying the fundamental adjacency relation, get mapped onto events in the spacetime which can be at arbitrarily large distances now as measured in the metric of the emerging spacetime.[29] Locality is notoriously tricky in GR, of course, but in globally hyperbolic relativistic spacetimes, a precise notion of locality is readily available. Given a possibly physically privileged foliation, a spatial metric is induced on the leaf containing the events, which are thus spatially related. This now permits an explication of locality e.g. in terms of convex spatial neighbourhoods of events. Thus, what is adjacent in the fundamental structure in general is not local or nearby in the emerging spacetime as judged by the latter's induced spatial metric.

From the perspective of the emerging spacetime, the spin networks generally get the locality structure wrong, or so one would expect. The expectation that these non-localities are generic arises from the fact that relation between spin networks and classical spacetimes—to the extent to which we understand it—is many-to-one.[30] In other words, there are in general many spin network states whose best classical approximation is the same relativistic spacetime. Since these spin networks are physically distinct, and one of the main ways in which they can differ is by their connectivity defined by the obtaining adjacency relations, spin networks with distinct topologies will be best approximated by one and the same spacetime. As spin networks that give rise to realistically large universes will consist of very many adjacent pairs of nodes, it seem natural to think that at least some of them will be non-local in the present sense. If this is right, then non-localities generically arise in

---

[28]Cf. e.g. [28].

[29]Cf. Figure 1 in [22].

[30]Cf. Section 4. Cf. also ([28] §2) who give a related reason.

spin networks, and we have a second deep departure of the latter from relativistic spacetimes.

These non-localities are suppressed in the low-energy approximation from the spin network to the relativistic spacetime. In fact, they must be suppressed, for otherwise they would have to be emulated by the emerging structure in the sense that these adjacency relations would re-occur in the spacetime in the form of neighbouring relations and thus not qualify as 'non-localities'. To repeat, 'non-localities' of the relevant sort are fundamental adjacencies with no vicinity-type counterpart in the emerging spacetime. If the course graining attendant to the emergence of spacetime from spin network states—of which more in §4—would not 'wash out' the non-local connections, they would have to be encoded in the emerging relativistic spacetime, perhaps as non-local 'wormholes'. If, however, their presence were so strong as to preclude essentially local physics at comparatively low energy scales, such as described by quantum field theory on relativistic spacetime backgrounds, then the corresponding theory, or at least model, would have to be considered empirically inadequate.[31] So we would expect those non-localities to be generically present, but suppressed in the coarse graining to macroscopic scales.

Relativistic spacetimes arguably differ in significant ways in how they conceptualize space and time from our intuitive concepts of space and time. But whatever differences these are, they do not suffice to call into question why we refer to the structures of GR as 'spacetimes', and justifiably so. Whatever the differences between intuitive space and time and spacetime in GR may be, it is clear that the departures of LQG from the manifest image run much deeper. Not only is the fundamental structure discrete and non-local, but as we have seen in §1, the problem of time in its different forms illustrated how our common concepts of time, change and dynamics and the way these concepts are standardly encoded in physical theories and their languages completely and utterly fails. Even though this failure was enunciated in §1 at the classical level already, it crucially depended on the particular non-standard, and inequivalent, formulation of GR necessary for the canonical programme to get going. If we could directly quantize GR from its standard formulation, the resulting theory's departure from classical spacetime physics might be milder. But alas, no promising strategy along these lines is known.

I conclude that we can safely assume that spacetime has been lost, at least in its traditional, relativistic sense, somewhere in the transition from GR to LQG. Now that the Babylonians of quantum gravity have removed spacetime from its sacred place, amid rampant speculation concerning its whereabouts, serious efforts have commenced to recover the lost spacetime and restore it to its lawful place. He or she who recaptures it may be blessed with wisdom—or be smitten, as the case may be.

---

[31] This does not entail that the fundamental non-localities could not have observable consequences, such as those proposed by [32].

3 What Emergence of Spacetime is Not

In order to honour the covenant—and to avoid being smitten—, then, let this section clarify what the emergence of spacetime could not be. First, §3.1 explains the difference between the standard concepts of 'emergence' as they figure in philosophy and physics, respectively, and states that it is the physicists' use that will be relevant for our purposes. Second, it will be argued in §3.2 that the use of the notion of 'unitary equivalence' will not serve to determine whether spacetime still maintains fundamental existence in LQG.

3.1 Non-reductive Relation

The concept of 'emergence' has a venerable history in philosophy: arguably stretching back to Aristotle and Galen, it attracted renewed interest in the nineteenth century, reflected in the work of George Henry Lewes, John Stuart Mill, and C D Broad in Britain, and Nicolai Hartmann on the continent. Despite some variation among them, authors in this tradition as well as contemporary philosophers use the term so as to imply a *non-reductive relation* between the emergent and the fundamental, presupposing that reality is somehow layered into different 'strata' and that the properties and relations attributed to entities at different levels in general differ from one another. The general spirit of the concept is well captured by Brian McLaughlin's definition in terms of supervenience:

Definition 3 (*Emergent property*) "If P is a property of w, then P is emergent if and only if (1) P supervenes with nomological necessity, but not with logical necessity, on properties the parts of w have taken separately or in other combinations; and (2) some of the supervenience principles linking properties of the parts of w with w's having P are fundamental laws." ([30] 39)

Definition 3 only gets traction if all the terms in the definiens are defined in their turn. Let us briefly discuss some of them. The first clause in the definition betrays the physicalist underpinnings of the version of emergentism which I assume here as standard. As ([30] §3) explains, the relevant notion of 'supervenience' in this context is based on the idea of a "required-sufficiency relationship" (ibid.), i.e., that the possessing of a higher-level property requires the possessing of a lower-level property which in turn suffices for the possessing of the higher-level property. This supervenience should not be forced by logic alone, but instead result from contingent laws of nature. To grasp the meaning and the role of the second clause, let me state the definition of 'fundamental law' as given by McLaughlin:

Definition 4 (*Fundamental law*) "A law L is a fundamental law if and only if it is not metaphysically necessitated by any other laws, even together with initial conditions." (ibid., 39)

The second clause is necessary; for without it, Definition 3 would be overly inclusive, as McLaughlin argues, in that reducible properties would often also qualify as emergent, against the stated intention of the emergentists. If the laws which codify the connections between the properties of the lower level entities with those of the higher level, or those of the parts with those of the whole, are fundamental, then they are in principle not reducible to other laws governing the properties of lower levels, thus ruling out that reducible properties qualify.[32]

It should be emphasized that in the context of the present study, and of much of the physics literature on the subject, 'emergent' should not be understood as the terminus technicus defined in Definition 3, where an emergent property (or, mutatis mutandis, an emergent entity) is not even weakly reducible. Rather, it is to be understood as a collective designation for broadly reductive relationships. Indeed, *that* is the point of the entire enterprise: to understand how classical spacetime and its properties reduce, or more neutrally *relate*, to the fundamental non-spatiotemporal structure. Reduction, as an inter-theoretic relation, can thus be regarded as a *working hypothesis* of the quest to regain spacetime.

3.2   Unitary Equivalence

Leaving behind the general philosophical literature, we find in the pertinent philosophy of physics a very specific criterion which has been proposed to determine whether or not in a quantum theory of gravity spacetime can still be regarded as fundamental or not. Almost as an aside, Craig Callender and Nick Huggett ([11] 21) use the criterion of *unitary equivalence* for exactly this purpose, and in the context of LQG! Unitary equivalence, here as elsewhere, is used as a sufficient condition for physical equivalence. Callender and Huggett state that if bases of spin network states and of (functionals of) three-metrics in quantum geometrodynamics are unitarily equivalent, then they would merely constitute different representations of the same objects—viz. space—, rather than of numerically distinct objects. Hence, if successful, unitary equivalence would establish a particularly direct (reductive) relation, at least concerning *space*. If the two bases turn out to be unitarily inequivalent, then the reductive relation will be more complex. To invoke unitary equivalence as a (necessary and sufficient) condition for physical equivalence is well motivated.[33] Despite qualms one might entertain regarding the equivalence of the equivalences, let us grant, for the sake of argument, that unitary equivalence and physical equivalence come together. It turns out, however, that the criterion is nevertheless unhelpful, for three reasons.

---

[32] For an up-to-date review on emergent properties, cf. [31].

[33] At least at the level of ordinary quantum mechanics; in relativistic quantum theories, matters become more subtle. Cf. ([38] §2.2).

Since unitary (in)equivalence is usually predicated of *representations*, not of *bases*, let us translate the condition into the language of bases of Hilbert spaces before we start listing the problems:

Definition 5 (*Unitary equivalence between bases*) Two bases $\{|a^{(k)}\rangle\}$ and $\{|b^{(l)}\rangle\}$ of two Hilbert spaces $\mathscr{H}$ and $\mathscr{H}'$, respectively, are *unitarily equivalent* just in case there is a unitary map $U : \mathscr{H} \to \mathscr{H}'$ such that $U|a^{(k)}\rangle = |b^{(k)}\rangle$ for all k.

Now, given this definition, and the orthonormality and the completeness of bases, it is easy to construct such a unitary map between Hilbert spaces of the same dimension: $U = \sum_k |b^{(k)}\rangle\langle a^{(k)}|$. For our discussion below, we need to put two theorems on the table. Here is the first one:

Theorem 1 (*[15], 3.11.3(a)*) *If $\mathscr{H}$ is an infinite-dimensional separable Hilbert space, then it is isomorphic to l^2, the space of square-summable sequences.*

Two Hilbert spaces are *isomorphic* just in case there is a unitary map that leaves the inner product invariant. Since being isomorphic is a transitive relation, *any* two infinite-dimensional separable Hilbert spaces are isomorphic. In other words, there is a unitary map between the bases of any two infinite-dimensional separable Hilbert spaces. This entails, of course, that for any two infinite-dimensional separable Hilbert spaces, we can find unitarily equivalent bases in the sense of Definition 5. In fact, we have the more general theorem:

Theorem 2 (*[20], §16*) *Any two Hilbert spaces $\mathscr{H}$ and $\mathscr{H}'$ are isomorphic iff* $\dim(\mathscr{H}) = \dim(\mathscr{H}')$.

An immediate consequence of this theorem is that any two Hilbert spaces of the same dimension will have unitarily equivalent bases. So our knee-jerk reaction right after Definition 5 stands vindicated. Quite generally, the theorem shows that Hilbert spaces of the same dimension are geometrically indistinguishable and can thus rightfully be considered identical as far as their physically salient structure is concerned.

Let us return to the proposal by Callender and Huggett [11] and discuss its problems. As announced above, there are three of them. *Primo*, in order for this criterion to get any traction, the relevant Hilbert spaces would have to be known—but they are not. We have already seen that the physical Hilbert space $\mathscr{H}$ of LQG has not yet been constructed, only its kinematic Hilbert space $\mathscr{H}_K$. The same is true for geometrodynamics, where the constraints are non-polynomial and so far defy solution. No Hilbert space, no basis. No basis, no checking for unitary equivalence. But let us proceed, again for the sake of argument, on the assumption that we had the relevant Hilbert spaces.

Secundo, the criterion, although perhaps necessary, is far removed from anything close to a sufficient condition, at least on its own. Consider the following three exhaustive possibilities. First, the physical Hilbert spaces of quantum geometrodynamics and LQG are both separable, i.e. they each have a countable basis. Second, one of them is separable, but the other is not. And third, both Hilbert spaces are non-separable, with either (a) their bases having the same cardinality, or (b) different cardinality.

In the first case, the criterion is trivially satisfied because two bases in *any* two (infinite-dimensional) separable Hilbert spaces are unitarily equivalent. In the second case, the criterion is trivially violated, for corresponding reasons. In the third case, if the bases of the two Hilbert spaces have the same cardinality, we are back to the first situation; if they do not, we find ourselves in the second case again. So either way, the criterion by itself is not very illuminating and clearly not sufficient. It would have much more bite—and that may be the unarticulated intention behind Callender and Huggett's proposal—if it were augmented by some additional condition such as the preservation of the characteristic algebraic relations among the operators (such as the canonical commutation relations) in the transformation from one to the other.

Tertio, the Callender-Huggett criterion gives the metric codification, which is used in quantum geometrodynamics, undue precedence over the connection codification, which is LQG's vantage point, in that it assumes that only the first captures the geometric essence of relativistic spacetimes. At least classically, both the metric and the connection descriptions are equally respectable ways of capturing the geometry of a spacetime and I see no reason to elevate one at the expense of the other. So we might, with equal justification, demand that a quantum theory of gravity offers a description of a quantum *spacetime* just in case a basis of its physical Hilbert space is unitarily equivalent to the connection basis of the physical Hilbert space of a quantum theory of gravity based on a connection representation.[34] Such a choice would be, of course, vulnerable to the same charge raised here.

Thus, unitary equivalence between a basis of the physical Hilbert space of a theory in question and the three-metrics basis of quantum geometrodynamics is certainly not sufficient to think that the fundamental structure proposed by the theory in question is still spacetime. Perhaps it is not even necessary. But even if the criterion were valuable, we would still be faced with a rather complete dissolution of the classical continuous and local spacetime structure into granular structure with odd non-localities, represented by labelled graphs. And the question would still naturally arise how come our world looks like it is well described at sufficiently large scales by relativistic spacetimes. This explanation would still be owed, even if we managed to convince ourselves that the fundamental structure still deserves to be called 'spacetime'.

4   Re-emergence of Spacetime

Before we venture into the enterprise of investigating how spacetime emerges from spin networks, one mistaken argument should be put to the side. I am thinking of a

---

[34]Strictly speaking, LQG basic variables are the holonomies and fluxes introduced in §2.1, which are not identical to the connection and the canonically conjugate electric field of the connection representation but are constructed from them.

Kantian who nonchalantly responds to the present situation of the fundamental loss of spacetime by declaring that spacetime is a 'pure form of intuition' and as such does not exist mind-independently anyway. So, the Kantian continues, we should not have expected to find spacetime as an ontological posit of a fundamental theory in the first place. But such a complacent 'told-you-so' reaction would be entirely misguided; assuming space and time to be pure forms of intuition does nothing to relieve us from the obligation to explicate how relativistic spacetimes emerge from what physics tells us is fundamental. On a Kantian perspective, the job of physics is to describe nature as it appears to us, not as it may be *an sich*. And the natural world surely appears to be spatiotemporally ordered, which is why (earlier) physical theories made the natural assumption that there are space and time. Since physical theories involving such postulations have been empirically very successful, any theory seeking to supplant a theory as successful as, e.g., GR, must explain why the latter was as successful as it was given that it is not true. In this sense, recovering spacetime from the fundamental structure becomes part of the task of justifying the fundamental theory. This aspect assumes great urgency in a field plagued by the lack of empirical data.

This justificatory task of understanding the emergence of spacetimes from fundamental structures, such as spin networks, is discharged by 'taking the classical limit' of the fundamental theory: one shows that the classical theory results from an appropriate mathematical procedure which is interpreted to physically explain why and how the proprietary effects of the fundamental theory are hidden behind the phenomena so well represented by the classical theory. To express the situation in Reichenbachian terms, taking the classical limit, and thus showing how relativistic spacetimes emerge from fundamental structures, constitutes, at least partially, the 'context of justification'. As indicated in Figure 2, the reverse process by which we arrived at the quantum theory of gravity from the classical theory is of course the quantization studied in §2 and can thus be understood as the 'context of discovery' (of the novel quantum theory). Understanding how classical spacetimes re-emerge is thus not only important to save the appearances and to accommodate common sense, but also a methodologically central part of the entire enterprise of quantum gravity.

Nota bene, the quantization procedure as outlined above lacks a unique implementation for which every step is well justified. At various steps, one can choose to follow different paths, all presumably leading to inequivalent quantum theories. Some may find the fact that the construction of the quantum theory does not proceed along more principled lines troublesome. Applying this Reichenbachian terminology also

Fig. 2 Quantization and the classical limit as 'inverse' tasks

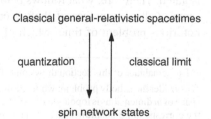

illustrates why this need not be a problem: the 'context of discovery' is dominated by creative elements which defy being bound by the narrow strictures of a research logic. On the other hand, the same traditional philosophy of science also urges that the other direction, the 'context of justification', be taken very seriously. Regardless of this traditional philosophy of science's merits, the urgency clearly applies to the case at hand.[35]

'Taking the classical limit' means establishing a mapping between, in some principled way, either individual models of the fundamental, 'reducing', theory to individual models of the higher level, 'reduced', theory, or 'generic' models of the reducing theory to 'generic' models of the reduced theory, or the totality or near-totality of models of the reducing theory to the totality or near-totality of models of the reduced theory. It will not suffice to just procure a merely mathematical expression of such a mapping; instead, any formal articulation of it will need to be supplemented by a demonstration of its 'physical salience' [22]. To start with the obvious, the map from the set of quantum states to the set of classical spacetimes should not be expected to be bijective, but many-to-one as there will be multiple distinct quantum states with the same classical limit.[36] Furthermore, there will be no classical analogue for some sets of quantum states. Also, the quantization of a classical theory might not guarantee the re-emergence of the classical structure from the resulting quantum theory, due to interpretational issues ([10], 80).

So far, the classical limit of LQG (and many other quantum theories of gravity) has resisted understanding. The difficulties tend to be of two disparate kinds. First, there are technical intricacies. Second, and of present interest, there are numerous conceptual and interpretational issues. This is where philosophers can hope to make contributions by helping to explore the conceptual landscape, to map possibilities, and, more concretely, bring the literature on emergence and reduction to bear on the problems at hand. To date, only few philosophers have ventured into this area. I hope that more will follow—and there are hopeful signs. But still, [9] and [10] constitute more or less the complete philosophical literature on emergence in canonical quantum gravity, together with my dissertation ([47], §9.2) on which the remainder of this section is based.

A caveat before we proceed to portray the emergence scheme proposed by Jeremy Butterfield and Chris Isham and articulate its application to LQG and hence to the emergence of the full spacetime, rather than just time, as Butterfield and Isham do. As we noticed above (in §2.1), LQG is not a complete theory in that the 'dynamics' is not well understood and in this sense the physical Hilbert space has not yet been isolated. Therefore, what follows below is limited to the kinematical level. This has some of the advantages of theft over honest toil, as we can thus circumvent the notorious problem of time, which of course Butterfield and Isham address. But it

---

[35]The remainder of this section draws on ([47] §9).

[36]Consider the n-body problem: while the phase space of states of an n-particle system in a physical space of m dimensions is topologically $\mathbb{R}^{2mn}$ and therefore finite-dimensional in classical mechanics, the corresponding quantum space of states is the infinite-dimensional Hilbert space $L^2(\mathbb{R}^{mn})$, the space of square-integrable functions on $\mathbb{R}^{mn}$.

brings with it the distinct disadvantage that the following remains preliminary and must thus be taken with a grain of salt.

4.1 The Butterfield-Isham Scheme

Let us then orient our conceptualization of the problem toward the extant literature on emergence in canonical quantum gravity. Similarly to my suggestion above, [9, 10] propose to regard quantization and emergence as two distinct, somewhat inverse, and *independent* strategies for solving the problem of quantum gravity. Butterfield and Isham consider various potentially helpful explications of the concept of emergence. As it turns out, all of them cast emergence as a reductive relation. As we have seen in §3.1, this usage is consonant with the physics literature, but dissonant with the one in general philosophy. Given the richness and diversity of the literature on reductive relations between theories, [9] conclude that this should be taken to sustain the conclusion that there may not be a single concept of reduction to fit all instances considered, not even if the analysis is confined to physics.[37]

Butterfield and Isham [9] distinguish three ways in which theories (or their concepts, entities, laws, or models) can stand in a reductive relation to one another: definitional extension, supervenience, and emergence. The first typically assumes a syntactic understanding of theories, i.e. it understands a theory as a deductively closed set of propositions. Applying Butterfield and Isham's definition of it to the case at hand, one could say that GR is a *definitional extension* of LQG iff it is possible to add to LQG definitions of all non-logical symbols of GR such that every theorem of GR can be proven in LQG thus augmented. The concept of definitional extension is attractive because it gives us a clear understanding of how two theories, one of which is a definitional extension of the other, relate to one another. Thus, definitional extension goes a long way to explain why the predecessor theory was as successful as it was and why it breaks down where in fact it does. However, we do not expect the relation between GR and LQG to be as clear-cut as it is between Newtonian mechanics and special relativity, where the concept of definitional extension admits a rather straightforward application. In order to determine whether or not GR is a definitional extension of LQG, one would need to know how to recover the classical limit. Unless there is at least some progress in the recovery of the classical limit of LQG, the concept of definitional extension cannot usefully be applied to the case at stake. One would expect, to be sure, that relating LQG to GR will involve approximations such that general-relativistic propositions only hold approximately in LQG, and only under certain conditions. More specifically, one first extends the definitions of LQG such as to make it conceptually sufficiently potent to be able to prove all theorems of an intermediate theory, from which GR can, in a well-understood way,

---

[37]No attempt shall be made to substantially consider the wider literature on the topic. Cf. [43] for an analysis of various proposals for reduction as an inter-theoretic relation, with a particular eye on the physical sciences.

be recovered as an approximation. This process of approximation can be defined as follows:

Definition 6 (*Approximating procedure*) An *approximating procedure* designates the process of either neglecting some physical magnitudes, and justifying such neglect, or selecting a proper subset of states in the state space of the approximating theory, and justifying such selection, or both, in order to arrive at a theory whose values of physical quantities remain sufficiently close to those of appropriately related quantities in the theory to be approximated.

But all of this goes beyond the concept of definitional extension and shall be discussed below when I will discuss approximation as a form of emergence.[38]

The second relation considered by Butterfield and Isham is supervenience. *Per definitionem*, GR *supervenes* on LQG iff all its predicates supervene on the predicates of LQG, with respect to a fixed set $\mathfrak{A}$ of objects on which both predicates of GR and of LQG are defined. The set of predicates of GR is said to *supervene* on the set of predicates in LQG, given a set $\mathfrak{A}$ of objects, iff any two objects in $\mathfrak{A}$ that differ in what is predicated of them in GR must also differ in what is predicated of them in LQG. The fact that supervenience requires a stable set $\mathfrak{A}$ of objects underlying both theories, i.e. an identical ontology on which the ideologies of both theories are defined, renders it rather useless in the present case. In a very rough way, the ontology of both theories of course contains the gravitational field. But the finer structure of the ontologies of both theories do not resemble each other: in LQG, one might perhaps find loops, or spin networks, or more generally the inhabitants of the physical Hilbert space in its ontology, while in GR, no such objects can be found. Hence, supervenience, at least as defined above, does not offer any help in understanding the relation between GR and LQG. Of course, the requirement that the set $\mathfrak{A}$ must underlie both theories can be relaxed: one could instead demand that the set $\mathfrak{A}$ of objects on which the sets $\mathfrak{P}_1$ and $\mathfrak{P}_2$ of properties figuring in the two theories are defined must be closed under compositional operations such as mereological sums or the formation of sets. The sets $\mathfrak{P}_1$ and $\mathfrak{P}_2$ would then be defined with respect to some base individuals, forming subsets $\mathfrak{A}_1$ and $\mathfrak{A}_2$ of $\mathfrak{A}$. Typically, these predications would induce some properties on the non-basic composite objects. Conceivably, this relaxation might be sufficient to overcome the disjointness of the sets $\mathfrak{A}_1$ and $\mathfrak{A}_2$.[39]

Consequently, we should not harbour any hope that GR either is a definitional extension of LQG or supervenes on LQG. However, if one admits a sufficiently liberal notion of emergence, hope resurges. The third broadly reductive relation proposed by Butterfield and Isham, and termed 'emergence' by them, fits the bill:

Definition 7 (*Emergence*) For Butterfield and Isham, a theory T_1 *emerges* from another theory T_2 iff there exists either a limiting or an approximating procedure to relate the two theories (or a combination of the two).

---

[38]The clause "appropriately related quantities in the theory to be approximated" in Definition 6 above occludes substantive work that must be completed to achieve such "appropriate relation". I am grateful to Erik Curiel for pushing me on this point—I most certainly deserve the pushing here.

[39]I wish to thank Jeremy Butterfield for suggesting this relaxation.

The definition of 'approximating procedure' was given in Definition 6; here is the one for 'limiting procedure':

Definition 8 (*Limiting procedure*) A *limiting procedure* is taking the mathematical limit of some physically relevant parameters, in general in a particular order, of the underlying theory in order to arrive at the emergent theory.

For it to have any prayer of sufficing to relate two theories, a limiting procedure as envisioned by Butterfield and Isham must be accompanied by a specification of a map between the theories that relates at least some of their algebraic or geometric structures.[40] For both technical and conceptual reasons, one should not expect that the emergence of GR from LQG can be understood only as a simple limiting procedure. Carlo Rovelli ([34] §6.7.1) delivers an account of how limiting procedures alone are incapable of establishing the missing link. He relates how loop quantum gravitists have not suspected that quantum space might turn out to have a discrete structure during the period from the discovery of the loop representation of GR around 1988 to the derivation of the spectra of the area and volume operators in 1995. He reminisces how during this period researchers believed that the classical, macroscopic geometry could be gained by taking the limit of a vanishing lattice constant of the lattice of loops. This limiting procedure was taken to run analogously to letting the lattice constant of a lattice field theory go to zero and thus define a conventional quantum field theory (QFT). With this model in mind, something remarkable happened when people tried to construct so-called weave states which are characterized as approximating a classical metric: when the quantum states were defined as the limit one gains when the spatial loop density grows to infinity, i.e. when the loop size is assumed to go to zero, it turned out that the approximation did not become increasingly accurate as the limit was approached. This can be taken as a clear indication that taking this limit was physically inappropriate. What was observed instead was that eigenvalues of the area and volume operators increased. This, of course, meant that the areas and volumes of the spatial regions under consideration also increased. In other words, the physical density of the loops did not increase when the 'lattice constant' was decreased. The physical density of loops, it turned out, remains unaffected by how large the lattice constant is chosen; it is simply given by a dimensional constant of the theory itself, Planck's constant. This result is interpreted to mean that there is a minimal physical scale. Or, in Rovelli's words, "more loops give more size, not a better approximation to a given [classical] geometry" (ibid.). The loops, it turns out, have an intrinsic physical size. Taking this limit, then, does not change the structure from discrete quantum states to smooth manifolds. It just does not change anything in the physics, except that we look at larger volumes. As some of the features of the classical geometry such as smoothness cannot be reduced to or identified with properties of the quantum states of the more fundamental theory, GR in toto does not reduce to LQG. Thus, a limiting procedure, at least if used in isolation, will just not do the trick.

---

[40]Thanks to Erik Curiel for holding me to task here.

On the conceptual side, a limiting procedure never eliminates superposition states, which of course are generic in a quantum theory. For this reason alone, a limiting procedure cannot succeed in recovering a classical theory from a quantum one. As argued by Klaas Landsman [26], the classical world only emerges from the quantum theory if some quantum states and some observables of the quantum theory are neglected, *and* some limiting procedure is executed. According to his view, to be discussed below, relating the classical with the quantum world thus takes both, the limiting as well as the approximating, procedures.

Turning to approximations then, a series of theories the last of which will mimic classical spacetimes via approximations needs to be constructed. First, let us consider what the 'approximandum', the classical theory to be approximated, should be. In GR, and in quantum theories based on it, one standardly, and perhaps somewhat unprincipledly, distinguishes between gravity and matter—a distinction routinely downplayed in particle-physics based approaches. They differ in their role and where they show up in the Einstein equations: gravity, the "marble" as Einstein called it, constitutes the left-hand side of the equations and determines the spacetime geometry; matter, the "low-grade wood", enters the stress-energy tensor on the right-hand side. In the quantization that led to LQG, no matter was assumed to be present: LQG results from a vacuum quantization of GR. It would seem, therefore, that states in LQG's physical Hilbert space should generically give rise to semi-classical states which yield emergent classical spacetimes that are vacuum solutions. But this expectation may be disappointed, and perhaps for a reason: it has been claimed that matter is implicitly built into LQG and that it would therefore be a mistake to think that no matter is present in spin network states. In particular, it may be that the very structure of the spin networks gives rise to matter in the appropriate low-energy limit. This means that it may be advisable not to be fixated on vacuum spacetimes.

Similarly, Hamiltonian GR is restricted to spacetime models with topology $\Sigma \times \mathbb{R}$. Should we thus expect that the procedure for recovering relativistic space-times would only yield spacetimes of such topology? While spacetimes with different topology may be suppressed and the generic result thus be concentrated on $(\Sigma \times \mathbb{R})$-spacetimes, the quantum structures with their combinatorial and topologically var-iegated connections may lead to spacetimes with more complicated topologies than those permitted by Hamiltonian GR.

In order to prepare the field for applying the Butterfield-Isham scheme, let us consider the major ways in which classical physics is typically held to relate to quantum physics, as listed and discussed, e.g., by Landsman [26]: (i) by a limiting procedure involving the limit $\hbar \to 0$ for a finite system, (ii) by a limiting procedure involving the limit $N \to \infty$ of a large system of N degrees of freedom while $\hbar$ is held constant, and (iii) either by decoherence or by a consistent histories approach. Landsman defends the point of view that while none of these manners is individually sufficient to understand how classicality emerges from the quantum world, they jointly suggest that it results from ignoring certain states and certain observables from the quantum theory.[41]

---

[41]For a more thorough discussion of Landsman's argument, cf. ([47] §9.2.1).

As Landsman shows, taking limits such as $N \to \infty$, albeit 'factual', i.e., pertaining to our world, and hence physically more reasonable, is mathematically just a special case of the 'counterfactual', and hence physically more problematic, limit $\hbar \to 0$. Regardless of their physical salience, these limits will in themselves not suffice because no such limit can ever resolve a quantum superposition state into a classical state. Thus, something more will be necessary, and that is where many think 'decoherence' will come into play. The main idea of the program of decoherence is that the generically assumed presence of interference in quantum states is suppressed by the system's interaction with the 'environment', such as is thought to occur in the measurement process.[42] Decoherence, then, is the phenomenon that pure quantum states, by virtue of their interaction with the system's 'environment', evolve, over very short time spans, from superposition states to 'almost' mixed states with classical probability distributions but 'almost' no quantum interference left. Roughly speaking, decoherence leaves the quantum system, to a high approximation, in an eigenstate of a macroscopically relevant operator; the classical probabilities of the resulting mixed states then only reflect our ignorance as to *which* eigenstate the system's in.

Given that the system at stake is the universe, and all of it, of course, the notion that 'environmental' degrees of freedom are those which decohere the system must be generalized so as to include 'internal' degrees of freedom of the system. This does not mean that the system is put in a mixed state from the beginning—that would be begging the question, as a referee correctly remarked—, but instead to 'coarse grain' and thereby 'wash out' many degrees of freedom, which then effectively act as the environment of the 'system' consisting of the remaining, physically salient degrees of freedom. This 'internal' environment then induces the decoherence of the originally pure state. We will return to this 'cosmological' problem below in the specific context of LQG.

The cosmological problem thus requires that we operate with a generalized notion of decoherence, which does not rely on a decohering system being embedded in an environment which is literally external to it. There is, however, a second issue that needs to be addressed. Decoherence is usually understood as a *dynamical* process of a system interacting with a large number of 'environmental' degrees of freedom. How should we conceive of a dynamical process in the general quantum-gravitational context in which time itself is part of the system at stake and, at least for canonical approaches, in which we face the nasty problem of time? Unfortunately, I have no solution to offer here, but can only note the puzzling problem and venture a guess as to the direction in which its resolution may have to go. In my view, the solution will come from a considered understanding of how dynamical processes such as decoherence can co-emerge with spacetime such that the emergence of the former facilitates the emergence of the latter, *and vice versa*, to let dynamics and spacetime mutually enable one another.

In sum, *if*—and only if—a theory of decoherence manages to give us a handle on how to identify the relevant degrees of freedom, and under what circumstances the

---

[42]For reviews of decoherence, see [7] and [40].

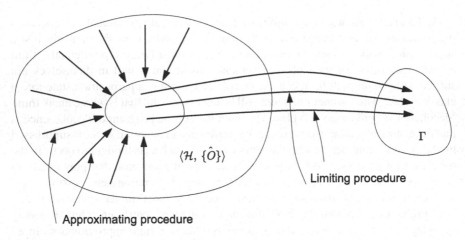

Fig. 3 The Butterfield-Isham scheme transposed to the present case

interaction between these degrees of freedom and those which were not picked out as 'environmental' leads to a suppression of interference, and how this suppression works in detail, particularly concerning its 'dynamics', *then* we will have a mechanism that 'drives the system' to the right sorts of semi-classical quantum states. In other words, such a mechanism would then justify the selection of the subset of states (and of a subset of physical magnitudes) we made in what Butterfield and Isham called the 'approximating procedure'.

In general outline, then, following Butterfield and Isham's proposal will lead to a two-step procedure, as illustrated in Figure 3. The first step consists of an approximating procedure, driving the generic quantum states, by some physical mechanism or other, into the semi-classical states, which are more closely related to classical states. The second step involves a limiting procedure relating these semi-classical states to states in the classical phase space, denoted in Figure 3 by Γ. Regardless of how the details of this story work out, one thing is clear: a whole host of issues known from the traditional problem of understanding the relation between the quantum and the classical world will arise.

4.2 Applying the Butterfield-Isham Scheme

The thesis—or should I say the 'promissory note'—to be suggested in the remainder of this essay asserts that at least to the extent to which LQG is a consistent and complete theory, (a close cousin of) GR can be seen to emerge from LQG if a delicately chosen ordered combination of approximations and limiting procedures is applied. This note is yet to be redeemed. All approaches to finding the semi-classical and classical limits of LQG are confined, to date, to using the kinematical Hilbert space $\mathcal{H}_K$ rather than the physical Hilbert space $\mathcal{H}$ as their starting point.

This raises the concern of both the viability and the meaningfulness of relating the kinematical states to corresponding classical spacetimes, or spaces. But concerns like these, although perhaps ultimately critical, should not keep us from attempting to get a grasp on what it means to draw the classical limit of the background-independent QFT as it stands now (and has been sketched above), as it may turn out to be an eminent help in the construction of the physical Hilbert space itself. To be sure, even the relationship between kinematic LQG and classical theories is ill-understood. Let me sketch, however, how preliminary work by physicists might bear out the Butterfield-Isham scheme.

The rough idea of constructing semi-classical states from the kinematical Hilbert space $\mathcal{H}_K$ is to find those kinematical states which correspond to almost flat three-metrics, i.e. to three-geometries where the quantum fluctuations are believed to be negligibly small. Two major approaches to construct semi-classical theories dominate the extant literature, the so-called 'weave state approach' and the ansatz using coherent states. The latter has been pioneered by Thomas Thiemann and Oliver Winkler.[43] Other proposals include Madvahan Varandarajan's 'photon Fock states' and generalizations thereof [1, 45], and the Ashtekar group's 'shadow states' [6].[44] The remainder of this essay shall be dedicated, however, to the most prominent approach of constructing semi-classical states, the so-called 'weave states'.

The idea of a *weave state* originally introduced by Ashtekar, Rovelli, and Smolin [5],[45] revolves around selecting spin network states that are eigenstates of the geometrical operators for the volume of a (spatial) region $\mathcal{R}$ with eigenvalues which approximate the corresponding classical values for the volume of $\mathcal{R}$ as determined by the classical gravitational field. Simultaneously, these selected spin network states are eigenstates of the geometrical area operator for a surface $\mathcal{S}$. More technically, consider a macroscopic three-dimensional region $\mathcal{R}$ of spacetime with the two-dimensional surface $\mathcal{S}$ and the three-dimensional gravitational field $e_a^i(\mathbf{x})$ defined for all $\mathbf{x} \in \mathcal{R}$. This gravitational field defines a metric field $q_{ab}(\mathbf{x}) = e_a^i(\mathbf{x})e_b^j(\mathbf{x})\eta_{ij}(\mathbf{x})$, where η_{ab} is the Minkowski metric, for which it is possible to construct a spin network state $|S\rangle$ such that $|S\rangle$ approximates the metric q_{ab} for sufficiently large scales $\Delta \gg \ell_{\text{Pl}}$, where ℓ_{Pl} is the Planck length, in a yet to be rigorously specified sense.[46] Classically, the area of a two-dimensional surface $\mathcal{S} \subset \mathcal{M}$ and the volume of a three-dimensional region $\mathcal{R} \subset \mathcal{M}$ with respect to a (sufficiently well-behaved)

---

[43] For a review, cf. [39] and ([44] §11.2). Thiemann's book also discusses weave states in §11.1 and the photon Fock states in §11.3.

[44] As ([44] §11) points out, there are deep connections between the various semi-classical programmes.

[45] For an intuitive introduction, see ([34] §6.7.1). The picture is that of the gravitational field like a (quantum cloud of) fabric(s) of weaves which appears to be smooth if seen from far but displays a discrete structure if examined more closely. Hence *weave states*.

[46] The 'upper case' spin network states $|S\rangle$ live in $\mathcal{K}^*$, the *pre-kinematical Hilbert space*, i.e. the Hilbert space containing all spin network states which solve the Gauss constraints, but not necessarily the spatial diffeomorphism constraints. Thus, the spin network states in $\mathcal{K}^*$ are not represented by abstract graphs, as are those in the full kinematical Hilbert space $\mathcal{H}_K$, but as embedded graphs on a background manifold. This choice is just conveniently following the established standard in the

fiducial gravitational field $^0e^i_a$ are given by ([34] §2.1.4)

$$\mathbf{A}[^0e, \mathscr{S}] = \int |d^2\mathscr{S}|, \tag{11}$$

$$\mathbf{V}[^0e, \mathscr{R}] = \int |d^3\mathscr{R}|, \tag{12}$$

where the relevant measures for the integrals are determined by $^0e^i_a$. This fiducial metric is typically, but not necessarily, chosen to be flat. The requirement that the spin network state $|S\rangle$ must approximate the classical geometry for sufficiently large scales is made precise by demanding that $|S\rangle$ be a simultaneous eigenstate of the area operator $\hat{\mathbf{A}}$ and the volume operator $\hat{\mathbf{V}}$ as mentioned above with eigenvalues equal to the classical values as given by (11) and (12), respectively, up to small corrections of the order of $\ell_{\mathrm{Pl}}/\Delta$:

$$\hat{\mathbf{A}}(\mathscr{S})|S\rangle = \left(\mathbf{A}[^0e, \mathscr{S}] + \mathscr{O}(\ell^2_{\mathrm{Pl}}/\Delta^2)\right)|S\rangle, \tag{13}$$

$$\hat{\mathbf{V}}(\mathscr{R})|S\rangle = \left(\mathbf{V}[^0e, \mathscr{R}] + \mathscr{O}(\ell^3_{\mathrm{Pl}}/\Delta^3)\right)|S\rangle. \tag{14}$$

If a spin network state $|S\rangle$ satisfies these requirements, then it is called a *weave state*. In fact, the length scale Δ, which is large compared to the Planck length ℓ_{Pl}, characterizes the weave states, which are for this reason sometimes denoted by $|\Delta\rangle$ in the literature. At scales much smaller than Δ, the quantum features of spacetime would become relevant, while at scales of order Δ or larger, the weave states exhibit a close approximation to the corresponding classical geometry in the sense that it determines the same areas and volumes as the classical metric q_{ab}. In this sense, the weave states are semi-classical approximations.

It should be noted that the correspondence between weave states and classical spacetimes is many-to-one. In other words, equations (13) and (14) do not determine the state $|S\rangle$ uniquely from a given three-metric q_{ab}. The reason for this is that these equations only put constraints on values averaged over all of $\mathscr{S}$ and $\mathscr{R}$, respectively, and we have assumed *ex constructione* that these regions are large compared to the Planck scale. Of course, there are many spin network states with these averaged properties, but only one classical metric which exactly corresponds to these averages values. The situation can be thought of as somewhat analogous to thermodynamics, where a physical system with many microscopic degrees of freedom has many different microscopic states with the same averaged, macroscopic properties such as temperature.[47]

(Footnote 46 continued)
literature on weave states; we will see below in Footnote 47 that this poses no problem as everything can be directly carried over to the spatially diffeomorphically invariant level.

[47] The weave states as introduced above have merely been defined at the pre-kinematic level, i.e. they are not formulated in terms invariant under spatial diffeomorphisms (cf. also Footnote 46). The reason for this choice lies mostly in that this is the canonical choice in the literature, but also because in this way, the weave states can be directly related to three-metrics, rather than equivalence classes

Apart from a serious difficulty in constructing semi-classical weave states corresponding to classical Minkowski spacetime,[48] it seems as if the notion of approximation as captured in Definition 6 and the Butterfield-Isham scheme might bear fruit in relating semi-classical weave states to classical spacetimes (or at least spaces). If the weave states are taken to be simultaneous eigenstates of the area and volume operators, as they are in (13) and (14), then some physical quantities must be neglected, viz. all those operators constructed from connection operators, since the 'geometrical' eigenstates are maximally spread in these operators, and the kinematical (weave) states must be carefully selected to only include those which are peaked around the geometrical values determined by the fiducial metric. It is at least questionable, however, whether the neglect of connection-based operators can be justified. If it cannot, then only semi-classical states which are peaked in both the connection and the triad basis, and are peaked in such a manner as to approximate classical states, should be considered. In this case, we would still only have a selection of states, but perhaps no operators, or no physically salient ones, which are being ignored.

None of this gives us just as yet a *physical mechanism* that drives generic kinematical states to the semi-classical weave states. Just as above in the general case, decoherence is widely assumed to offer such a mechanism in the context of weave states. But this brings what I termed above the 'cosmological problem' back into the fold: how should such a story possibly apply to the present context where the spin network states are supposed to be the quantum account of space—and all of it. If we thus think of an 'environment' as something external to the system for which it is an environment, then relying on such an environment in our story implies that there must be something outside of space. But this is clearly incoherent. Not all hope is lost, however, as there are at least two ways to escape the incoherence. First, as in the general case above, one might conceive of decoherence not in terms of external, environmental degrees of freedom which interact with the system, but instead as interactions among different degrees of freedom of the system itself. This will presuppose a partition of the system's degrees of freedom into 'salient' ones and mere 'background'; but there is no reason that this could not be done in a principled fashion.

Second, we may reconceptualize LQG's subject matter. We may, more specifically, conceive of areas and volumes as local properties of the quantum gravitational field, just as these geometrical properties were local in GR. As was explicated in §2.1, given a region $\mathscr{R}$ of quantum space, e.g. a chunk of space in our laboratory, each node of the spin network state represents a grain of such a space as it contributes to

---

(Footnote 47 continued)
of three-metrics. This, however, does not constitute a problem whatsoever, as the characterization of weave states carries over into the context of diffeomorphically invariant spin network states in $\mathscr{H}_K$, as follows. If we introduce a map $P_{\mathrm{diff}} : \mathscr{K}^* \to \mathscr{H}_K$ which projects states in $\mathscr{K}^*$ related by a spatial diffeomorphism unto the same element of $\mathscr{H}_K$, then the state $\mathscr{H}_K \ni |s\rangle = P_{\mathrm{diff}}|S\rangle$ is a *weave state* of the classical three-geometry $[q_{ab}]$, i.e., the equivalence class of three-metrics q_{ab} under spatial diffeomorphisms, just in case $|S\rangle$ is a weave state of the classical three-metric q_{ab} as defined above.

[48] For details, cf. ([47] 181).

the eigenvalue of the volume operator. Similarly, each link from a node within $\mathscr{R}$ to a node outside of $\mathscr{R}$, i.e. each link which intersects the boundary $\mathscr{S}$ of $\mathscr{R}$, contributes to the eigenvalue of the area operator. If we had measurement devices at our disposal with Planck-scale accuracy, we could, in principle, measure the volume and the surface area of a region of space(time) given in our lab. Such a measurement would essentially amount to counting (and weighing) the nodes within a region as well as counting (and weighing) the links which leave the region. If the region $\mathscr{R}$ considered does not encompass all of space, but only a delimited piece of it, then of course finding an environment for such a 'mid-sized' region is straightforward and the cosmological problem dissolves. In fact, it would arguably also resolve the dynamical problem, as the lab frame would offer a context in which dynamical processes unravel. It could thus be the case that if we performed an area or volume measurement on surface $\mathscr{S}$ or region $\mathscr{R}$, respectively, then we would find the quantum state of this 'mid-sized' region decohered into an eigenstate of the relevant operator, and thus into a weave state.

Once we have completed this stage, and we have found semi-classical states which approximate classical states, then a limiting procedure can be executed. Such a limiting procedure will involve taking the limit $\ell_{Pl}/\Delta \to 0$, which will make the small corrections in (13) and (14) disappear. This limit can be performed by either having Δ go to infinity, or ℓ_{Pl} go to zero (or both). The first choice corresponds to letting the size of the spatial region $\mathscr{R}$ grow beyond all limits, and thus resembles the 'factual' limit $N \to \infty$ as discussed above. The second choice, letting the Planck size go to zero, corresponds, accordingly, to the 'counterfactual' case $\hbar \to 0$. With the second choice, but arguably not the first, we leave the realm of the quantum theory and arrive at a strictly classical description of the spatial geometry.

It should be noted that none of this solves the measurement problem. Only a full solution of the measurement problem will ultimately give us complete comprehension of the emergence of classicality from a reality which is fundamentally quantum. But to solve this problem is hard in non-relativistic quantum mechanics, harder still if special relativity must be incorporated, and completely mystifying once we move to fully relativistic quantum theories of gravity. In light of this, I submit that we would have reason to uncork our champagne even if we only managed to articulate a complete and consistent quantum theory of gravity with a well-understood approximation to semi-classical states and a somewhat rigorous limiting procedure connecting these semi-classical states to classical states of the gravitational field.

5 Conclusion

We have seen how classical space and time 'disappear' in quantum gravity and considered a sketch of how they might re-emerge from the fundamental, not obviously spatiotemporal structure. Even though the situation is technically and conceptually more demanding overall and even though a *case* must be made for the applicability of a traditional measurement concept more specifically, I hope the reader has also recognized that the way in which classicality emerges from the quantum theory does not radically differ from ordinary quantum mechanics, at least along some dimensions of comparison.

The project of analyzing the emergence of spacetime, and hence of classicality, from quantum theories of gravity, which often deny at least some aspects of spatiotemporality, is relevant for two reasons. First, important foundational questions concerning the interpretation of, and the relation between, theories are addressed, which contributes to the conceptual clarifications in the foundations of physics arguably necessary to achieve a breakthrough. Not only philosophers of physics will contribute to this project, of course. They are not even likely to shoulder the lion's share, which will still fall on the physicists. But they can nevertheless bring their unique skill set to the table, to the benefit, it is hoped, of the entire dinner party. Second, and conversely, quantum gravity is rich with implications for specifically philosophical, and particularly metaphysical, issues concerning not just space and time, but also causation, reduction and even modality. Quantum gravity thus turns out to be a very fertile ground for the philosopher. Altogether, I take it, there is no reason for philosophers to keep aloof from these exciting developments in the foundations of physics.

References

1. Abhay Ashtekar and Jerzy Lewandowski. Relation between polymer and Fock excitations. *Classical and Quantum Gravity*, 18:L117–L128, 2001.
2. Abhay Ashtekar and Jerzy Lewandowski. Quantum theory of geometry I: Area operators. *Classical and Quantum Gravity*, 14:A55–A81, 1997.
3. Abhay Ashtekar and Jerzy Lewandowski. Quantum theory of geometry II: Volume operators. *Advances in Theoretical and Mathematical Physics*, 1:388–429, 1998.
4. Abhay Ashtekar and Jerzy Lewandowski. Quantum field theory of geometry. In Tian Yu Cao, editor, *Conceptual Foundations of Quantum Field Theory*, pages 187–206, Cambridge University Press, Cambridge, 1999.
5. Abhay Ashtekar, Carlo Rovelli, and Lee Smolin. Weaving a classical metric with quantum threads. *Physical Review Letters*, 69:237–240, 1992.
6. Abhay Ashtekar, Stephen Fairhurst, and Joshua L Willis. Quantum gravity, shadow states, and quantum mechanics. *Classical and Quantum Gravity*, 20:1031–1062, 2003.
7. Guido Bacciagaluppi. The role of decoherence in quantum theory. In Edward N. Zalta, editor, *Stanford Encyclopedia of Philosophy*, 2012. URL http://plato.stanford.edu/entries/qm-decoherence/.
8. Gordon Belot and John Earman. Pre-Socratic quantum gravity. In Craig Callender and Nick Huggett, editors, *Physics Meets Philosophy at the Planck Scale*, pages 213–255. Cambridge University Press, Cambridge, 2001.
9. Jeremy Butterfield and Chris Isham. On the emergence of time in quantum gravity. In Jeremy Butterfield, editor, *The Arguments of Time*, pages 111–168. Oxford University Press, Oxford, 1999.
10. Jeremy Butterfield and Christopher Isham. Spacetime and the philosophical challenge of quantum gravity. In Craig Callender and Nick Huggett, editors, *Physics Meets Philosophy at the Planck Scale*, pages 33–89. Cambridge University Press, Cambridge, 2001.

11. Craig Callender and Nick Huggett. Introduction. In Craig Callender and Nick Huggett, editors, *Physics Meets Philosophy at the Planck Scale*, pages 1–30. Cambridge University Press, Cambridge, 2001.
12. Yvonne Choquet-Bruhat and James W York, Jr. The Cauchy problem. In Alan Held, editor, *General Relativity and Gravitation: One Hundred Years After the Birth of Albert Einstein*, pages 99–172. Plenum Press, New York, 1980.
13. Erik Curiel. Against the excesses of quantum gravity: A plea for modesty. *Philosophy of Science*, 68(3):S424–S441, 2001.
14. Erik Curiel. General relativity needs no interpretation. *Philosophy of Science*, 76(1):44–72, 2009.
15. Lokenath Debnath and Piotr Mikusiński. *Introduction to Hilbert Spaces with Applications*. Academic Press, San Diego, 1999.
16. John Earman. Tracking down gauge: An ode to the constrained Hamiltonian formalism. In Katherine Brading and Elena Castellani, editors, *Symmetries in Physics: Philosophical Reflections*, pages 140–162. Cambridge University Press, Cambridge, 2003.
17. John Earman. The implications of general covariance for the ontology and ideology of spacetime. In Dennis Dieks, editor, *The Ontology of Spacetime*, pages 3–23. Elsevier, Amsterdam, 2006.
18. John Earman. *Bangs, Crunches, Whimpers, and Shrieks: Singularities and Acausalities in Relativistic Spacetimes*. Oxford University Press, New York, 1995.
19. Helmut Friedrich and Alan Rendall. The Cauchy problem for the Einstein equations. *Lecture Notes in Physics*, 540:127–224, 2000.
20. Paul R Halmos. *Introduction to Hilbert Space and the Theory of Spectral Multiplicity*. Chelsea Publishing Company, New York, 1951.
21. Marc Henneaux and Claudio Teitelboim. *Quantization of Gauge Systems*. Princeton University Press, Princeton, NJ, 1992.
22. Nick Huggett and Christian Wüthrich. Emergent spacetime and empirical (in)coherence. *Studies in the History and Philosophy of Modern Physics*, 44:276–285, 2013.
23. Nick Huggett, Tiziana Vistarini, and Christian Wüthrich. Time in quantum gravity. In Adrian Bardon and Heather Dyke, editors, *A Companion to the Philosophy of Time*, pages 242–261. Wiley-Blackwell, Chichester, 2013.
24. Claus Kiefer. Quantum gravity: whence, whither? In Felix Finster, Olaf Müller, Marc Nardmann, Jürgen Tolksdorf, and Eberhard Zeidler, editors, *Quantum Field Theory and Gravity: Conceptual and Mathematical Advances in the Search for a Unified Framework*, pages 1–13. Springer, Basel, 2012.
25. Vincent Lam and Michael Esfeld. A dilemma for the emergence of spacetime in canonical quantum gravity. *Studies in the History and Philosophy of Modern Physics*, 44:286–293, 2013.
26. Nicolaas P Landsman. Between classical and quantum. In Jeremy Butterfield and John Earman, editors, *Handbook of the Philosophy of Science. Vol. 2: Philosophy of Physics*, pages 417–553. Elsevier B.V., Amsterdam, 2006.
27. John Byron Manchak. What is a physically reasonable space-time? *Philosophy of Science*, 78:410–420, 2011.
28. Fotini Markopoulou and Lee Smolin. Disordered locality in loop quantum gravity states. *Classical and Quantum Gravity*, 24:3813–3823, 2007.
29. Tim Maudlin. Thoroughly muddled McTaggart: Or how to abuse gauge freedom to generate metaphysical monstrosities. *Philosophers' Imprint*, 2(4), 2002.
30. Brian McLaughlin. Emergence and supervenience. *Intellectica*, 2:25–43, 1997.
31. Timothy O'Connor and Hong Yu Wong. Emergent properties. In Edward N. Zalta, editor, *Stanford Encyclopedia of Philosophy*, 2015. URL http://plato.stanford.edu/entries/properties-emergent/.
32. Chanda Prescod-Weinstein and Lee Smolin. Disordered locality as an explanation for the dark energy. *Physical Review D*, 80:063505, 2009.
33. Carlo Rovelli. Partial observables. *Physical Review D*, 65:124013, 2002.
34. Carlo Rovelli. *Quantum Gravity*. Cambridge University Press, Cambridge, 2004.

35. Carlo Rovelli. A new look at loop quantum gravity. *Classical and Quantum Gravity*, 28:114005, 2011.
36. Carlo Rovelli and Lee Smolin. Discreteness of area and volume in quantum gravity. *Nuclear Physics B*, 442:593–622, 1995a. Erratum: *Nuclear Physics B*, **456**:734.
37. Carlo Rovelli and Lee Smolin. Spin networks and quantum gravity. *Physical Review D*, 52:5743–5759, 1995b.
38. Laura Ruetsche. *Interpreting Quantum Theories: The Art of the Possible*. Oxford University Press, Oxford, 2011.
39. Hanno Sahlmann, Thomas Thiemann, and Oliver Winkler. Coherent states for canonical quantum general relativity and the infinite tensor product extension. *Nuclear Physics B*, 606:401–440, 2001.
40. Maximilian Schlosshauer. Decoherence, the measurement problem, and interpretations of quantum mechancis. *Reviews of Modern Physics*, 76:1267–1305, 2004.
41. Chris Smeenk and Christian Wüthrich. Time travel and time machines. In Craig Callender, editor, *The Oxford Handbook of Philosophy of Time*, pages 577–630. Oxford University Press, Oxford, 2011.
42. Lee Smolin. Generic predictions of quantum theories of gravity. In Daniele Oriti, editor, *Approaches to Quantum Gravity: Toward a New Understanding of Space, Time and Matter*, pages 548–570. Cambridge University Press, Cambridge, 2009.
43. Marshall Spector. *Concepts of Reduction in Physical Science*. Temple University Press, Philadelphia, 1978.
44. Thomas Thiemann. *Modern Canonical Quantum General Relativity*. Cambridge University Press, Cambridge, 2007.
45. Madvahan Varadarajan. Fock representations from $U(1)$ holonomy algebras. *Physical Review D*, 61:104001, 2000.
46. Robert M Wald. *General Relativity*. The University of Chicago Press, Chicago, 1984.
47. Christian Wüthrich. *Approaching the Planck Scale from a Generally Relativistic Point of View: A Philosophical Appraisal of Loop Quantum Gravity*. PhD thesis, University of Pittsburgh, 2006.
48. Christian Wüthrich. When the actual world is not even possible. *Manuscript*, 2015.
49. Christian Wüthrich. A la recherche de l'espace-temps perdu. In Soazig Le Bihan, editor, *Précis de philosophie de la physique*, pages 222–241. Vuibert, Paris, 2013.

Printed in the United States
By Bookmasters

Printed in the United States
By Bookmasters